Mine and Water

Mine and Water

Ruiping Liu
Youning Xu

Basel • Beijing • Wuhan • Barcelona • Belgrade • Novi Sad • Cluj • Manchester

Editors
Ruiping Liu
Xi'an Center of China
Geological Survey
Ministry of Natural Resources
Xi'an
China

Youning Xu
Xi'an Center of China
Geological Survey
Ministry of Natural Resources
Xi'an
China

Editorial Office
MDPI AG
Grosspeteranlage 5
4052 Basel, Switzerland

This is a reprint of articles from the Special Issue published online in the open access journal *Water* (ISSN 2073-4441) (available at: www.mdpi.com/journal/water/special_issues/JA4K6M845H).

For citation purposes, cite each article independently as indicated on the article page online and as indicated below:

Lastname, A.A.; Lastname, B.B. Article Title. *Journal Name* **Year**, *Volume Number*, Page Range.

ISBN 978-3-7258-2544-8 (Hbk)
ISBN 978-3-7258-2543-1 (PDF)
doi.org/10.3390/books978-3-7258-2543-1

Cover image courtesy of Changguang Satellite Technology Co., Ltd

Contents

About the Editors

Ruiping Liu

Liu Ruiping (1979~) Xi'an Geological Survey Center of China Geological Survey, is a member of the Mine Safety Professional Committee of the China Mining Federation, member of the Xi'an Science and Expert Database, Women's Pioneer of the Xi'an Petrochemical Industry Union, and member of the China Geological Survey Technology Expert Database. She is also an editorial board member of the journals "*China Geology*" and "*Geology and Resources*". Since 2004, she has been engaged in the investigation, research, and prevention of mine geological environmental issues. Since 2001, she has hosted and participated in more than 40 geological environmental research and projects. She has published 50 articles as the first author, including 13 SCI articles. She has published five monographs and obtained nine national invention patents. She has won four second prizes in science and technology from the Provincial and Ministerial Science and Technology Awards. She has high research capabilities and an abundance of work experience in the evaluation, prevention, and control of mine geological environmental issues.

Youning Xu

Xu Youning (1963) is a project consultant and academic leader; has a Ph.D.; is a second-level researcher; a deputy engineer at the Xi'an Geological Survey Center of the China Geological Survey; and a doctoral supervisor at the Chinese Academy of Geological Sciences. He is a review expert for the National Science and Technology; a peer review expert for the International Science and Technology Cooperation Key Projects of the Ministry of Science and Technology; a referee for the National Natural Science Foundation; an expert in mine environmental protection for the Ministry of Natural Resources; an expert in soil pollution investigation for the Ministry of Ecology and Environment; a member of the Heavy Metal Pollution Control Committee of the Environmental Science Society; a member of the Landslide and Debris Flow Committee of the Chinese Society of Soil and Water Conservation; and a member of the Quaternary Ge Academic Committee of the Geological Society of China. Since 2001, he has been engaged in the investigation, research, and prevention and control of mine environments and has compiled seven standards for the investigation, monitoring, and evaluation of mine geological environments. Since 2001, he has presided over thirteen projects for the investigation and research of the mine geological environment of the China Geological Survey; one general project of the National Natural Science Foundation; two special industry projects of the Ministry of Land and Resources; and one key research and development project of Shaanxi Province. He has published more than 50 articles as the first author, published 4 monographs, and obtained 7 national invention patents. He has won four second prizes in the Science and Technology Awards of the Ministry of Land and Resources and Shaanxi Province (ranked first). He has high research capabilities in the investigation and evaluation of mine geological environment and has rich working in the theory and technology of prevention and control.

water

MDPI

Editorial

Mine and Water

Ruiping Liu [1,2,3,*] and Youning Xu [1,2,3]

1 Xi'an Center, China Geological Survey, Ministry of Natural Resources, Xi'an 710054, China; xyouning@mail.cgs.gov.cn
2 Observation and Research Station of Environmental Geology of Typical Mines in Shaanxi Province, Department of Natural Resources, Xi'an 710054, China
3 Key Laboratory for Geo-Hazards in Loess Area, Ministry of Natural Resources, Xi'an 710054, China
* Correspondence: lrp1331@163.com; Tel.: +86-13279487642

1. Introduction to the Special Issue

Environmental geology is a specialized field that examines the interaction between natural geological processes and human activities, particularly in the context of mineral resource extraction. The study of the geological and environmental impacts of mining focuses on how mining operations affect and are influenced by the geological environment, including potential negative impacts such as environmental pollution and geological hazards. Environmental geologists apply scientific theories and methods to promote sustainable development in the mining industry by managing resources responsibly and implementing effective strategies to minimize and mitigate the environmental impacts of mining activities [1]. The accelerated development of science and technology has spurred unprecedented levels of large-scale, high-intensity mineral resource exploitation. This intense geological activity serves as a catalyst for intensified changes within a mining region's geological environment. The resulting modifications exert a profound influence on the local geological landscape, increasing the likelihood of sudden geological disasters. Additionally, the cumulative accumulation of heavy metals poses a significant environmental threat, disrupting the delicate balance between human activities and the natural environment. These consequences ultimately jeopardize the long-term sustainability of human existence and development [2]. In this context, water is an indispensable resource for both human life and economic progress and intimately intertwined with mining activities. Geothermal water is classified as a valuable mineral resource and serves as a crucial component of the mining environment, and it can pose significant environmental risks when extracted and utilized improperly [3,4]. The process of water extraction and its subsequent use can result in water pollution and depletion [5]. Moreover, mining-related operations can contaminate groundwater sources with harmful pollutants, including heavy metals, acids, and sediments [6]. This contamination has far-reaching implications for human health, agricultural practices, and the delicate balance of ecosystems. Water is also a contributing factor to other geological mining problems [7].

This Special Issue on mines and water aims to provide comprehensive guidance on hydrogeology, environmental geology, mining environments, disaster geology, and health geology. By delving into the intricate relationship between water and mining operations, this publication seeks to foster a deeper understanding of the role that mining processes play in mine monitoring and the development of innovative prevention strategies. This knowledge is instrumental in promoting the sustainable development of green mines and ensuring the harmonious coexistence of human activities and the natural environment [8]. Following a rigorous peer review process, twelve original research papers (Contributions 1–12) were accepted for publication in November 2022. These papers encompass a diverse range of topics, including the exchange of mineral water with other environmental media, the pollution and enrichment of mineral water elements, the evolution of mineral water composition, and associated human health risks. Additionally,

Citation: Liu, R.; Xu, Y. Mine and Water. *Water* **2024**, *16*, 3022. https://doi.org/10.3390/w16213022

Received: 12 October 2024
Accepted: 16 October 2024
Published: 22 October 2024

this Special Issue features research and development in simulation technology, studies utilizing unmanned aerial vehicle (UAV) technology for the identification of mine wastewater, strategies for the prevention and reuse of polluted mine water, and an exploration of the relationship between mine water and geological disaster risk. To provide a more comprehensive understanding of this Special Issue, we have summarized the key findings of the published papers below.

2. Overview of the Contributions to This Special Issue

The pollution, enrichment, evolution and human health risk of elements in water of mining area. Rui-ping Liu and colleagues investigated the factors influencing fluoride levels in geothermal water within mining regions. Their research revealed that water–rock interactions are the primary drivers of fluoride enrichment. Furthermore, the mixing of geothermal waters with near-surface cold water can also contribute to the concentration of fluoride. The hydrogeochemical factors influencing fluoride enrichment in confined geothermal water primarily include natural conditions such as pH, ion exchange, and mineral saturation. Although the groundwater in the study area is slightly alkaline, its fluoride levels exceed the drinking water quality standards of both China and the World Health Organization (WHO). Specifically, in the guide basin, the confined water shows elevated fluoride concentrations ranging from 0.43 to 5.7 mg/L, whereas the phreatic and surface waters contain insufficient fluoride levels for drinking. Therefore, to address this issue, the Department of Water Resources Management is advised to mix confined geothermal water with phreatic and surface water in appropriate proportions to ensure that the fluoride content in the resulting drinking water falls within safe limits. Moreover, excessive fluoride consumption from confined geothermal water poses significant health risks to both adults and children. Ultimately, this study highlights that fluoride is an element with potentially harmful long-term health consequences.

Focusing on the exchange between water and other environmental media, Wei-kang Sheng and colleagues conducted a comprehensive case study in Chehe Town, Guangxi, a region that is characterized by integrated metals mining and smelting operations. In order to investigate the geochemical distribution, migration, and transformation of cadmium (Cd) and other heavy metals, the researchers systematically collected samples from various surface media. These included atmospheric dust, surface water, stream sediments, ores, tailings, mine drainage, soil, and crops, both within and around the mining area. By analyzing these diverse samples, they aimed to gain a thorough understanding of the environmental impacts associated with mining activities in the region.

Focusing on simulation technology research and development, Marcos Sanz-Ramos and colleagues investigated the potential risks associated with mine tailing pond ruptures. Their research highlighted that, depending on the characteristics of the stored material, such ruptures can lead to the fluidization and release of hyper-concentrated flows, which exhibit non-Newtonian fluid behavior. To simulate these dynamics, numerical modeling tools solve mass and momentum conservation equations, incorporating rheological models to accurately capture the unique properties of these fluids. In this study, the researchers extended Iber, a two-dimensional hydrodynamic numerical tool, to enable the simulation of non-Newtonian shallow flows, particularly those associated with mine tailings. The performance of the numerical tool was rigorously evaluated through both benchmark tests and real-world case studies. The results demonstrated excellent agreement with analytical and theoretical solutions in the benchmark tests, thereby confirming the tool's accuracy. Furthermore, the numerical tool proved to be well suited for simulating both dynamic and static phases under realistic conditions. Consequently, the outputs generated by this tool offer valuable insights, allowing researchers to assess flood hazards and risks associated with mine tailing spill propagation scenarios.

UAV technology can be utilized for the identification of mining wastewater. Acidic mine drainage (AMD) from mining operations can severely contaminate surrounding rivers and lakes, leading to significant ecological problems. While traditional field surveys have been the primary method for monitoring mining environments, Xiaomei Kou and colleagues demonstrated that UAV technology offers several advantages, including the ability to capture real-time imagery, enhanced security, and improved image accuracy. By leveraging UAV technology, it is possible to mitigate the shortcomings of traditional methods and significantly improve the efficiency of environmental surveys of mines. As a result, UAV technology has become an increasingly essential tool for monitoring mining environments.

Working towards the prevention and reuse of polluted water in mining areas, Li Zhizhong and colleagues proposed an innovative in situ leaching mining technique for deep bauxite exploration. This technique is particularly well suited to the geological conditions of the Ordos Basin, as it ensures that the leaching solution flows within the bauxite ore bed without any leakage. To prevent the leaching solution from contaminating the main aquifer surrounding the basin, the researchers suggested several preventive measures, including the use of a polymer resin filling, plugging to control well pattern seepage, and the establishment of comprehensive monitoring systems. These measures are designed to enhance the effectiveness of the technique while safeguarding the surrounding environment. Furthermore, the findings of this study offer a valuable theoretical foundation for the adoption of environmentally sustainable mining practices and for effectively mitigating water pollution associated with deep bauxite exploration.

To examine the ways in which water both causes and is impacted by geological hazards in mining areas, Wei-cui Ding and colleagues developed a comprehensive methodology for assessing and managing geological hazards and risks in complex mountainous towns. Using Longlin Town in Gansu Longnan, China, as a case study, they focused on the Quanjia Bay debris flows and Panping Village landslides. The proposed methodology comprises six stages: risk identification, investigating the hazard and disaster model, risk analysis, vulnerability assessment, risk evaluation, and risk management and control measures. Based on this framework, the researchers analyzed geological hazards under various precipitation frequencies (5%, 2%, and 1%). The results indicated that 75.23% of the regions remained at low risk levels, while 24.38% exhibited an increase in risk levels as the precipitation frequency decreased. Furthermore, only 0.39% of the regions were classified as extremely high-risk areas across all precipitation frequency conditions. For the specific cases of the Quanjia Bay debris flows and Panping Village landslides, the study explored how to control the sources of geological hazards and risks, along with management and control technologies. Additionally, both engineering and non-engineering disaster prevention and control measures were tailored to urban disasters and specific hazard zones. Ultimately, this research provides valuable technical support and guidance for disaster prevention and mitigation, as well as territorial spatial planning.

3. Conclusions

The Guest Editors hope that the papers published in this Special Issue will be of interest to researchers, designers, and practitioners engaged in water-related monitoring, control, and management within the context of mineral resource development. These contributions are expected to stimulate further research and exploration in this field. Moreover, we encourage readers to investigate water and issues surrounding it from the perspectives of integrated monitoring of the sky, earth, and human factors, along with numerical simulations and experimentation. By applying these insights to learn from prevention and control projects, the research presented in this Special Issue offers both intellectual interest and practical significance. The findings and methods outlined in this collection, including the status of mining water pollution, the transport and enrichment of substances in multiple media, the evolution of these processes, and the associated human health risks, as well as strategies for water pollution prevention and reuse, highlight the pivotal role of water in driving geological hazards and ecological restoration in mining areas. These important contributions hold substantial research value and can significantly enhance our understanding of the intricate relationship between mining and water.

Author Contributions: Conceptualization, R.L.; methodology, Y.X.; validation, R.L.; formal analysis, R.L.; resources, Y.X.; data curation, R.L.; writing—original draft preparation, R.L.; writing—review and editing, R.L.; visualization, R.L.; supervision, R.L.; project administration, R.L.; funding acquisition, R.L. All authors have read and agreed to the published version of the manuscript.

Funding: This study was funded by a ministry–province cooperation-based pilot project entitled A Technological System for Ecological Remediation Evaluation of Open-Pit Mines, initiated by the Ministry of Natural Resources in 2023(2023ZRBSHZ003); the survey projects were initiated by the Land and Resources Investigation Project ([2023]06-03-04, 1212010634713), Key R & D projects in Shaanxi Provincein 2023 "2023-ZDLSF-63".

Acknowledgments: Thank you to the *Water* editorial team for their excellent editing work. Thank you also to Jiangang Jiao and Refaey M. EI-Wardany for polishing and proofreading the manuscript.

Conflicts of Interest: The authors declare no conflicts of interest.

List of Contributions

1. Baloyi, J.; Ramdhani, N.; Mbhele, R.; Ramutshatsha-Makhwedzha, D. Recent Progress on Acid Mine Drainage Technological Trends in South Africa: Prevention, Treatment, and Resource Recovery. *Water* **2023**, *15*, 3453. https://doi.org/10.3390/w15193453.
2. Li, B.; Yu, T.; Ji, W.; Liu, X.; Lin, K.; Li, C.; Ma, X.; Yang, Z. Geochemical Response of Surface Environment to Mining of Sn-Pb-Zn Sulfide Deposits: A Case Study of Dachang Tin Polymetallic Deposit in Guangxi. *Water* **2023**, *15*, 1550. https://doi.org/10.3390/w15081550.
3. Kou, X.; Han, D.; Cao, Y.; Shang, H.; Li, H.; Zhang, X.; Yang, M. Acid Mine Drainage Discrimination Using Very High Resolution Imagery Obtained by Unmanned Aerial Vehicle in a Stone Coal Mining Area. *Water* **2023**, *15*, 1613. https://doi.org/10.3390/w15081613.
4. Liu, R.; Liu, F.; Xu, Y.; Zhu, H.; Jiao, J.; El-Wardany, R.M. Hydrogeochemical Characteristics and Human Health Risk Assessment of Fluoride Enrichment in Water in Faulted Basins of Qinghai-Tibet Plateau—A Case Study of Sanhe Plain in Guide Basin. *Water* **2023**, *15*, 1968. https://doi.org/10.3390/w15101968.
5. Sheng, W.; Hou, Q.; Yang, Z.; Yu, T. Spatial Distribution, Migration, and Ecological Risk of Cd in Sediments and Soils Surrounding Sulfide Mines—A Case Study of the Dabaoshan Mine of Guangdong, China. *Water* **2023**, *15*, 2223. https://doi.org/10.3390/w15122223.
6. Li, Z.; Zhang, Y.; Luo, T.; Xia, P.; Mu, H.; Sun, P.; Wang, X.; Wang, J. In-Situ Leaching Mining Technique for Deep Bauxite Extraction and the Countermeasures for Water Pollution Prevention: An Example in the Ordos Basin, China. *Water* **2023**, *15*, 2381. https://doi.org/10.3390/w15132381.
7. Chen, H.; Zhao, A.; Xu, Y.; Zhang, J.; Yang, M. The Impact of Molybdenum Mining on Cd Pollution along Wenyu Stream in Qinling Mountains, Northwest China. *Water* **2023**, *15*, 2779. https://doi.org/10.3390/w15152779.
8. Ding, W.; Wang, G.; Yang, Q.; Xu, Y.; Gao, Y.; Chen, X.; Xu, S.; Han, L.; Yang, X. Risk Assessment and Control of Geological Hazards in Towns of Complex Mountainous Areas Based on Remote Sensing and Geological Survey. *Water* **2023**, *15*, 3170. https://doi.org/10.3390/w15183170.
9. Duan, R.; Chang, L.; Gu, X.; Li, X.; You, X.; Zhang, Q.; Wang, Q. The Hydrogeochemical Processes of Groundwater in the Bieletan Area, the Western Potash Production Region in China. *Water* **2024**, *16*, 1833. https://doi.org/10.3390/w16131833.
10. Wang, X.; Sun, K.; Ma, W.; Peng, J.; Liu, R.; Chen, J.; Zhang, K.; Gao, S.; Li, C.; Zhang, P. Hydrochemical Characteristics and Water Quality Evaluation of Groundwater in the Luohe Formation of Binchang Mining Area, China. *Water* **2024**, *16*, 1913. https://doi.org/10.3390/w16131913.
11. Sanz-Ramos, M.; Bladé, E.; Sánchez-Juny, M.; Dysarz, T. Extension of Iber for Simulating Non–Newtonian Shallow Flows: Mine-Tailings Spill Propagation Modelling. *Water* **2024**, *16*, 2039. https://doi.org/10.3390/w16142039.
12. Huang, Y.; Liu, S.; Wang, Q.; Huang, G.; Zhang, X.; Liu, Y. Dechlorination of Hexachlorobenzene by Ni/Fe Bimetallic Nanoparticles and the Influence of Co-Existing Heavy Metal Ions. *Water* **2024**, *16*, 2855. https://doi.org/10.3390/w16192855.

References

1. Department of Geological Environment. *Geological Environment Management Manual*; Ministry of Land and Resources of the People's Republic of China, China Dadi Press: Beijing, China, 2002; pp. 37–38.
2. Xu, Y. Mine environmental geology and mine geological environment. *Northwest. Geol.* **2005**, *38*, 108–113.

3. Huang, Y.; Lei, H.; Na, J.; Yuan, Y.; Tian, H. Investigations of the impact of geothermal water reinjection on water-rock interaction through laboratory experiments and numerical simulations. *Appl. Geochem.* **2024**, *175*, 106180. [CrossRef]
4. Wang, M.; Zhang, H.; Liang, L.; Zhu, Z.; Zhang, A. Constraining the properties of the heat sources of high-temperaturehydrogeo thermal systems: Evidence from the lithium concentrations ofgeothermal waters. *J. Hydrol.* **2021**, *610*, 131696. [CrossRef]
5. Ruppen, D.; Runnalls, J.; Odermatt, D. Optical remote sensing of large-scale water pollution in Angola and DR Congo caused by the Catoca mine tailings spill. *Int. J. Appl. Earth Obs. Geoinf.* **2023**, *118*, 103237. [CrossRef]
6. Bhuiyan, M.A.H.; Islamc, M.A.; Dampare, S.B.; Parvez, L.; Suzuki, S. Evaluation of hazardous metal pollution in irrigation and drinking water systemsin the vicinity of a coal mine area of northwestern Bangladesh. *J. Hazard. Mater.* **2010**, *179*, 1065–1077. [CrossRef] [PubMed]
7. Al Heib, M.; Franck, C. A methodology for multi-hazard interaction assessment of abandoned mines. *J. Ind. Saf.* **2024**, 100018. [CrossRef]
8. Xiong, Y.; Kong, D.; Song, G. Research hotspots and development trends of green coal mining: Exploring the path to sustainable development of coal mines. *Resour. Policy* **2024**, *92*, 105039. [CrossRef]

water

MDPI

Article

In-Situ Leaching Mining Technique for Deep Bauxite Extraction and the Countermeasures for Water Pollution Prevention: An Example in the Ordos Basin, China

Zhizhong Li [1,2], Yi Zhang [3,4,*], Tengyue Luo [4,*], Peng Xia [5,6], Huayi Mu [1], Pingping Sun [7], Xin Wang [1] and Jianhua Wang [8]

1 Xi'an Center of Geological Survey, China Geological Survey, Xi'an 710054, China
2 Huawei Mineral Exploration Technologies Co., Ltd., Xi'an 710076, China
3 College of Earth Sciences, Jilin University, Changchun 130012, China
4 Research Institute of Shaanxi Yachang Petroleum (Group) Company Ltd., Xi'an 710075, China
5 China Geological Survey, Beijing 100037, China
6 Faculty of Engineering, China University of Geosciences, Wuhan 430074, China
7 School of Human Settlement and Civil Engineering, Xi'an Jiaotong University, Xi'an 710054, China
8 Aerospace Information Research Institute, Chinese Academy of Sciences, Beijing 100101, China
* Correspondence: zhangyi.allright@163.com (Y.Z.); luotengyue@sxycpc.com (T.L.)

Abstract: As the second most significant metal following steel, aluminum plays a vital role in the advancement of both strategic emerging industries and national economic development. The existing oil and gas drilling data indicate that the deep bauxite deposits is abundant around the Ordos Basin in China, at the depths ranging from several hundred meters to several kilometers. Based on the geological and hydrogeological characteristics analysis, it is found that deep bauxite deposits in the basin have distinct electrical characteristics, characterized by *four highs and two lows*. While there is scarcity of prior research on the exploration topic for the technique limitation. In this paper, a logging interpretation model has been developed, which allows the evaluation of bauxite deposits. An efficient technology was proposed for deep bauxite exploration, utilizing an in-situ leaching mining technique. This technology is well-suited to the geological conditions of the Ordos Basin, ensuring that the solution flows within the bauxite ore bed without any seepage loss. To prevent the leaching solution from seeping into and polluting the main aquifer around the basin, several measures have been proposed. These include filling with polymer resin, well pattern seepage control plugging, and establishing monitoring systems. The results of this study provide a theoretical basis for the adoption of environmentally sustainable mining techniques and the mitigation of water pollution in deep bauxite exploration.

Keywords: bauxite; occurrence state; microstructure; nuclear magnetic resonance; water pollution prevention and control; in-situ leaching technique

Citation: Li, Z.; Zhang, Y.; Luo, T.; Xia, P.; Mu, H.; Sun, P.; Wang, X.; Wang, J. In-Situ Leaching Mining Technique for Deep Bauxite Extraction and the Countermeasures for Water Pollution Prevention: An Example in the Ordos Basin, China. *Water* **2023**, *15*, 2381. https://doi.org/10.3390/w15132381

Academic Editor: Catherine N. Mulligan

Received: 14 May 2023
Revised: 13 June 2023
Accepted: 16 June 2023
Published: 28 June 2023

1. Introduction

Bauxite, the primary source of metallic aluminum, is composed of aluminum-bearing minerals, iron-bearing minerals, and small amounts of silicates, titanates, sulfates, and carbonates [1–3]. In-situ leaching mining is a novel mining method. It utilizes the physical and chemical properties of minerals to selectively dissolve, leach, and recover useful mineral components by injecting water, chemical solvents, or microorganisms into the deposit or heap [4,5]. This technology is commonly employed in the mining of salt, uranium, and rare earth ores. For instance, water-soluble mining of salt mines can be traced back over 1400 years. It remains in use today due to its efficiency and cost-effectiveness. In 1961, Soviet hydrogeological engineers designed an in-situ pumping test for a sandstone uranium mine that achieved a 77% uranium recovery rate by 1978. Similarly, the United

States conducted a series of successful in-situ leach uranium mining experiments during this period. It leads to the mainstream adoption of this process in the U.S. uranium mining industry by 1992 [6]. While attempts at in-situ leaching mining of other non-ferrous metals were made as early as the 16th century in Spain, large-scale application was not feasible due to industrial limitations at the time. In 1995 a project study of in-situ copper mining was conducted in Santa Cruz, Arizona, USA. Copper ore below 370 m was successfully extracted from the ground through field trials with acid dissolution. The project delivered 25 gallons of saturated solution per minute and yielded 1 ton of copper output per day [7], demonstrating the commercial viability of this technology. Recently, Kongar (2023) describes the possibility of leaching of metals from geo-raw materials. The sequentially-selective extraction of metals from raw materials by changing the alkalinity (pH factor) of the agent is proposed, which significantly increases the extraction of the valuable component, reduces the agent consumption and increases the economic efficiency of the leaching method used [8]. Researchers studied the leaching process and achieved the maximum extraction of the valuable component from the geo-raw materials [9]. The studies on the extraction of the valuable component from the geo-raw material by leaching were carried out with different activation times and irrigation of the working solution with hydrochloric acid of differential concentration.

Throughout the history of in-situ leaching technology, its most mature and widely used applications is in the mining of uranium and rare earth elements [10,11]. The use of this technology for the extraction of other metal minerals has been less prevalent. In recent years, gas has been obtained from the bauxite layer during oil and gas exploration, challenging previous assumptions that the dense bauxite layer could only serve as a regional cap layer [12]. Furthermore, it has been confirmed that bauxite is an effective reservoir due to the development of microfractures and dissolved pores, which create a well-connected pore network. Motivated by the results of previous research, this study introduces the conception of the in-situ leaching mining technique for extracting deep bauxite deposits. The approach is grounded in a comprehensive analysis of the geological attributes and petrophysical properties of bauxite deposits located within the Ordos Basin.

The Ordos Basin is one of China's most significant energy reserves, boasting abundant resources such as oil, gas, coal, bauxite, uranium, rock salt, oil shale, cement limestone, mirabilite, and gypsum. The proven reserves of natural gas, coalbed methane, coal, uranium, and rock salt within the basin rank first in China, while its proven petroleum geological reserves rank fourth. The basin contains rich bauxite resources with proven reserves of 770 million tons. These resources are primarily distributed across Shaanxi, Shanxi, Inner Mongolia and Gansu provinces [13,14]. Numerous scholars have investigated the distribution, genesis, classification, and geological characteristics of typical bauxite minerals in the periphery of the Ordos Basin [15–17]. However, research on deep bauxite deposits within the basin remains limited. Oil and gas drilling data indicate that the basin contains rich bauxite resources at depths ranging from several hundred meters to several kilometers. The exploration and exploitation of deep bauxite could greatly alleviate the national shortage of bauxite resources.

By 2015, six in-situ leach uranium mines were constructed in China, primarily utilizing acid leaching methods. In an effort to reduce the pollution of shallow and groundwater, new types of neutral leaching and microbial leaching have been gradually implemented in recent years [18]. Researchers focuse on minimizing the impact of productive brines on the environment. The authors investigated the penetration of these brines through the fracture system into the mass, established the radius of penetration and proposed a way to isolate the sources of contamination [19]. There are also research formulates a multifactorial mathematical model of degradation of environmental ecosystems as a result of the impact of waste [20]. Considering the geological profile of the Ordos Basin, this paper examines the implementation of a barrier layer to inhibit leachate infiltration into the aquifer. This is achieved through the utilization of well network seepage control and the injection of polymer resin. Furthermore, the establishment of a dynamic monitoring system at the mining site is proposed, to continuously track and assess groundwater conditions, thereby preventing and mitigating water pollution. The research aims to provide a theoretical foundation for the exploration and environmentally sustainable extraction of deep bauxite deposits within the Ordos Basin.

2. Characteristics of Deep Bauxite in the Ordos Basin

2.1. The Geological Characteristics

2.1.1. Formation of Orebearing Rock Series

Bauxite within the Ordos Basin is found in aluminiferous rock series in the middle and lower portions of the late Carboniferous Benxi Formation. The Benxi Formation consists of early Late Paleozoic sediments deposited on a basement geomorphology shaped by Caledonian Ordovician erosion and weathering. Sedimentation primarily occurred through filling and supplementing low and concave paleogeomorphology on the weathering surface. Sedimentary thickness, which generally ranges from 10 to 80 m, is mainly controlled by paleogeomorphology and tends to be thicker in the east and thinner in the west.

The rock types of the Benxi Formation are mainly bauxite of weathering products, clastic rock of littoral and shallow Marine facies, tidal flat limestone of littoral marsh facies, coal rock and carbonaceous mudstone, which are characterized with ferro-aluminite assemblages containing pyrite and siderite nodules or bands. Vertically, the Benxi Formation can be divided into Member 1 and Member 2 (Figure 1). The lithology of the upper part of Member 1 is mainly dark gray micritic limestone and coal seam, mixed with gray coarse sandstone and pebbled coarse sandstone. The natural gamma is low and massive, while the deep and shallow lateral resistivity is high and massive. The lower part of the Member 1 is mainly dark gray mudstone, interbedded with thin gray sandstone and limestone. The natural gamma is medium-high, micro-dentate and local finger. The lithology of the upper part of the Member 2 is gray, light gray medium-coarse sandstone and dark gray mudstone, partially interbedded with thin micritic limestone. The natural gamma is medium-high value, dentate, partially massive, and the resistivity is mainly low value. The lithology of the middle part of the Member 2 is dark gray mudstone, light gray conglomerate and pebbly coarse sandstone. The natural gamma is dominated by high value, weakly dentate, locally low value and massive, and the resistivity is dominated by medium and low value. The lower part of the Member 2 is high gamma-aluminum mudstone, whose thickness is generally 3~6 m, and the lower part is Majiagou Formation limestone or dolomite.

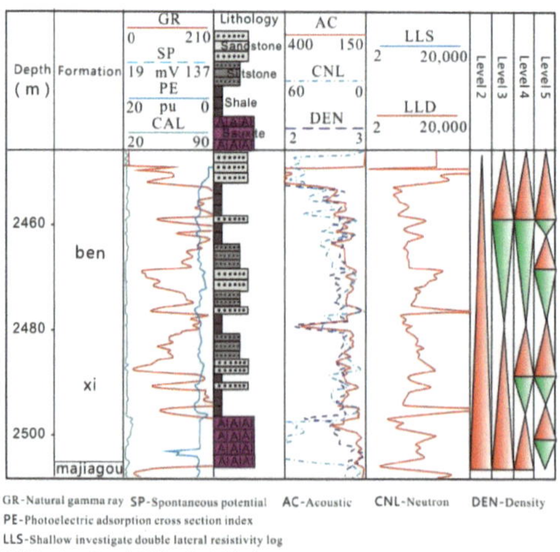

Figure 1. Sedimentary characteristics of Benxi Formation.

2.1.2. Structure and Thickness Variation of Bauxite Layer

On the paleoerosion surface of Ordovician carbonate rocks, the vertical aluminiferous rock series exhibits a sandwich structure. Specifically, bauxite is often found between top and bottom layers of clay rock or shale. The aluminiferous rock system developed on an unconformity surface where the basement rock is typically carbonate. The bottom layer consists of aluminum iron bauxite and bauxitic mudstone with poor physical properties and a thickness of 2–4 m. The middle layer consists of bauxite and porous bauxite with better physical properties and an average thickness of approximately 6.3 m. The upper layer consists of black carbonaceous mudstone with poor physical properties and a thickness of 1–2 m (Figure 2).

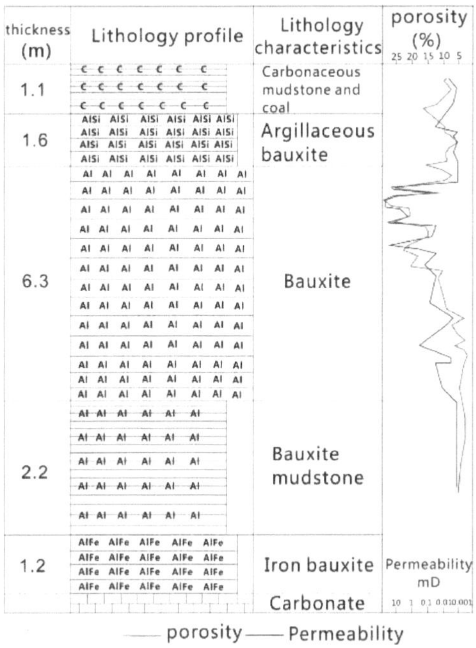

Figure 2. Sedimentary Sequence and Vertical Physical Properties of Bauxite in Benxi Formation.

The distribution of the bauxite is controlled by karst palaeogeomorphology. The palaeoterrain is flat, the orebodies are layered or quasi-layered, and the palaeoterrain is uneven and mostly lens-shaped. On the plane, the bauxite is mainly deposited in low and concave paleogeomorphic areas, with a banded or lamellar distribution (Figure 3). In the same strip or area, bauxite beds are stably distributed with good continuity (Figure 4). The average thickness of the ore beds ranges from 1.5 m to 3 m, which is positively correlated with the thickness of ore-bearing rock series, with a correlation coefficient of 0.38–0.76.

2.1.3. Electrical Characteristics of Bauxite

Bauxite has obvious electrical characteristics [21], which has been used as an important marker layer at the bottom of Benxi Formation in oil and gas exploration, with unusually high natural gamma (GR), generally above 300 API. The electrical resistivity (LLD, LLM) is generally about 35 Ω/m, with a maximum value of 100 Ω/m. The compensated neutron (CNL) porosity is high, usually about 60%. The density (RHOB) is high, generally around 2.74 g/cm^3, and the acoustic time difference (AC) is low, generally around 170 $\mu s/m$ (Figure 5).

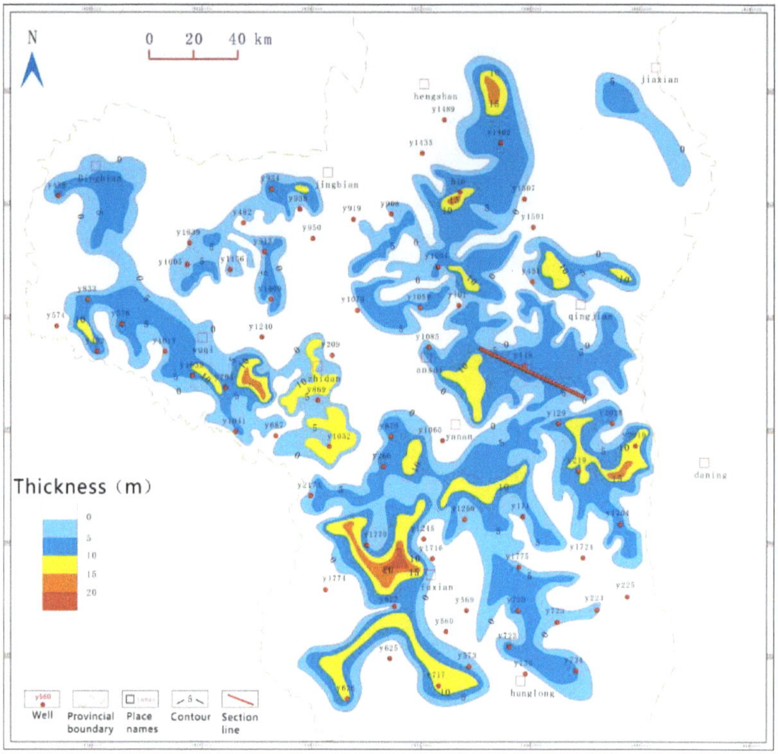

Figure 3. Distribution of bauxite in southern Ordos Basin.

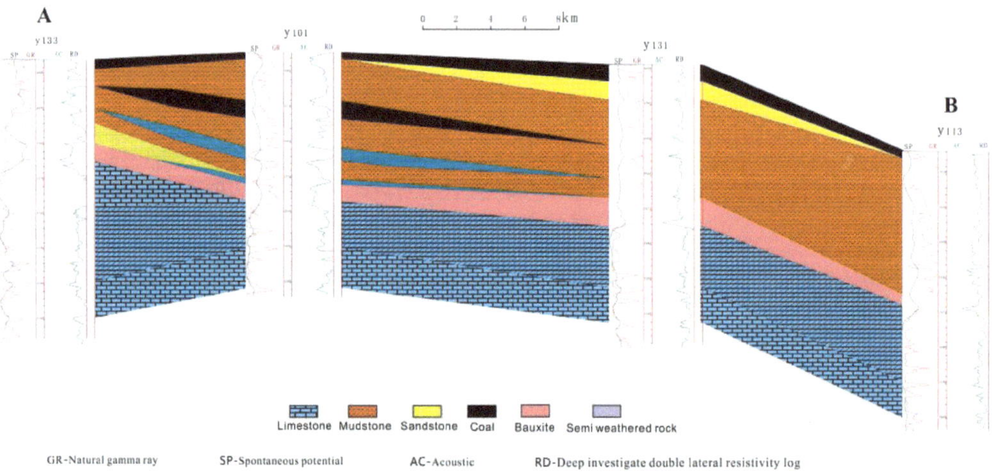

Figure 4. Lithological Structure of Top and Bottom of Bauxite Deposits.

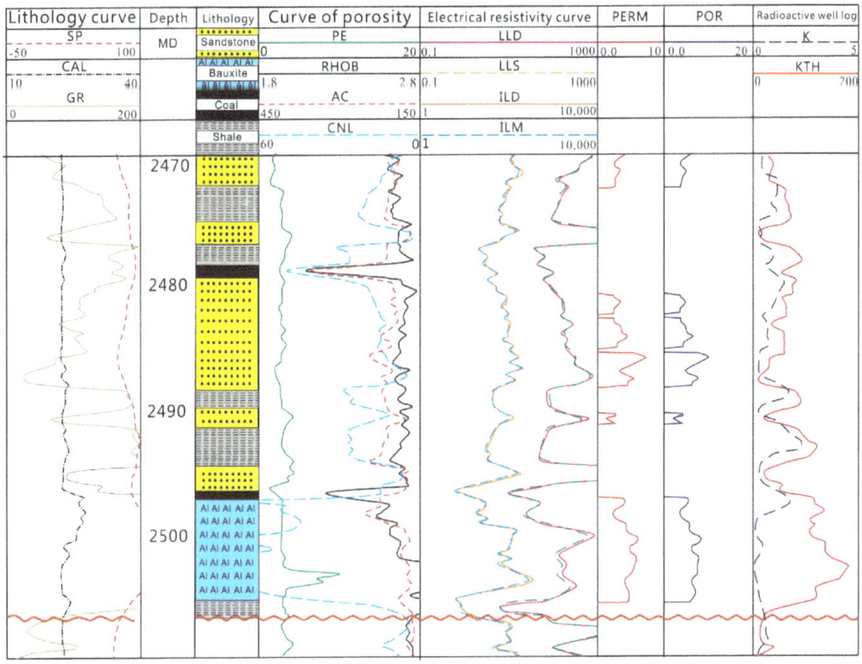

GR-Natural gamma ray SP-Spontaneous potential AC-Acoustic CNL-Neutron DEN-Density K-potassiurm

KTH-gamma ray without uranium PE-Photoelectric adsorption cross section index POR-porosity

PERM-permeability MD-Measuring depth LLS-Shallow investigate double lateral resistivity log

LLD-Deep investigate double lateral resistivity log

Figure 5. Logging response characteristics of bauxite (taking Yan x well as an example).

Therefore, with the help of core calibration logging, conventional logging and special logging, bauxite identification standard and bauxite logging parameter model can be established, and bauxite evaluation indexes such as diaspore content and aluminum-silicon ratio can be calculated, which can complete the evaluation of the bauxite in great buried depth.

2.2. Hydrogeological Characteristics of Basin and Water Bearing of Bauxite Layer

2.2.1. Hydrogeological Characteristics of the Basin

Ordos Basin is a giant groundwater basin with multiple aquifer systems with different characteristics that are overlaid or linked laterally, cut and connected with each other in different degrees in space [22]. These aquifer systems are composed of a variety of different types of rocks. From bottom to top, they are: Cambrian-Ordovician carbonate karst aquifer system, Carboniferous—Jurassic-Cretaceous sandstone fractured aquifer system, Quaternary loess pore aquifer system and Cenozoic loose rock pore aquifer system (Figure 6).

The groundwater in the Cambrian and Ordovician carbonate strata belongs to the karst-fractured water type (karst water for short), and the Cambrian crystalline rock at the bottom and the carboniferous aluminite shale at the top are the regional water barrier layer. The carbonate rocks in the middle of the basin are buried at a depth of 4000 m. The research shows that the karst water in the carbonate rocks cannot penetrate the basin for deep circulation, and the groundwater stagflation sealing zone is formed in the middle of the basin. The circulation of the modern karst water only occurs in the karst body at a certain depth around the basin (generally buried within 800~1200 m), which is the main aquifer with large water volume and good water quality. It is mainly controlled by climate, topography, lithology and lithofacies, structure, buried depth and karst development.

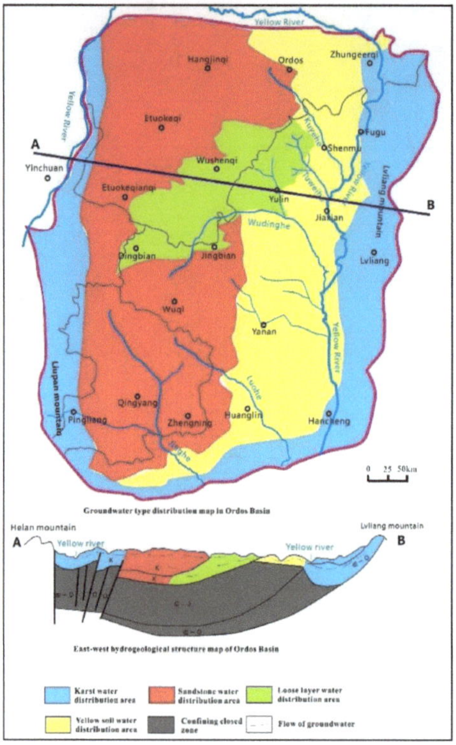

Figure 6. Types and distribution characteristics of groundwater in Ordos Basin.

The main aquifers of loose rocks include Quaternary aeolian and alluvial sand layers, alluvial (flood) sand and gravel layers, and part of loess. The pores of these rock strata play a leading role in water storage and water conduction. The water-bearing medium is relatively uniform, and the occurrence of groundwater is relatively uniform. In the shallow part of the basin, an incomplete continuous plateau pore aquifer system with obvious difference between north and south is formed. Generally speaking, the water quantity is large, and the water quality is good. The basin has a certain mining value, and often become an important water supply source for the severe water shortage areas.

2.2.2. The Bauxite Layer Is Water-Bearing

It is believed that Carboniferous aluminite shale is a regional water-proof layer. The permeability of pure bauxite is (6.50~14.50) $\times$ 10^{-8} μm^2, the permeability of aluminite mudstone is (0.83~6.40) $\times$ 10^{-8} μm^2, and the breakthrough pressure is greater than 5 MPa. The aluminite shale has high expansibility, and it is often associated with argillite. The diagenetic process is not easy to produce fractures, which is an ideal water barrier for gas reservoirs. However, in 2015, Changqing oilfield drilled 9.3 m into the bauxite rock layer of Benxi Formation in Well Shaan 464 in the eastern Ordos Basin. The low-yielding gas flow of 1849 m^3 per day was obtained through gas test, and the bauxite gas reservoir of Benxi Formation was discovered for the first time. In the same year, well D66-172 in the bauxite section of Benxi Formation obtained daily gas of 1.0375 $\times$ 10^4 m^3 and daily water of 2.9 m^3. In 2020, Changqing Oilfield conducted a large number of gas tests in Longdong area. Most of the gas wells produced gas and water, with the highest daily water production of 33.6 m^3 and the average daily water production of 18.6 m^3, proving that the local Carboniferous Benxi Formation bauxite has a certain porosity and permeability, as well as a certain water content. The analysis shows that the formation water of Carboniferous Benxi Formation bauxite is mainly confined and retained during the deposition period.

2.3. *Microstructure of Bauxite*

To gain a comprehensive understanding of the microstructure characteristics of Bauxite, this study focuses on outcrop Bauxite located in the southern region of the Ordos Basin. Utilizing a range of techniques, including casting thin section, X-ray diffraction, scanning electron microscopy, conventional physical properties, and constant pressure mercury injection, the mineral composition and pore structure of the bauxite ore bed were systematically analyzed. The instruments employed in this research include the D8 Focus X-ray diffractometer from Bruker Company (Hanau, Germany), the CMS300 pore permeability tester from Corlab Company (Denver, CO, USA), the 9520 II constant pressure mercury intrusion meter, the QuantaFEG 450 field emission scanning electron microscope from FEI Company (Alhambra, CA, USA), and the 4500P polarization microscope from Leica Company (Wetzlar, Germany).

2.3.1. Mineral Composition and Occurrence State

X-ray diffraction analysis showed that the mineral composition of bauxite ore was mainly diaspore or boehmite, with a content of 40–77%. The secondary clay minerals are mainly kaolinite, with a content of 3–37%. Trace minerals include quartz, feldspar, hematite, anatase, goethite, etc., with a content of 1–11%. Some of the pores are filled with calcite and (iron) dolomite, with a content of 1–6% (Figure 7). Diaspore presents plate-like, needle-like and columnar crystals, and flake and scale forms can also be seen. The flake and scale forms of diaspore are filled between beans and oolitic grains with interstitium. Kaolin is distributed in the core of bean oolitic grains in fine sheets and fibers or mixed with diaspore to form the belt of bean grains and oolitic grains. Kaolin is also used as interstitial material between grains. Calcite and (iron) dolomite are mostly used as interstitial material, occasionally with ingrain cores or ring structures (Figure 8).

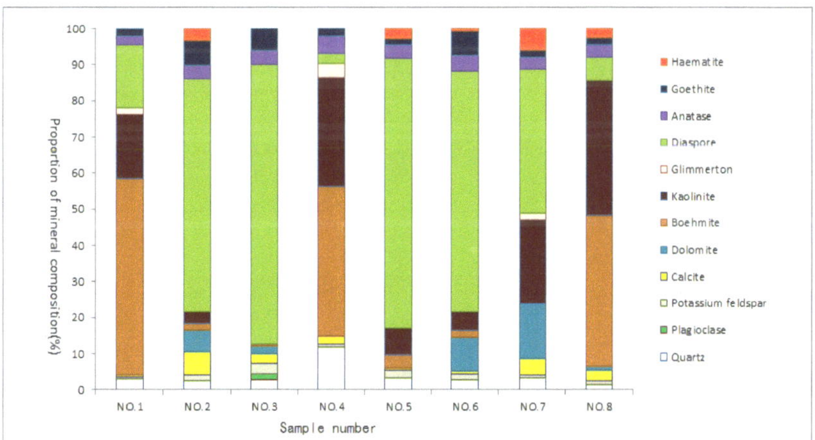

Figure 7. X-ray diffraction mineral composition map of a bauxite deposit in Ordos Basin.

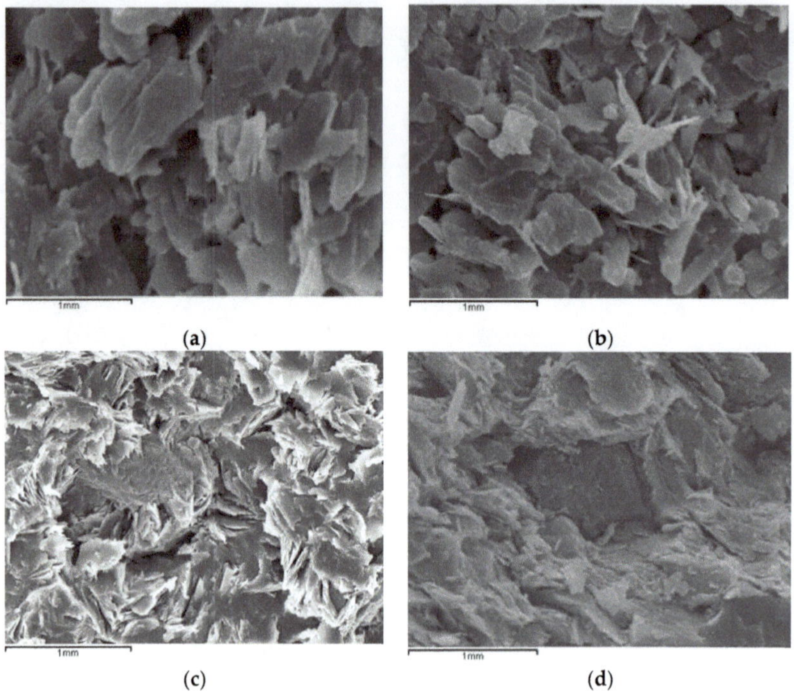

Figure 8. Microstructure of bauxite under scanning electron microscope. (**a**) Compact tabular diaspore. (**b**) Diaspore, anatase. (**c**) Scaly diaspore. (**d**) Fine sheets of kaolin.

The petrochemical components of the bauxite are mainly Al_2O_3, SiO_2, Fe_2O_3, TiO_2, and a small amount of CaO, MgO, K_2O, Na_2O, etc. (Figure 9).

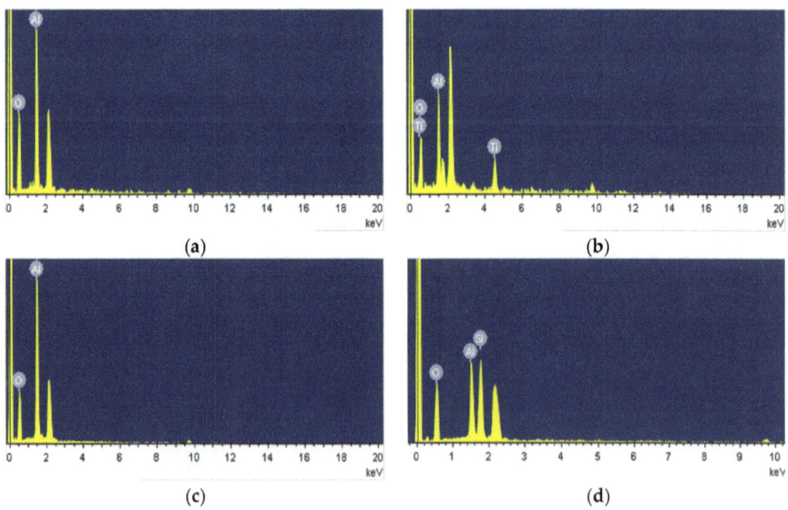

Figure 9. EDS analysis of bauxite. (**a**) EDS analysis of compact tabular diaspore. (**b**) EDS analysis of diaspore and anatase. (**c**) EDS analysis of scaly diaspore. (**d**) EDS analysis of fine sheet kaolin.

2.3.2. Physical Property and Pore Structure Characteristics of Ore Body

According to the test data of scanning electron microscope and pore permeability analysis of bauxite, the bauxite develops pores and fractures, and the reservoir has showed good physical properties. The porosity ranges from 1.3 to 6.1%, and the permeability varies from 0.042 to 5.25 mD.

Scanning electron microscopy showed that the bauxite and aluminum-soil mudstone were mainly characterized by intercrystalline pores of diaspore (maximum pore size up to 1.2 μm), intergranular and intrgranular dissolution pores, and microfractures (Figure 10).

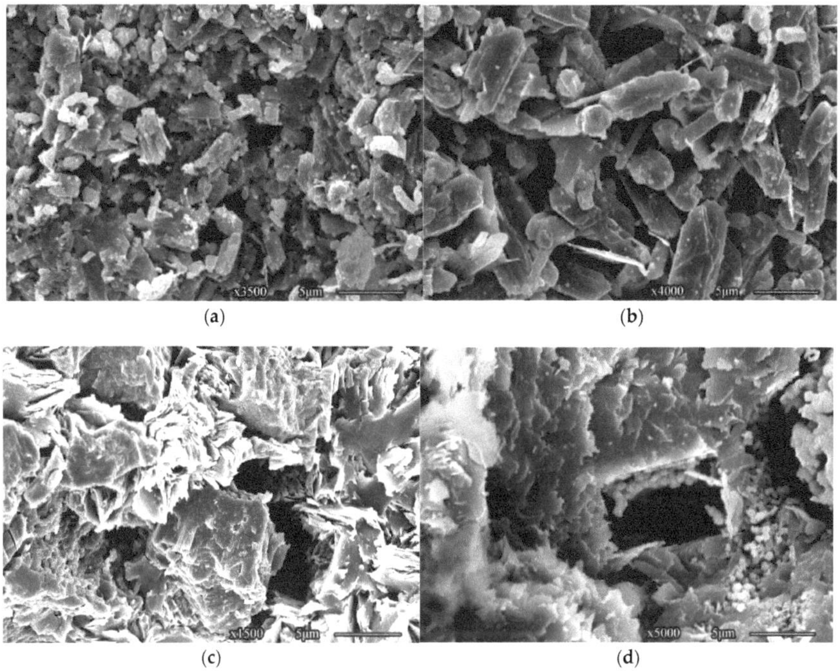

Figure 10. Characteristics of micro pore throat structure of bauxite. (**a**) Closely packed bauxite, semi-autogenous -autogenous pseudohexagonal sheet, bauxite intercrystalline pores 3500×. (**b**) Bauxite, semi-autogenous—autogenous false hexagonal sheet, bauxite intercrystalline pores 4000×. (**c**) Bauxite, mostly semi-self-shaped false hexagonal sheet or other shape, poor crystallization, bauxite intercrystal pores micro-cracks 1500×. (**d**) Amorphous silicon and aluminum colloidal particles, irregular, granular pores and intergranular pores 5000×.

The constant mercury injection into the high permeability reservoir section indicates a good pore connectivity, a drainage pressure between 0.21~0.42 MPa, and the maximum connected pore-throat radius between 1.81~5.22 μm. The median pressure is between 5.90 and 6.11 MPa, and the average throat radius is generally between 0.09 and 0.67 μm, which is a common characteristic of dissolved pore type reservoir.

2.4. Evaluation of the Deep Bauxite

In oil and gas exploration, well logging shows that the bauxite is developed at the bottom of Benxi Formation, and the buried depth is from several hundred meters to several thousand meters. Previously, this set of bauxite has been used as the marker of Benxi Formation and the underlying Majiagou Formation in stratigraphic correlation, and has been regarded as the regional cap of weathering crust gas reservoir in oil and gas explorations.

As a marker, the bauxite layer has obvious electrical characteristics, which are characterized by "four highs and two lows", namely high gamma, high thorium, high uranium content, and high compensating neutrons, and low resistance and low acoustic time difference. The GR is abnormally high. Due to the finer particles and larger specific surface of bauxite, it is easy to adsorb radioactive substances such as uranium, thorium and titanium. The natural gamma is positively correlated with the content of diaspore. When the bauxite content exceeds 75%, the GR ray content is greater than 450.0 API, therefore, natural NG mark is the most key of logging parameters for identification of bauxite. Dual lateral resistivity (LLD, LLM) is 35 Ω/m or so commonly, a maximum of 100 Ω/m. The resistivity corresponding to the high gamma interval generally decreases, and the positive difference between the log curves of the double lateral resistivity is obvious. Due to the high gamma section of high contents of bauxite, bauxite is crisp, easy to form micro fractures and solution pores, resulting in the decrease of resistivity [23].

The parameter characterization of bauxite layer is the basis of bauxite identification, evaluation and reserve calculation. On the basis of X-ray diffraction analysis of whole rock minerals, ore physical property measurement, scanning electron microscopy and other data, a parameter interpretation model of bauxite layer is established based on core scale. According to the mineral types of the classical rock volume model and the mineral analysis results of the whole rock by X-ray diffraction, the content of diaspore was determined by natural gamma ray and density fitting. The porosity model was established by intersection of core analysis porosity and acoustic time difference. According to the correlation between porosity and permeability, the permeability of bauxite layer was obtained by using porosity [24]. Finally, deep bauxite logging evaluation can be completed, and bauxite identification, evaluation and reserve estimation can be realized.

Special logs such as ECS, NMR, dipole acoustic and acoustic imaging can also be used to evaluate bauxite deposits. ECS element logging can obtain the content of formation elements and rock mineral composition, which can satisfy the evaluation of various properties of the formation, obtain the physical parameters of the formation, and calculate the content of clay minerals. Dipole acoustic logging can provide the time difference of longitudinal and shear waves, and can perform anisotropy analysis and processing to determine the direction of the maximum horizontal formation stress, calculate the maximum and minimum formation stress horizontally, and obtain the Poisson's ratio, Young's modulus, and shear modulus of the rock. All these important rock mechanical parameters are used to meet the requirements of establishing the calculation model of rock and guide the fracturing transformation. Acoustic and electric imaging logging has the characteristics as high resolution, high borehole coverage and good visibility, and it is of great significance to identify fracture types to guide reconstruction and evaluate development effects [25].

Based on the above results, the bauxite resources in the south of Ordos Basin are evaluated. The bauxite is widely distributed in the Ordos Basin, which has a good correlation with the Ordovician karst paleogeomorphology. The bauxite is controlled by the Ordovician microgeomorphology, distributed in strips and lenses, and has great exploration potential.

3. Conseptation of the In-Situ Leaching Mining Technique and the Applicable Conditions

3.1. Conception of In-Situ Mining of Deep Bauxite

In-situ leaching mining is the best way to effectively exploit the deep bauxite resources in the basin. The sedimentary stability, roof and floor conditions, orebody permeability and other aspects of bauxite in the Ordos Basin are well matched with in-situ leaching mining.

Bauxite is a sedimentary rock with certain porosity and permeability, and presents a sandwich structure, which is often surrounded between the top and bottom clay rocks. It has the geological conditions of in-situ leaching mining, especially according to the stress sensitivity tests, new pores or cracks will appear with the increase of stress in bauxite, which can improve the permeability of the ore layer. Based on the in-situ leaching mining technology of uranium ore and rare earth ore, the in-situ leaching mining of bauxite can be considered.

In-situ leaching mining uses a liquid injection well to inject specific leaching solution into the target orebody, leaching useful metals through a series of physical and chemical reactions, and pumping the leaching solution out of the surface through the pumping well to achieve the process of extraction of useful metals in the surface factory [26]. As shown in Figure 11, the main implementation steps of in-situ leaching mining of bauxite include:

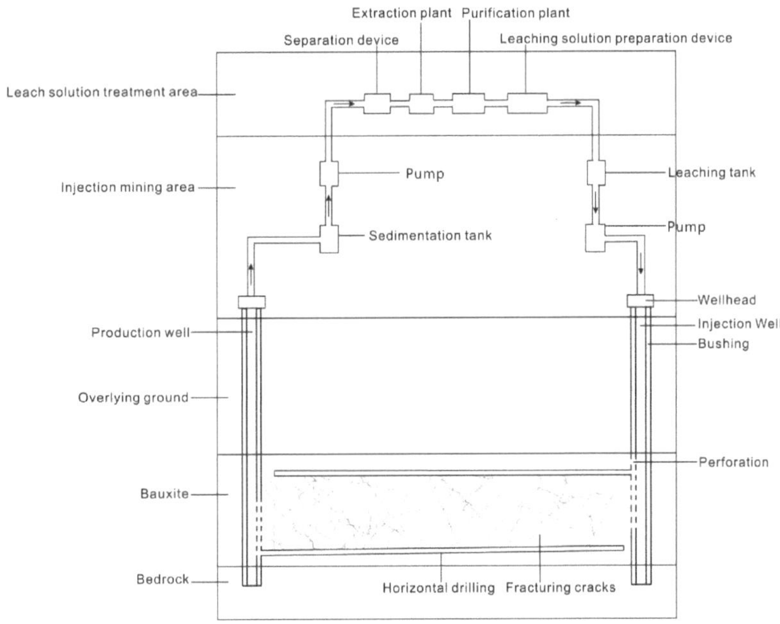

Figure 11. In-situ leaching of bauxite mining concept map.

(1) Drilling: cluster well are adopted, with the inverted nine-point method, the inverted seven-point method or the inverted five-point method to distribute the injection and production well pattern, including the liquid injection well and the drainage well. The liquid injection well is located in the middle position, and the drainage well is located in the surrounding position. The liquid injection well is a vertical well, and the drainage well is a directional well (as shown in Figure 12). The well was drilled through the overlying bauxite to 20 m of the underlying bauxite bedrock and completed with casing.

(2) Reconstruction of the ore layer: The perforation section is set at the top of the bauxite bed in the liquid injection well, and the drainage well is set at the bottom of the bauxite bed. The horizontal borehole is drilled on the top of the bauxite bed with the liquid injection well as the starting point and the drainage well as the ending point. The horizontal borehole extends horizontally from the liquid injection well to the drainage well parallel to the roof of the ore bed, where the first horizontal borehole is not connected with the drainage well. Then, a horizontal borehole is drilled at the bottom of the bauxite layer with the drainage well as the starting end and the liquid injection well as the end point. The inclination angle of the horizontal borehole from the drainage well to the liquid injection well is upward inclined. The horizontal drilling hole is not connected to the liquid injection well. The liquid injection well is fractured or blasted by hydraulic power to increase the permeability to form fracturing cracks in the bauxite layer and establish a seepage channel.

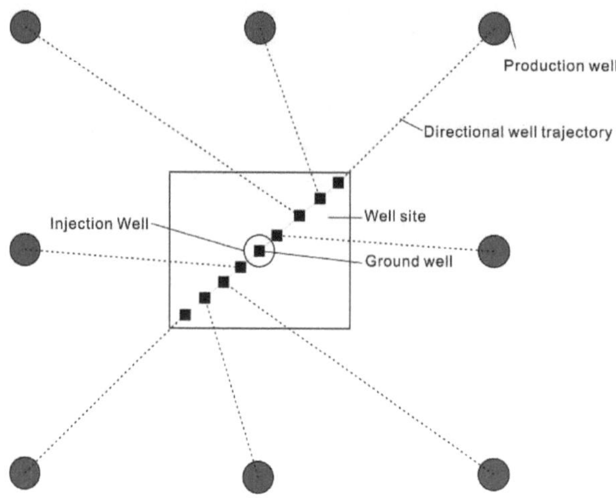

Figure 12. In situ leaching of bauxite mining well pattern diagram.

(3) Leaching operation: leaching solution is injected into the liquid injection well, and it enters the bauxite layer through the perforation section of the liquid injection well and the horizontal borehole at the top. The bauxite is dissolved through the downward seepage of the fracturing fracture, and then the dissolving liquid containing bauxite is extracted to the ground through the seepage of the horizontal borehole at the bottom and the perforation section of the drainage well.
(4) Ground operation: After settling the leaching solution containing bauxite, the leaching solution containing bauxite can be separated and extracted, and then reused after purification treatment.

3.2. Conditions for In-Situ Leaching of Deep Bauxite Deposits

In-situ leaching mining is a very special mining method, and its application conditions are very strict. Not all deposits can be mined by in-situ leaching, and those that can be mined may not be technically and economically feasible. Only bauxite with certain geological-hydrogeological conditions has the possibility of in-situ leaching technology for economic exploitation. The investigations show that the feasibility evaluation of in-situ leaching is a very important and difficult subject. Therefore, it is very important to study the selection of in-situ leaching mining for bauxite. The bauxite selection that can be in-situ leaching mining needs to solve the following problems: (1) whether it is permeable to ensure that the leaching solution can flow in the ore layer; (2) whether it is soluble, which can ensure the leaching rate, the content of beneficial minerals in the leaching solution and the consumption of leaching agent; (3) whether there is a closed top and bottom conditions, can ensure that the leaching solution does not leak, does not pollute the formation water, to ensure the recovery rate of leaching solution. On the basis of considering the above problems, the conditions of bauxite deposits suitable for in-situ leaching mining were sorted out (Table 1).

Table 1. Deposit conditions of in-situ leaching bauxite.

Category of Conditions	Parameter
Condition of geology	Lithology and lithofacies characteristics Degree of cementation and diagenesis of ore The mineral composition of an ore and its percentage Thickness of roof and floor Continuity of roof and floor Grade of ore Distribution characteristics of ore beds Thickness of ore bed The depth of the mine
Hydrogeological conditions	Reservoir permeability (experimental data) Reservoir effective porosity Roof and floor permeability Reservoir permeability (data from pumping tests) Top bottom aquifer
Ore leaching process conditions	Beneficially metal leaching rate of ore sample The leaching solution has beneficial mineral content Unit metal leach agent consumption Time required for leaching Cleaning process test ore sample to cleaning water consumption

4. Countermeasures for Water Pollution Control In Situ Mining of Deep Bauxite

Compared with traditional mining and open-pit mining, the in-situ leaching mining has achieved green and pollution-free mining to a large extent, but the complex and changeable underground ore bodies, surrounding rock, roof and floor, permeability and solubility of ore bodies are difficult to predict, and it is easy to form leaching agent leakage, which will not only reduce the efficiency of in-situ leaching mining, but also cause water pollution [27]. To prevent and control this potential problem, the seepage control strategy of deep orebody leaching solution is put forward, including polymer resin filling and well pattern seepage control.

4.1. Seal with Polymer Resin

Injection of polymer resin plugging refers to the injection of polymer resin insoluble in leaching agent into the top and bottom of the mineral layer before the injection of leaching agent, forming a polymer material barrier layer to prevent leaching solution through the top and bottom into the aquifer (Figure 13). According to the hydrogeology of Ordos Basin, the Carboniferous Benxi Formation bauxite layer is covered by Ordovician carbonate rocks, and its groundwater belongs to karst water, which is the main aquifer in the periphery of the basin. Therefore, the injection of polymer resin between bauxite layer and carbonate rock can effectively form plugging and prevent the pollution of karst water by leaching agent. Leakage points are identified through a combination of drilling, logging, geophysical exploration, and tracer water injection testing. Subsequently, polymer resin is injected into the borehole to seal these points and prevent the leakage of leaching solution into the aquifer.

4.2. Well Pattern Seepage Control and Plugging

Well pattern seepage control plugging refers to controlling the flow direction of the underground leaching solution by setting the well pattern to prevent the leaching agent from entering the aquifer (Figure 14). Radial horizontal holes are drilled at the top of the ore layer to inject the leaching solution, and radial horizontal holes are drilled at the bottom of the ore layer to drain the leaching solution. To prevent the leaching solution from running through the floor, multiple radial horizontal drilling holes can be drilled at the bottom of the ore layer in the drainage well to collect the leaching solution, and the flow direction of the leaching solution can be controlled by the pressure difference between the horizontal drilling hole of the injection well and the horizontal drilling hole of the drainage well to prevent leaching solution from leaking and contaminating karst water at the bottom of the ore layer.

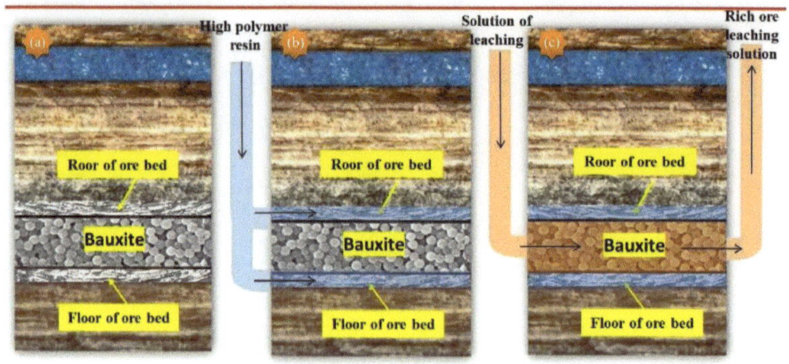

Figure 13. Schematic diagram of water pollution prevention with polymer resin infusion. (**a**) The stratigraphic procedure; (**b**) Schematic diagram of polymer resin filling; (**c**) In-Situ Leaching Mining after resin filling.

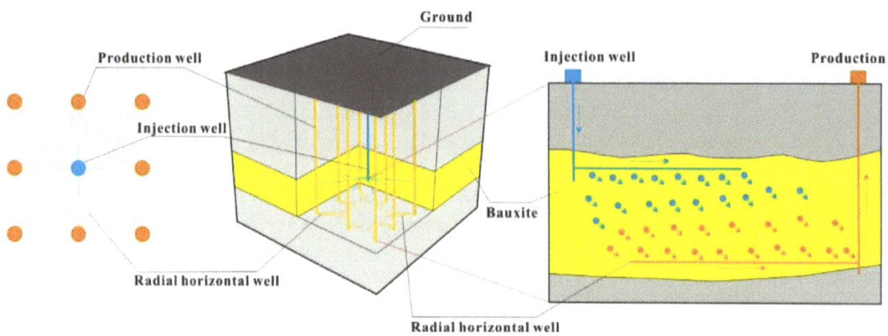

Figure 14. Schematic diagram of seepage control of well pattern to prevent water pollution.

4.3. Establish Production Dynamic Monitoring System

Based on the hydrogeological research of the mining area, the dynamic production monitoring system is designed to find out the geological conditions of the mining area, deploy monitoring wells around the leaching area, track the groundwater situation of the mining area in real time, investigate whether there are objects that need to be targeted for pollution prevention, treat environmental problems from the root, and build a dynamic production monitoring system [28]. The dynamic monitoring system includes the global underground seepage monitoring system, groundwater index monitoring system, and three-dimensional visualization system of leaching mining area, which can monitor and prevent potential problems in real time, and deal with the problems in time, so as to comprehensively prevent and control the occurrence of water pollution events.

5. Conclusions

(1) Geological research, including drilling and logging data, has revealed that the deep bauxite deposits in the Ordos Basin exhibit a sandwich structure. This structure is characterized by layers of claystone or shale above and below the bauxite deposit. Furthermore, analysis of the bauxite using techniques such as Scanning Electron Microscopy (SEM) and pore seepage analysis has shown that the bauxite contains pores and fractures, indicating favorable reservoir physical properties.

(2) The use of in-situ leaching mining technology in the extraction of metal solid minerals has certain prerequisites. Sedimentary bauxite ore exhibits inherent characteristics that make it suitable for this technology, including solubility, stability of ore body deposition, favorable top and bottom plate conditions, and permeability of the ore body. These characteristics provide a foundation for the application of in-situ solution leach mining to deep bauxite ore in the Ordos Basin, and the concept and steps for this mining method are proposed.
(3) This paper discusses how to establish barrier layer by means of well pattern seepage control and polymer resin infusion to prevent leaching solution from entering aquifer, and establish dynamic monitoring system in mining area to track and monitor groundwater situation in real time and prevent water pollution.

Author Contributions: Z.L.: Methodology, Investigation, Data curation, Funding acquisition, Writing—original draft, Visualization. T.L.: Methodology, Supervision, Validation, Resources, Writing—review and editing. Y.Z.: Methodology, Supervision, Validation, Resources, Writing—review and editing. P.X.: Supervision, Resources, Writing—review and editing. H.M.: Supervision, Writing—review and editing. P.S.: Methodology, Supervision, Writing—review and editing. X.W.: Supervision, Writing—review and editing. J.W.: Supervision, Writing—review and editing. All authors have read and agreed to the published version of the manuscript.

Funding: This study was the financially supported the China Geological Survey (No. DD20221774).

Data Availability Statement: All data included in this study are available upon request by contacting the corresponding author.

Conflicts of Interest: The authors declare no conflict of interest.

References

1. Zhang, H.; Hu, P.; Jiang, J.; Cheng, X.; Wang, J.; Liu, J.; Xiang, P. Distribution, genetic types and current situation of exploration and development of bauxite resources. *Geol. China* **2021**, *48*, 68–81.
2. Mameli, P.; Mongelli, G.; Oggiano, G.; Dinelli, E. Geological, geochemical and mineralogical features of some bauxite from Nurra (Western Sardinia, Italy). *Int. J. Earth Sci.* **2007**, *96*, 887–902. [CrossRef]
3. Bogatyrev, B.A.; Zhukov, V.V.; Tsekhovsky, Y.G. Formation conditions and regularities of the distribution of large and superlarge bauxite deposits. *Lithol. Miner. Resour.* **2009**, *44*, 135–151. [CrossRef]
4. Zhao, Y.S.; Liang, W.G.; Feng, Z.J. Science, technology and engineering of in-situ modified mining by fluidization. *J. China Coal Soc.* **2021**, *46*, 25–35.
5. Zhao, Y.S.; Xu, S.G. Theoretical study of in-situ solution mining. *J. Taiyuan Univ. Technol.* **2012**, *43*, 382–387.
6. Benes, V.; Boitsov, A.V.; Fuzlullin, M.; Hunter, J.; Mays, W.; Novak, J.; Slezak, J.; Stover, D.E.; Tweeton, D.; Underhill, D.H. *Manual of Acid in Situ Leach Uranium Mining Technology*; International Atomic Energy Agency: Vienna, Austria, 2001; pp. 1–283.
7. Yan, C.W. Potential of deep drilling in in-situ mining. *Geotech. Drill. Eng.* **1997**, *5*, 45–46.
8. Kongar-Syuryun, C.B.; Aleksakhin, A.V.; Eliseeva, E.N.; Zhaglovskaya, A.V.; Klyuev, R.V.; Petrusevich, D.A. Modern Technologies Providing a Full Cycle of Geo-Resources Development. *Resources* **2023**, *12*, 50. [CrossRef]
9. Golik, V.I.; Klyuev, R.V.; Martyushev, N.V.; Brigida, V.; Efremenkov, E.A.; Sorokova, S.N.; Mengxu, Q. Tailings Utilization and Zinc Extraction Based on Mechanochemical Activation. *Materials* **2023**, *16*, 726. [CrossRef] [PubMed]
10. Xiao, W.G.; Huang, K.L.; Zhu, J.L. Application of production and exploration results of ionic rare ore in in-situ leaching mining design. *China Metal Bull.* **2020**, *12*, 27–28.
11. Chi, R.A.; Liu, X.M. Prospect and development of weathelution-deposited rare earth ore. *J. Chin. Soc. Rare Earths* **2019**, *37*, 129–140.
12. Fu, J.H. A study of the sealing properties of the Palaeozoic caprocks in Erduosi Basin. *Nat. Gas Ind.* **1991**, *11*, 6–11.
13. Gao, L.; Wang, D.H.; Xiong, X.Y.; Qi, S.J.; Y, C.W.; Jia, S.H. Minerogenetic characteristics and resource potential analysis of bauxite in China. *Geol. China* **2015**, *42*, 853–863.
14. Lu, Y.L.; Lin, Y.; Yi, J.N. The current situation of mineral resources in the Ordos Basin and suggestions for exploration and development. *China Min.* **2015**, *24*, 15–32.
15. Sun, S.L. Division of bauxite metallogenic belt and characteristics of ore bearing rock series in Baode Xingxian County, Shanxi Province. *Land Resour. North. China* **2018**, *6*, 13–14.
16. Du, Y.S.; Yu, W.C. Subaerial leaching process of sedimentary bauxite and the discussion on classifications of bauxite deposits. *J. Palaeogeogr.* **2020**, *22*, 812–826.
17. Wang, Q.F.; Deng, J.; Liu, X.F.; Zhang, Q.Z.; Li, Z.M.; Kang, W.; Cai, S.H.; Li, N. Review on research of bauxite geology and genesis in China. *Geol. Explor.* **2012**, *48*, 430–448.

18. Sun, Z.X.; Asghar, F.; Zhao, K.; Zhou, Y.; Li, G.; Xu, L. Review and prospect of uranium mining and metallurgy in China. *Nonferrous Met. (Extr. Metall.)* **2021**, *8*, 1–8.
19. Golik, V.I.; Gashimova, Z.A.; Liskova, M.Y.; Kongar-Syuryun, C.B. Improvement of the occupational safety by radical isolation of pollution sources during underground ore mining. *Bezop. Tr. V Promyshlennosti* **2021**, *2021*, 7–12.
20. Rybak, Y.; Khayrutdinov, M.; Kongar-Syuryun, C.; Tyulyayeva, Y. Resource-saving technologies for development of mineral deposits. Sustain. *Dev. Mt. Territ.* **2021**, *13*, 406–415.
21. Liu, K.K.; Fu, X.; Rong, W.; Wang, J.Q.; Lin, J.; Jiang, Y.P. Analysis of bauxite reservoir in X area of Ordos Basin. *J. Xi'an Shiyou Univ. (Nat. Sci. Ed.)* **2022**, *37*, 25–31.
22. Wang, D.Q.; Liu, Z.Z.; Yin, L.H. Hydro-Geological characteristics and ground water systems of the Ordos Basin. *Quat. Sci.* **2005**, *25*, 6–14.
23. Meng, W.G.; Li, X.G.; Wu, B.W.; Gong, Z.C.; Dong, D.S.; Liu, Y.Y.; Xian, X.M. Research on gas accumulation characteristics of aluminiferous rock series of Taiyuan Formation in Well Ninggu 3 and its geological significance, Ordos Basin. *China Pet. Explor.* **2021**, *26*, 79–87.
24. Nan, J.X.; Liu, N. Characteristics and formation mechanism of bauxite reservoir in Taiyuan Formation, Longdong area, Ordos Basin. *Nat. Gas Geosci.* **2022**, *33*, 288–296.
25. Liu, S.L.; Lu, H.S. Evaluation methods and characteristics of log evaluation technology in shale gas. *Well Logging Technol.* **2011**, *35*, 112–116.
26. Zhao, Y.S.; Liang, W.G.; Feng, Z.J. *Introduction to In-Situ Modified Fluid Mining*; Science Press: Beijing, China, 2019; pp. 1–5.
27. Chen, F.H.; Zhao, J.F.; Chang, B.C.; Gao, Y.J. A preliminary analysis and assessment of hydrogeological conditions for in-situ leach mining of sandstone-type uranium deposit in northern Ordos basin. *Uranium Geol.* **2006**, *22*, 163–167.
28. Wang, L.M.; Luo, Y.K.; Yin, S.H. Exploration study of synergistic mining between the fluidized leaching process enhancement of deep metal mines and geothermal energy development. *Chin. J. Eng.* **2022**, *44*, 1694–1708.

water

MDPI

Article

Hydrogeochemical Characteristics and Human Health Risk Assessment of Fluoride Enrichment in Water in Faulted Basins of Qinghai-Tibet Plateau—A Case Study of Sanhe Plain in Guide Basin

Ruiping Liu [1,2,3,4,5], Fei Liu [6], Youning Xu [1,2,3], Hua Zhu [1,2], Jiangang Jiao [7,*] and Refaey M. El-Wardany [7,8]

1 Xi'an Center, China Geological Survey, Ministry of Natural Resources, Xi'an 710054, China; lrp1331@163.com (R.L.)
2 Observation and Research Station Environmental Geology of Typical Mines in Shaanxi Province, Ministry of Natural Resources, Xi'an 710054, China
3 Key Laboratory for Geo-Hazards in Loess Area, Ministry of Natural Resources, Xi'an 710054, China
4 Observation and Research Station of Groundwater and Ecology in Shaanxi Yulin, Ministry of Natural Resources, Yulin 719053, China
5 Key Laboratory for Groundwater and Ecology in Arid and Semi-Arid Areas, China Geological Survey, Xi'an 710054, China
6 Dalian Ocean and Fishery Comprehensive Administration Supervision Lochus, Dalian 116000, China
7 School of Earth Science and Resources, Chang'an University, Xi'an 710054, China
8 Faculty of Science, Al-Azhar University, Assiut 71524, Egypt
* Correspondence: jiangang@chd.edu.cn

Abstract: Fluoride (F) is an essential element of drinking water for human health, especially for bone development and enamel creation. However, if the fluoride content in drinking water is higher than 1.5 mg/L or lower than 0.5 mg/L, it will cause endemic diseases, such as dental fluorosis. There are two main hydrogeological characteristics: the properties of the water-bearing rocks and groundwater conditions controlled the groundwater in guide basin. The geothermal water can be divided into fracture convection and sedimentary basin geothermal water according to its geological environment and heat transfer mode. Inductively coupled plasma spectrometry is a significant tool for groundwater quality analysis. The geochemical factors of fluoride enrichment in confined geothermal water mainly include pH, ion exchange, and mineral saturation. Both groundwater samples are slightly alkaline, while the phreatic water and surface water record pH values of 8.5, 7.78, and 7.8, respectively. The salinity of groundwater water is not high, but for confined geothermal water, phreatic water, and surface water measures 706.0, 430.1 and 285.9 mg/L respectively. The higher the pH of groundwater, the more beneficial it is to the enrichment of fluoride. In contrast, the main cations in phreatic water and surface water are calcium ions and magnesium ions. The anions in groundwater and surface water mainly include SO_4^{2-} and HCO_3^-, followed by Cl^-, indicating that the groundwater and surface water here is mainly leaching. Fluoride was shown to be positively correlated with sodium and bicarbonate. Moreover, the results indicate that F^- enrichment is usually associated with high HCO_3^- and Na^+ concentrations in water, while a high Ca^{2+} concentration tends to lower the F^- concentration in water. This means that the ion exchange between calcium ions and sodium ions may lead to fluoride enrichment in natural water. As mentioned above, high-sodium and low-calcium water are favorable for fluoride enrichment. Moreover, saturation indices of fluorite, gypsum, dolomite, and calcite, as well as the saturation index of fluorite, represent a vital method to understand the fluoride enrichment. According to this study, fluoride as a pollutant poses great risks to human health overall, whether lower than or higher than the drinking water limit. Children face higher health risks than adults caused by confined geothermal water drinking intake. This study suggests that groundwater treatment should be conducted to reduce fluoride concentration in drinking water. It is suggested that when confined geothermal water is used as drinking water, it should be mixed with phreatic water and surface water in a certain proportion to make the fluoride in groundwater reach the range of safe drinking water.

check for updates

Citation: Liu, R.; Liu, F.; Xu, Y.; Zhu, H.; Jiao, J.; El-Wardany, R.M. Hydrogeochemical Characteristics and Human Health Risk Assessment of Fluoride Enrichment in Water in Faulted Basins of Qinghai-Tibet Plateau—A Case Study of Sanhe Plain in Guide Basin. *Water* **2023**, *15*, 1968. https://doi.org/10.3390/w15101968

Academic Editor: Maurizio Barbieri

Received: 2 March 2023
Revised: 3 May 2023
Accepted: 17 May 2023
Published: 22 May 2023

Keywords: Qinghai-Tibet plateau; Sanhe Plain in guide basin; confined geothermal water; fluoride; hydrogeochemistry; human health risks

1. Introduction

Fluoride (F) is an essential element of drinking water for human health, especially for bone development and enamel creation. A small amount of fluoride in drinking water is highly beneficial to dental health, capable of preventing dental caries [1,2]. Nevertheless, a high fluoride concentration in drinking water is one of the most critical problems in the world. It is known that a fluoride content in drinking water is higher than 1.5 mg/L or lower than 0.5 mg/L will cause endemic diseases, e.g., dental fluorosis. Approximately 200 million people suffer from chronic fluorosis in 28 developed and developing countries, including India, Brazil, the African continent, Pakistan, and China [3]. The high fluoride content in groundwater is closely related to the geothermal distribution area in terms of geochemical and hydrogeochemical characteristics, especially the geothermal water highly affected by the water–rock interaction through the migration process. On the other hand, the effects of the natural conditions and human activities on the physical and chemical features of groundwater rise day after day. On the other hand, the most significant and main priority is to understand the influence of water–rock interaction on element concentration to face and resolve these challenges [4]. Many authors have focused on the fluoride concentration and the factors which control and affect the behavior of fluoride ions in the groundwater. Liu [2] and others pointed out that the essential factors which control the enrichment of the fluoride ions include the alkaline environment, temperature, cycle depth, Lithology, and dissolution of fluoride-bearing minerals (fluorite or fluorspar apatite, cryolite, etc.), in addition to the cation exchange, as the main geochemical process which controls the hydrochemical characteristics of high-fluoride geothermal water [2,5–8]. Generally, the environmental studies and their issues, especially green energy, have global attention regarding to its potential development and utilization value. China has been considered the sconed richest country globally in terms of geothermal resources and reserves. The development and utilization of geothermal energy are increasing at an annual rate of 10%. accompanied by many environmental problems, such as air pollution, groundwater level decline, land subsidence, etc., posing a hidden danger to the safety of the local drinking water.

The Qinghai-Tibet Plateau, known as the "Asia Water Tower", is the main birthplace of major rivers in China and Asia. In addition, the groundwater of the Qinghai-Tibet Plateau represent an important part of China's water resources, with a high abundance of underground geothermal resources. Since 1978, there have been more than 20 artesian wells with a depth of 200~600 M in this area. Up to now, more than 10 hot water wells have been used as a drinking water source for humans and animals for a long time. An estimated 59,000 people are at risk from water-borne fluorosis and arsenic poisoning [9]. Shi [9] and Zhang [10] believed that fluoride in hot water exists in a confined artesian hot water in the Guide Group, and is controlled by structure, depth, and temperature [9,10]. The chemical types of geothermal water in the Guide basin are SO_4.Cl-Na. Instead, the local crustal partial melting or magma chamber at 15~35 km beneath the Gonghe basin is characterized by a relatively high geothermal background induced by the continental collision between the Indian and Eurasian plates [11]. δD and $\delta^{18}O$ indicated that the meteoric water is the most likely source of the geothermal water in the Zhacang field, with depleted δD and $\delta^{18}O$ in deeper water indicative of recharge via snow melt and/or from cooler climates [12]. In contrast, precipitation is the main recharge source for unconfined and confined groundwater in the study area, where the chemical composition of unconfined groundwater mainly derives from the weathering of silicate and the dissolution of carbonate in the aquifer, while that of confined groundwater is mainly controlled by the dissolution of evaporite, which increases along the flow path. The chemical composition of unconfined groundwater is variable seasonally, while the chemical composition of confined groundwater is temporally constant [13]. Dai [14] studied the convective geothermal water from five faults in the Guide Basin as the research object and comprehensively utilized the hydrochemical characteristics and the evolution of geothermal water in Guide Area, Qinghai, etc., suggesting that the water–rock interaction degree in Guide Basin is high, and the geothermal water comes from atmospheric precipitation [14]. Sun et al. [15] concluded that the enrichment degree of fluoride ions in the water of the hydrochemical type Na in Tibet geothermal water is higher than that of the water of the type of Ca. The enrichment mechanism of high fluoride geothermal water mainly includes the upwelling mixing of the deep geothermal fluid and the leaching of fluoride silicate minerals and fluorite during the downward seepage of make-up water [15]. Zhang et al. [16] suggested that the typical harmful elements in geothermal water in Tibet have an impact on river water quality [16]. The control function, hydrogeochemistry, and human health risks of basin-type high-fluoride groundwater have not been discussed in detail by the scientists in the context of the Qinghai-Tibet Plateau. This paper will discuss the hydrogeochemical effects of the spatial distribution of fluoride ion content on geothermal water and human health in the Guide Basin as an example, in order to study the occurrence and the mechanism of faulted basins in Qinghai-Tibet Plateau and the various risks caused by drinking water to provide a basic reference data for the development and utilization of geothermal energy in fluoride-rich areas.

The guide basin is located in the southeast of Hainan Tibetan Autonomous Prefecture, Qinghai province, approximately 97 km from Xining, and the basin is surrounded by mountains (the Lagrange in the north, and Chamah in the south). The terrain is high in the north and south, low in the middle, with an average elevation of about 3100 m. The water system is developed in the basin, where steep slopes, low mountains, and hills are common. The Yellow River enters from the Longyangxia Gorge in the western part of the basin. Tributaries flow into the northern and southern sides of the river, forming multi-level river and hilly landforms, where the two main largest tributaries flow into the Yellow River on the east and west sides of Guide County's seat. The confluence of these three tributaries form a wide valley plain called the "Sanhe plain" (Figure 1). On the other hand, the region is considered semi-arid continental climited, with an average annual temperature of 8.5 °C, with large diurnal variations. The annual average rainfall is 275.8 mm, and the precipitation mainly concentrates from June to September. Moreover, the mean annual potential evaporation is 1417.1 mm.

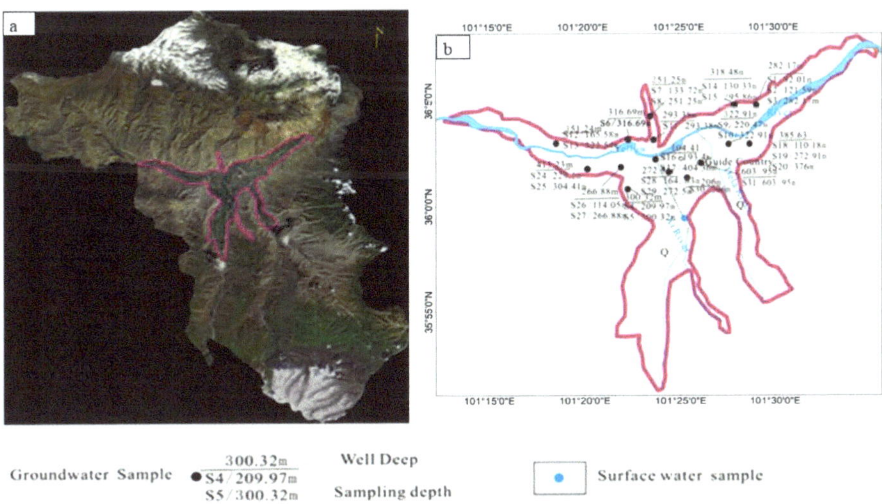

Figure 1. (**a**,**b**) Groundwater and surface water sampling distribution map, (**a**)—Remote sensing image map of the Guide Country.

2. Geological Setting

Proterozoic and Palaeozoic strata are exposed around the basin, Paleocene-Oligocene Xining Group, and Miocene-Pliocene Guide Group, which are in turn from bottom to top. Notably, the Neogene Guide Group and Quaternary have the widest distribution area and greatest thickness. The Neogene Guide Group not only includes a confined artesian water, but also confined artesian hot water (Figure 2) [17].

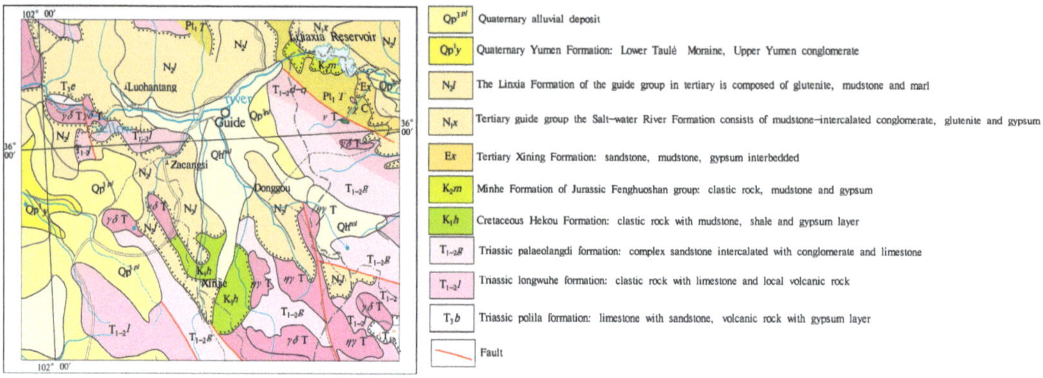

Figure 2. Geological map of Guide County.

Xining Group is composed of brownish-red mudstone, sandy mudstone, gray-green, gray-white gypsum rock, and sandy argillaceous gypsum rock interbedded with sandstone and siltstone convex mirror, which is a lacustrine gypsum-containing red clastic rock deposit in a dry climate.

Guide Group is a set of a stratigraphic sequence composed of khaki, brownish yellow mudstone, sandy mudstone, grayish white fine-medium conglomerate, gravelly sandstone, sandstone, and marlstone, and this sedimentary environment is composed of salt lakeside-brackish water or freshwater lakeside facies. This set of strata is widely distributed in the Guide Basin. Except for the local areas along the fault zone in the basin edge belt, it inclines slightly from the basin edge to the center of the basin. On the other hand, the upper part of the Guide area is covered by the Lower Pleistocene of Quaternary sediments.

The Neogene Guide Group was formed in the environment of saltwater lakeside-brackish water or freshwater lakeside in an arid and hot climate and is mainly composed of fine-grained lacustrine strata, which are rich in organic matter. Rock cutting statistics show that the sources of fluoride and arsenic in the strata are mainly Archean to Lower Proterozoic metamorphic rocks and volcanic rocks in the Riyueshan and Lajishan areas in the northern part of the basin [10]. The modern semi-enclosed topography and hydrogeochemical environment of underground hot water are the main promoting factors to form drinking water endemic fluorosis and arsenic poisoning areas.

The Guide Basin is dominated by fault structures, followed by fold structures. There are mainly NWW-oriented structures in the north, NW-oriented structures and NNW-oriented structures in the south-central part, and SN-oriented structures in the central and eastern margins, with the development of EW-oriented and NE-oriented structures.

3. Hydrogeological Characteristics

Hydrogeological characteristic can be divided into two main categories, the first part related to the type and properties of the water-bearing rocks and the second part related to the main groundwater conditions.

3.1. The Water-Bearing Rock Includes Four Essential Groups

(1) Loose Rock Pore Group:

This group is mainly distributed in the plain area of Sanhe Valley in the basin and emerges as pores of Quaternary loose alluvial proluvial. The loose sediment is thick, and the water quantity is controlled by the irregular recharge source. The variation in diving water level is affected by rainfall, river recharge, and burial depth.

(2) Clastic Rock Fissure Pore Group:

This group is mainly distributed in the red bed, hilly area, and presents in the fissures and pores of Paleogene clastic rocks. The lithology of this aquifer is mainly sand and conglomerate, with pressure bearing poor to good water quality.

(3) Granite Fissure Group:

This group is mainly distributed in the Zhongshan area in the west of the survey area and occurs in bedrock fissures or fault zones in Mesozoic clastic rocks and Indosinian intrusive rocks, with poor to good water quality.

(4) Frozen Layer of Water:

The frozen layer of water is mainly distributed in mountainous areas above 3800 m above sea level at the edge of the basin, and there are two modes of occurrence. The first mode occurs in moraine and ice-water deposits of the middle and late Pleistocene and Holocene, while the second occurs in Triassic sand slate and granite.

3.2. Groundwater Conditions (Recharge, Diameter, and Discharge)

The formation and circulation of groundwater in the study area are influenced by stratum lithology, geological structure, hydrometeorology, and other factors. According to the different types of groundwater, there are four main types of water in the study area:

(1) Porewater (Groundwater):

The groundwater is mainly recharged by rivers and a lateral supply of bedrock fissure water in mountainous areas. The runoff condition is controlled by topography, with horizontal runoff as a main factor, while the underground runoff is sluggish. Instead, the main ways of discharge are evaporation and discharge to the Yellow River.

(2) Pore fissure water:

The hilly area is a distribution area of clastic rock fissure pore water, with strong terrain cutting and well-developed valleys. Pore fissure water is laterally replenished by atmospheric precipitation and bedrock mountain areas, with discharge usually to the surface forming some springs in the valleys after a short run-off. The clastic fissure-confined artesian water occurs usually in a fault contact with bedrock mountain areas as a complex structure and extends to the Neogene water storage structure after being guided by a fault zone.

(3) Fissure water:

The mountainous area exhibits high elevation, strong weathering of rocks, and steep terrain with abundant precipitation, which is conducive to infiltration and replenishment of atmospheric precipitation and the formation of bedrock fissure water. During the process of groundwater migration and runoff along its fissures, there is some discharge into the valleys in the form of springs, whereas the residual is supplied to the fissure pore water of clastic rocks in the hilly area in a hidden way through stratum contact parts or pore water of Quaternary loose rocks.

(4) Frozen layer water

Frozen soil (rock) is widely developed in Zhongshan and Gaoshan areas about 3800 m above sea level and contains a frozen layer of water. Frequently, the groundwater in this area present seasonal phase changes of liquid and is solid in shallow depths of 1~20 m [14].

4. Occurrence of Geothermal Water

In general, the study area can be divided into fracture convection geothermal water and sedimentary basin geothermal water according to its geological environment and heat transfer mode. Meanwhile, the code for the geological exploration of geothermal resources (GB-T-11615-2010) stipulates that the temperature of cold water is <25 °C and that of hot water is ≥25 °C.

4.1. Fracture Convection Geothermal Water

Convective geothermal water is often distributed in the well-developed fault zone around the basin in the form of hot springs. The Waligong Mountain fault zone in the west is the most active geothermal area in the whole basin, with many hot springs, such as Qunaihai, Zhacang Temple, and Xinjie Hot Springs, with the highest temperature reaching the local boiling point of 93 °C. Furthermore, hot springs, such as Lancai, are also developed in the Duohemao fault zone on the east side.

4.2. Sedimentary Basin Geothermal Water

The geothermal water in the sedimentary basin is exposed as a geothermal well, which are distributed in the Sanhe Plain basin, with a width of approximately 6 km from north to south and a length of approximately 20 km from east to west. There are two NS-trending compressive fractures and one EW-trending tensile fracture in the plain. These fractures are good heat conduction channels. On the other hand, there are no hot springs exposed. The thermal reservoir in Sanhe Plain can be divided into three thermal reservoirs:

(1) The Neogene thermal reservoir cap is approximately 160~240 m thick, while the thermal reservoir section is approximately 180 m thick. The water content mostly is good, and the orifice temperature is approximately 18.5~28 °C, reaching 34.6 °C. In contrast, the geothermal gradient is generally approximately 6.67~9.70 °C/100 m, which is the low-temperature thermal reservoir.

(2) The buried roof depth of the Paleogene thermal reservoir cap is between 1200 and 1500 m, and the thickness of the thermal reservoir section is approximately 600~800 m, with good water-rich properties and high temperature.

(3) The buried top plate depth of Cretaceous and Jurassic thermal reservoirs is 2700~3400 m, and the thickness of thermal reservoirs is 1000~1600 m. The poor property of water and the low temperature of the thermal fluid can be attributed to the lack of water and heat [17]. More than 20 exploratory and mining combined hydrogeological boreholes with a depth of 200~600 m have been established out in the Guide area. The revealed confined artesian water of fissures and pores of clastic rocks in the basin mainly occurs in the Guide Group of Neogene, and the water-bearing rock group is composed of multiple aquifers, with complex distribution and water abundance. The confined artesian water mainly occurs in the siltstone of the Zhongyan Formation [9].

5. Methodology

The groundwater samples were obtained from the anysis data in some previous works [9] combined with the test results of "Investigation on Mineral Resources Development and Geological Environment Impact of Qinghai-Tibet Plateau" from 2012 to 2015 by Xi'an Geological Survey Center of China Geological Survey. The groundwater samples were taken in layers and preserved at 4 °C, and then to the lab for analysis within 4 days. An inductively coupled plasma spectrometer was used for groundwater quality analysis, and the contains of Na^+, SO_4^{2-}, Cl^-, Ca^{2+}, Mg^{2+}, and HCO_3^- were measured by double indicator-neutralization titration. The content of F^- in groundwater was analyzed by ion chromatography. All analysis was performed by the Qinghai Hydrogeological Engineering Geological Environment Geological Survey Institute. Hydrochemical components were compared with Chinese national groundwater quality guidelines [18,19] to assess their suitability for drinking water. Therefore, only drinking water intake was considered in this study. Fluoride is a non-carcinogenic risk factor. Therefore, the non-carcinogenicity evaluation risk model recommended by the US EPA [20] was used to evaluate the fluorine content in groundwater in the study area for health risks.

Based on the assessment results, F^- was selected to perform the non-carcinogenic health risk assessment using the model recommended by the U.S. Environmental Protection Agency (USEPA) [21], with parameter values suitable for the study area. In terms of exposure pathways of fluoride, drinking water intake was considered. The assessment model of intake-induced risks to health is as follows:

$$HQ = CD_i/RFD \tag{1}$$

where HQ denotes the hazard quotient of noncarcinogenic health risks, and $HQ > 1$ means high potential health risks that are unacceptable for adults and children.

The daily average exposure doses per unit weight (CD_i) through drinking water are expressed using CD_{id} and CD_{if}, respectively. They can be calculated as follows:

$$CD_{if} = (C_i \times Y_{fi})/BW \tag{2}$$

$$CD_f = \Sigma CD_{if} \tag{3}$$

$$CD_{id} = (C_i \times IR \times EF \times ED)/(BW \times AT) \tag{4}$$

The calculation parameters are shown in Table 1.

Table 1. Parameters in the health risk assessment model.

Parameter	Meaning	Children	Adults	Reference Standards
C_i	Contaminant concentration in water			
IR	Daily intake of drinking water	1.0 L/d	0.6 L/d	
EF	Exposure frequency	365 days	365 days	
ED	Exposure duration	12 years	25 years	
BW	Body weight	15.9 kg	56.8 kg	USEPA 2017
AT	Average time for noncarcinogenic effect	4380 days	9125 days	
CD_i	Quotient			
RFD	non-carcinogenic reference dose of F^- through oral intake	0.03 mg/(kg·d)	0.06 mg/(kg·d)	
HQ	Non-carcinogenic risk			

6. Results and Discussion

6.1. Hydrochemistry and Water Quality Assessment

According to the statistical data of water samples listed in Table 2, both groundwater and surface water samples are slightly alkaline, with average pH values of confined geothermal water, while the phreatic water and surface water measure 8.5, 7.78, and 7.8, respectively. The salinity of groundwater and surface water is not high, but for confined geothermal water, phreatic water, and surface water, 706.0, 430.1, and 285.9 mg/L, respectively. Therefore, groundwater and surface water in Guide Basin are freshwater resources. The cations of underground hot water are mainly sodium and potassium ions followed by calcium and magnesium ions. The cations in phreatic water and surface water are mainly calcium ions and magnesium ions. The anions in groundwater and surface water mainly include SO_4^{2-} and HCO_3^- followed by Cl^-, indicating that the groundwater and surface water here is mainly leaching. The fluoride concentration in hot water under pressure is 0.43–5.7 mg/L, which is relatively high, while the fluoride content in diving water and surface water cannot be detected. The fluoride content in most of the confined hot water in the study area at 160–240 m exceeds the standard, except in the upper reaches of Xihe River and Donghe River. From an epidemiological point of view, excessive fluoride in drinking water will increase the risk of dental fluorosis and skeletal fluorosis (Figure 3b). Consequently, the drinking water must be treated to reduce fluoride concentration.

Table 2. Statistical analysis of physicochemical indices of samples against drinking water guidelines.

Index		TDS (mg/L)	pH	T	$Na^{+} + K^{+}$ (mg/L)	Ca^{2+} (mg/L)	Mg^{2+} (mg/L)	Cl^{-} (mg/L)	SO_4^{2-} (mg/L)	HCO_3^{-} (mg/L)	F^{-} (mg/L)
Confined Geothermal water	Min.	379.8	7.35	18.5	113.1	4.4	0.3	15.5	44.8	82	0.4
	Max.	1393.2	8.9	41.4	368.1	48.8	5.3	406.1	901	411.5	5.7
	Mean	706.0	8.5	29.7	211.1	9.60	1.79	143.7	132.73	199.79	2.7
Phreatic water		430.1	7.78	9.5	2.8	21.5	74.6	13.1	26.4	49.4	0
Surface water		285.9	7.8	14.0	7.4	7.4	51.4	9.2	7.2	24.6	0
Chinese guidelines		1000	6.5–8.5		200	/	/	250	250	/	1
WHO guidelines		1000	6.5–8.5		200	/	/	250	250	/	1.5

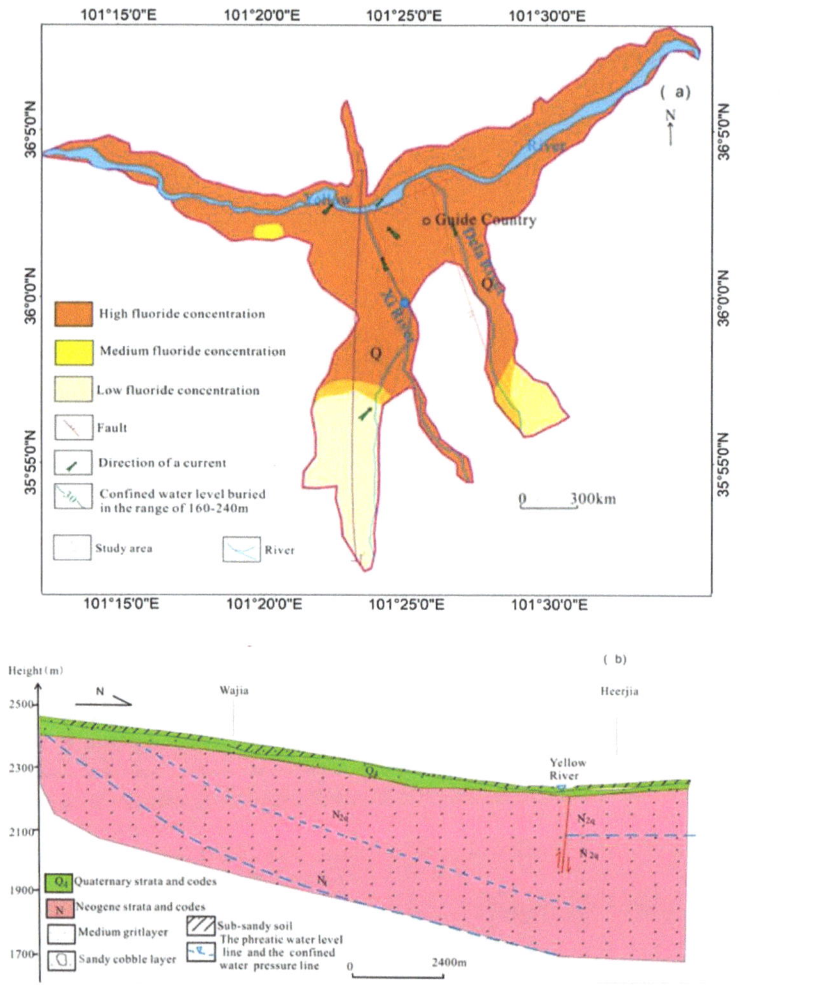

Figure 3. Spatial distribution of fluoride in groundwater. (**a**) Spatial distribution of fluoride in groundwater in Sanhe plain of Guide Country (The depth is 160–240 m under confined Geothermal water) (**b**) Hydrogeological section.

6.2. Hydrogeochemical and Their Influencing Factors

6.2.1. Dominant Zones of Hydrogeochemical Process

The Gibbs diagram was used to further analyze the fluoride sources of water samples (Figure 4). All water samples fell within the dominant zone of rocks, indicating that the chemical properties of the groundwater and surface water in the study area are controlled by weathering or the rock and water–rock interactions. In contrast, the groundwater samples which fall in the partially balanced or mixed water area indicate that the geothermal water may be mixed with cold water near the surface (Figure 5).

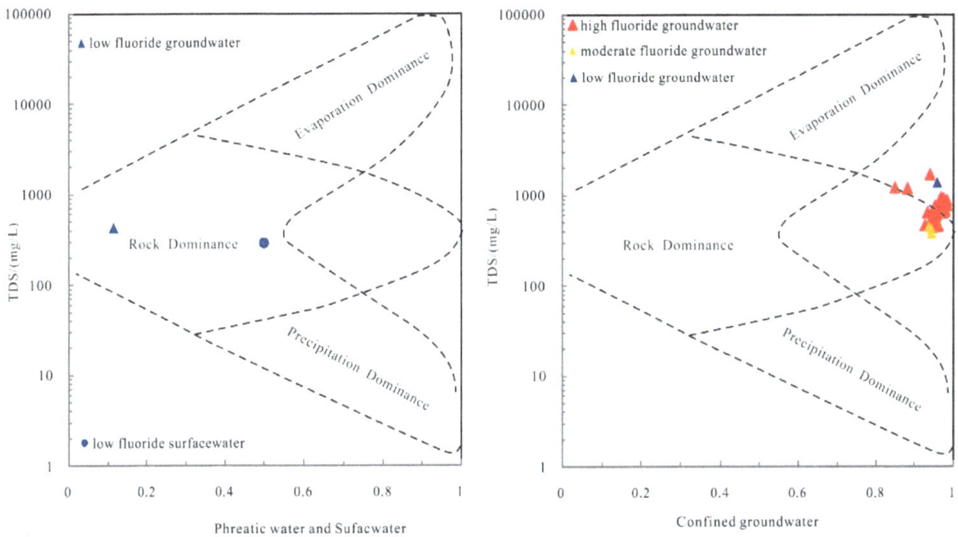

Figure 4. Gibbs diagram showing the control mechanisms of natural water chemistry.

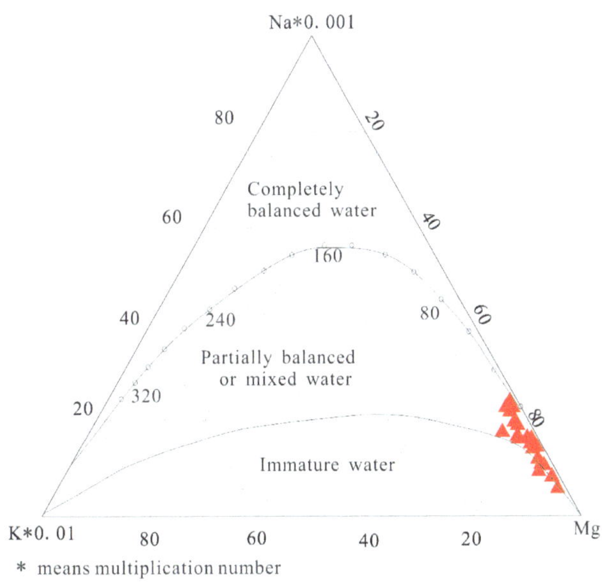

Figure 5. Geokit diagram showing the mixing mechanism of pressurized geothermal water and other types of water.

6.2.2. Impacts of pH

In geothermal water, fluorite (CaF_2) is one of the most vital minerals because it will react to produce fluoride ion (F^-), which can react with calcium ion Ca^2 [22]. The reaction formula is as follows:

$$CaF_2 \rightarrow 2F^- + Ca^{2+} \tag{5}$$

Groundwater with high pH can dissolve aluminosilicate minerals, so the alkali metal hydrolysis will increase the alkalinity of water and promote the dissolution of fluorosilicate minerals, e.g., fluorite and fluorosilicate minerals, then release fluoride to increase the fluoride content in groundwater. Therefore, the higher the pH of groundwater, the more beneficial it is to the enrichment of fluoride [23]. According to the investigation, the pH value of pressurized geothermal water in the study area is between 7.35 and 8.9, and the pH value of surface water is 7.8, which can facilitate the dissolution of fluoride-containing minerals. On the other hand, a Pearson correlation matrix showed that fluoride is positively correlated with sodium ($r = 0.583$) and bicarbonate ($r = 0.629$) (Table 3). Consequently, it can be inferred from Pearson correlation matrix that the fluoride-rich groundwater is highly alkaline and low in hardness [24]. The range of water hardness in the study is 5–51.2 mg/L, which is below the national drinking water standard of 450 mg/L. Furthermore, in alkaline environments, hydroxyl can replace fluoride-containing minerals, such as biotite and muscovite [25], which leads to an increase of the fluoride content in the groundwater. The main rock type in Zhacangsi geothermal field is granite, which mainly contains quartz, mica, and feldspar [26]. Moreover, the Guide County Basin is a geothermal anomaly area with characteristics of both structural fault-type thermal anomaly and structural basin-type thermal anomaly. The heat source of the Piedmont Zone in the southwestern part of the basin is Indosinian Granodiorite [27]. There are a lot of biotite/muscovite/chrysotile in granite. The exchange between hydroxyl and biotite/muscovite/chrysotile asbestos is as follows:

$$KMg_3(AlSi_3O_{10})F_2 + 2OH^- \rightarrow KMg_3(AlSi_3O_{10})(OH)_2 + 2F^- \tag{6}$$

$$KAl_2\ (AlSi_3O_{10)}F_2 + 2OH^- \rightarrow KAl_2(AlSi_3O_{10})(OH)_2 + 2F^- \tag{7}$$

$$Na_3A1F_6 \rightarrow 6F^- + 3Na^+ + A1^{3+} \tag{8}$$

Table 3. Pearson correlation matrix of hydrogeochemical parameters of confined geothermal water in guide basin.

	pH	K^+	Na^+	Ca^{2+}	Mg^{2+}	Cl^-	$SO_4{}^{2-}$	$HCO_3{}^-$	F^-
pH	1	−0.664 **	−0.669 **	−0.737 **	−0.316	−0.737 **	−0.079	−0.163	−0.151
K^+		1	0.616 **	0.823 **	0.364 *	0.772 **	−0.005	0.110	0.194
Na^+			1	0.411 *	0.469 **	0.699 **	0.004	0.635 **	0.583 **
Ca^{2+}				1	0.106	0.772 **	0.198	−0.315	−0.127
Mg^{2+}					1	0.410 *	−0.208	0.526 **	0.125
Cl^-						1	0.169	0.037	0.126
$SO_4{}^{2-}$							1	−0.215	−0.384 *
$HCO_3{}^-$								1	0.629 **
F^-									1

Notes: ** Correlation is significant at the 0.01 level (2-tailed). *: Correlation is significant at the 0.05 level (2-tailed).

6.2.3. Hydrochemical Environment

The enrichment of fluoride in water is controlled by a specific hydrochemical environment. A Piper diagram (Figure 6) was plotted to understand the impacts of the hydrochemical facies on the fluoride enrichment in water, as well as the high-fluoride groundwater mainly distributed in areas with Na^+ contents (meq) > 80%; and $HCO_3{}^-$

contents (meq) > 60%. Hydrochemistry analysis of geothermal fluids shows that the high-temperature thermal water is mainly of $SO_4 \cdot Cl$–Na type, and the low-temperature thermal water is mainly composed of SO_4–Na, $SO_4 \cdot HCO_3$–Na. The piper diagram indicates that F^- enrichment is usually associated with high HCO_3^- and Na^+ concentrations in water (Figure 5), while a high Ca^{2+} concentration tends to lower the F^- concentration in water. Therefore, fluorite (CaF_2) can be dissolved and F^- is released into groundwater with Na-bearing salts, or alternatively F^- can react with Ca^{2+} to form fluorite (Figure 7).

Fundamentally, the concentrations of major ions are the most affected factor in the hydrochemical environment. As shown in the Pearson correlation matrix (Table 3), HCO_3^- is significantly correlated with Mg^{2+} (r = 0.526) and the saturation index is lower than 0, indicating that carbonate mineral dissolution is an essential process in the groundwater system. This is consistent with the result of Wang Zhen's research, which promotes the increase of bicarbonate chemical compositions of confined groundwater. This process is controlled mainly by the dissolution of evaporite, which commonly increases along the flow path [13]. Moreover, the ion exchange between calcium ions and sodium ions may lead to fluoride enrichment in natural water. As mentioned above, high-sodium and low-calcium water are favorable for fluoride enrichment, and thus the changes in sodium and calcium concentrations caused by ion exchange will inevitably affect fluoride behavior. In this study, the binary graph of Na^+-Cl^- vs. (Ca^{2+} + Mg^{2+})-(HCO_3^- + SO_4^{2-}) is utilized to explain the cation exchange process [2]. This shows that there is a linear relationship between Na^+-Cl^- and (Ca^{2+} + Mg^{2+})-(HCO_3^- + SO_4^{2-}), with a slope of −0.279 (r = 0.767). The slope is far less than 1, indicating that the ion exchange is not too strong, and most of the samples are located in the cation exchange control area rather than the anion exchange area, which indicates that the exchange of Na^+ and Ca^{2+}/Mg^{2+} may occur in the study area (Figure 7a). Indeed, this is good evidence that the cation exchange plays an important role in regulating the chemical evolution of natural water. Additionally, the positive correlation between F^- and Na^+/Ca^{2+} ratio with a slope of 0.002 (r = 0.563) indicates that the cation exchange between Na^+ and Ca^{2+} affects the fluoride enrichment in the study area (Figure 7b).

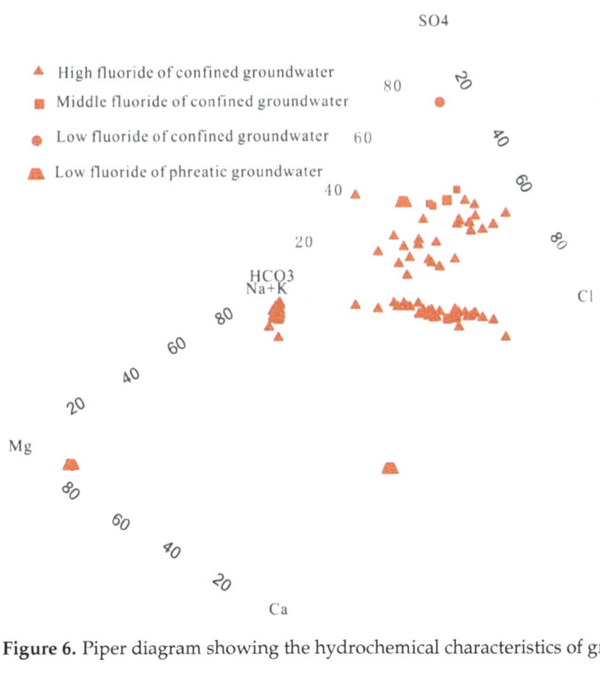

Figure 6. Piper diagram showing the hydrochemical characteristics of groundwater.

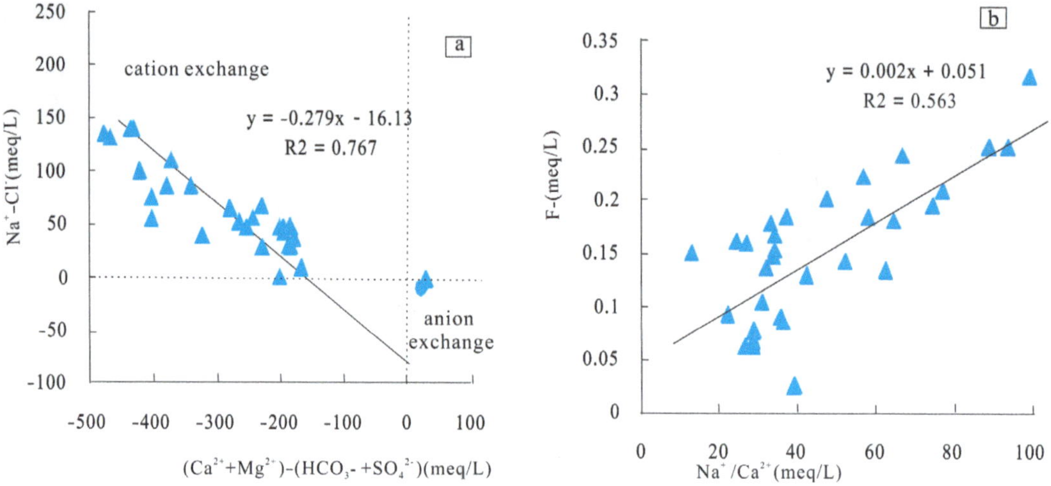

Figure 7. Bivariate plots showing ion exchange and its impacts on fluoride enrichment. (**a**) The relationship between major chemical components in groundwater. (**b**) The relationship between Na/Ca and fluoride ion.

6.2.4. Saturation Index of Minerals

The sedimentary lithology of the confined aquifer in the Guide basin is mainly composed of dark gray mudstone, earthy red clay, and dark gray fine sand. Furthermore, the main minerals in the composition are quartz, aragonite, mica, and albite. In addition, the Xining Group (Exn) is composed of reddish-brown mudstone, sandy mudstone, lime-green gypsum, lime-white gypsum, sandy argillaceous gypsum, and siltstone convex body. The rock composition of each stratum is mainly a set of flysch clastic rocks, siliceous rocks, some carbonate rocks, and a few volcanic rocks, which constitute the main provenance of Cenozoic basins in the southern and northern margin of the basin. The dissolution or precipitation of minerals depends on the saturation indices of minerals, which are used to denote the saturation state of water relative to minerals and indicate the dissolution or precipitation behavior of minerals. On the other hand, the impacts of mineral saturation on fluoride enrichment can be understood by calculating the saturation indices of fluoride bearing minerals, such as dolomite, calcite, gypsum, and fluorite. The dissolution behavior of these minerals was notably interdependent since they all contain calcium ions. The low solubility of gypsum will lead to the high solubility of fluorite [28]. Furthermore, the saturation index of calcite varied from −3.53 to 0.74, and that of dolomite ranged from 6.03 to 0.3, indicating that dolomite and calcite are unsaturated or close to saturation. The saturation index of gypsum was −3.62~−1.6, indicating that gypsum is also unsaturated. This implies that calcite, dolomite, and gypsum can continue to dissolve and release Ca^{2+} into groundwater, which will affect the fluoride enrichment in natural water (Figure 8). The saturation index of fluorite ranged from −3.37 to 0.01, indicating that fluorite is unsaturated. Meanwhile, fluorite dissolution contributes to an increase in fluoride content. This can be confirmed by the logarithmic positive correlation between fluoride and fluorite saturation index (R^2 = 0.851).

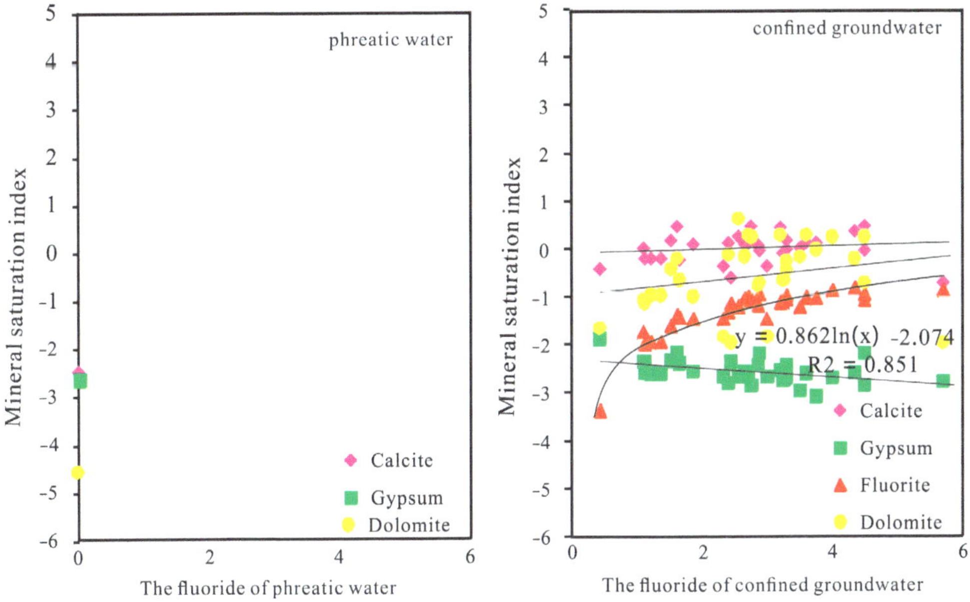

Figure 8. Relationships between F− and saturation indices F-bearing minerals (calcite, dolomite, gypsum, and fluorite).

6.3. Health Implication of Fluoride in Water

As proposed by the WHO, fluoride mainly affects human health in three ways, namely drinking water, skin contact, and food intake. The health risk of fluorine exposure through skin and air is very small [29]. This study focuses on the risks to the health of adults and children caused by excessive fluoride in drinking water. Table 4 shows that the HQ values of fluoride in confined geothermal water are between 0.54–7.17 (average: 4.12) and 0.25–3.35 (average: 1.62), respectively for children and adults. While the HQ values of fluoride in phreatic water and surface water were both 0. This indicates that the risks of fluoride through drinking confined geothermal water intake are unacceptable.

Table 4. Fluoride contents of drinking water and their non-carcinogenic health risks for adults and children.

Water	Sample Number	Statistic Parameter	Health Risks for Children	Health Risks for Adults
Confined Geothermal water	31	Average Maximum Minimum	4.12 7.17 0.54	1.62 3.35 0.25
phreatic water	1	concentration	0	0
surface water	1	concentration	0	0

Normally, if the fluoride content in water is lower than 0.5 mg/L, children will easily get dental caries after drinking for a long time. However, the formula developed by the USEPA [26] is imperfect and only aims at a design higher than the safety threshold. The authors think that the relative risk to human health can be adopted. Specifically, the fluoride concentration is expressed as the difference between the test concentration and the safety threshold concentration. Additionally, the calculated risk value is the relative risk to human health. More precisely, if the fluoride content is lower than 0.5 mg/L, there will be risks. However, there will be no risk within the safety thresholds.

Table 5 shows that the HQ values of fluoride in phreatic water and surface water for children and adults respectively were 0.63 and 0.29, respectively, while the HQ values of fluoride in confined geothermal water are between 0.00–5.28 (average: 1.67) and 0.00–2.46 (average: 0.78).

Table 5. Fluoride contents of drinking water and their modified non-carcinogenic health risks for adults and children.

Water	Sample Number	Statistic Parameter	Health Risks for Children	Health Risks for Adults
Confined Geothermal water	31	Average	1.67	0.78
		Maximum	5.28	2.46
		Minimum	0.00	0.00
phreatic water	1	concentration	0.63	0.29
surface water	1	concentration	0.63	0.29

As for the prevention and control measures in respect of fluoride in groundwater. The groundwater treatment should be conducted to reduce fluoride concentration in drinking water. It is suggested that when confined geothermal water is used as drinking water, it should be mixed with phreatic water and surface water in a certain proportion to make the fluoride in groundwater reach the range of safe drinking water.

7. Conclusions

High fluoride geothermal water has become a major public health problem in many areas of China, especially in the Qinghai-Tibet Plateau areas. In this study, the reasons for fluoride enrichment were revealed according to the hydrochemical characteristics of natural water in the confined geothermal water in faulted basins in the Qinghai-Tibet Plateau, and the potential human health risks of fluoride in drinking water were quantified. The conclusions are as follows.

(1) As indicated by the analysis of the Gibbs diagram concerning fluoride, the fluoride in groundwater is mainly related to the water–rock interaction. These geothermal waters may be mixed with near-surface cold water.
(2) The hydrogeochemical factors of fluoride enrichment in confined geothermal water mainly include specific natural factors, such as pH, ion exchange, and mineral saturation.
(3) The groundwater in the study area is slightly alkaline compared with the drinking water quality standards of China and the WHO. The confined water in the Guide basin presents high fluoride concentration (0.43–5.7 mg/L), while phreatic water and surface water present fluoride levels that are too low to drink. It is suggested that for Department of Water Resources Management that when confined geothermal water is used as drinking water, it should be mixed with phreatic water and surface water in a certain proportion to make the fluoride in groundwater reach the range of safe drinking water.
(4) Excessive fluoride in drinking confined geothermal water will cause health risks in adults and children. According to this study, fluoride is an element that causes great risks to human health over time in general. The HQ highest value of fluoride in confined geothermal water was 5.28. Meanwhile, children face higher health risks than adults caused by water drinking intake. Therefore, measures should be taken to ensure the health and safety of residents. Human health requires a groundwater fluoride concentration of less than 0.5 mg/L, which offers greater protection. This provides a reference basis for water management.

Author Contributions: R.L., Y.X. and H.Z. carried out the investigation, experiments, and sampling. R.L. wrote the manuscript. J.J. and F.L. helped R.L. Conceivably the original idea. R.M.E.-W. polished English manuscripts. All authors have read and agreed to the published version of the manuscript.

Funding: This study was funded by Key R & D projects in Shaanxi Provincein 2023 (2023-ZDLSF-63), the survey projects initiated by the Land and Resources Investigation Project ([2023]06-03-04), and Basic Research Project of Scientific Research Business Fee of Central University in 2016 (Open Fund) 310829161128. Investigation on geological environment effect of Qinghai mining industry develop ment (121201011000150022). The Qinghai-Tibet scientific research project "Geological Background Investigation of Freeze-thaw Disaster Development in Permafrost Regions of Qinghai-Tibet Plateau (2019QZKK0905)", Xi'an ecological Safety Investigation Evaluation and Risk Management Demonstration (DD20201317), and Xi'an Multi-factor Urban Geological Survey (DD20189220), Key Research and Development Program of Shaanxi (Program No. 2021ZDLSF05-01).

Data Availability Statement: The data is unavailable due to privacy can not be Published.

Acknowledgments: The authors are grateful to the experiment and testing lab of Xi'an Center, China Geological Survey for statistical data analyses. Thank you very much S.J. Assisted Sampling who coming from Qinghai Institute of Hydrogeology and Engineering geology and Environmental geology.

Conflicts of Interest: The authors declare no conflict of interest.

References

1. Liu, R.; Zhu, H.; Kang, M.; Jiao, J.; Yang, B.; Qiao, G.; Zhao, A.; Liu, J. Hydrogeochemistry of the fluoride in groundwater in the Dali area of the Guanzhong Basin. *J. Groundw. Sci. Eng.* **2009**, *5*, 84–93, (The Chinese Abstract Is English). [CrossRef]
2. Subba Rao, N. Controlling factors of fluoride in groundwater in a part of South India. *Arab. J. Geosci.* **2017**, *10*, 524. [CrossRef]
3. Wang, Z.; Guo, H.M.; Xing, S.P.; Liu, H. Hydrogeochemical and geothermal controls on the formation of high fluoride groundwater. *J. Hydrol.* **2021**, *598*, 126372. [CrossRef]
4. Chidambaram, S.; Sarathidasan, J.; Srinivasamoorthy, K.; Thivya, C.; Thilagavathi, R.; Prasanna, M.V.; Singaraja, C.; Nepolian, M. Assessment of hydrogeochemical status of groundwater in a coastal region of Southeast coast of India. *Appl. Water Sci.* **2018**, *8*, 27. [CrossRef]
5. Liu, R.P.; Zhu, H.; Yang, B.C.; Zhao, A.; Ke, H.L.; Qiao, G. Occurrence Pattern and Hydrochemistry Cause of The Shallow Groundwater Fluoride in the Dali County, Shaanxi Province. *Northwestern Geol.* **2008**, *41*, 0134–0142, (The Chinese Abstract Is English).
6. Ou, H.; Lu, G.P.; Hu, X.N. Fluoride enrichment in geothermal waters in Xinyi-Lianjiang region, Guangdong. *Environ. Chem.* **2019**, *38*, 1128–1138.
7. Liu, R.-P.; Zhu, H.; Liu, F.; Dong, Y.; El-Wardany, R.M. Current situation and human health risk assessment of fluoride enrichment in groundwater in the Loess Plateau, China: A case study of Dali County, Shaanxi Province, China. *China Geol.* **2021**, *4*, 487–497. [CrossRef]
8. Yuan, R.X.; Wang, G.L.; Liu, F.; Zhang, W.; Cao, S.W. Study on the indication of fluorine of the low—Medium temperature convective geothermal system in Northeastern Hebei Province. *Geol. Rev.* **2021**, *67*, 218–230.
9. Shi, W.D.; Guo, J.Q.; Zhang, S.Q.; Ye, C.M.; Li, J.; Ma, X.H. the distribution and geochemistry of geothermal groundwater bearing Fand as in the guide basin. *Hydrogeol. Eng. Geol.* **2010**, *37*, 0036–0041. [CrossRef]
10. Zhang, F.X. Study on environmental hydrogeology in endemic areas of drinking water in Qinghai Plateau—Preliminary study on the sources and occurrence laws of pathogenic elements in groundwater in endemic areas of Guide County. *China Ventur. Cap.* **2012**, *27*, 93–95.
11. Zhang, C.; Huang, R.; Qin, S.; Hu, S.; Zhang, S.; Li, S.; Zhang, L.; Wang, Z. The high-temperature geothermal resources in the Gonghe-Guide area, northeast Tibetan plateau: A comprehensive review. *Geothermics* **2021**, *97*, 102264. [CrossRef]
12. Jiang, Z.; Xu, T.; Owen, D.D.R.; Jia, X.; Feng, B.; Zhang, Y. Geothermal fluid circulation in the Guide Basin of the northeastern Tibetan Plateau: Isotopic analysis and numerical modeling. *Geothermics* **2018**, *71*, 234–244. [CrossRef]
13. Wang, Z. Distribution and Genesis Mechanism of HighArsenic Groundwater in the Guide Basin, Qinghai, China. Ph.D. Thesis, China University of Geosciences (Beijing), Beijing, China, 2019; pp. 1–128.
14. Dai, W. The Hydrogeochemical Characteristics and the Evolution of Geothermal Water in Guide area, Qinghai. Master's Thesis, China University of Geosciences (Beijing), Beijing, China, 2020; pp. 1–75.
15. Sun, H.; Ma, F.; Liu, Z.; Liu, Z.; Wang, G.; Nan, D. The distribution and enrichment characteristics off fluoride in geothermal active area in Tibet. *China Environ. Sci.* **2015**, *35*, 251–259.
16. Zhang, Q.; Tan, H.B.; Qu, T.; Zhang, W.J.; Zhang, Y.F.; Kong, N. Impacts of typical harmful elements in geothermal water on river water quality in Tibet. *Water Resour. Prot.* **2014**, *30*, 23–29.
17. Zhang, C.; Jiang, G.; Shi, Y.; Wang, Z.; Wang, Y.; Li, S.; Jia, X.; Hu, S. Terrestrial heat flow and crustal thermal structure of the Gonghe-Guide area, northeastern Qinghai-Tibetan plateau. *Geothermics* **2018**, *72*, 182–192. [CrossRef]
18. *GB/T 14848-2017*; General Administration of Quality Supervision, Inspection and Quarantine of China, Standardization Administration of China, Standards for Groundwater Quality. Standards Press of China: Beijing, China, 2017. (In Chinese)
19. WHO. *Guidelines for Drinking Water Quality: Fourth Edition Incorporating the First Addendum*; World Health Organization: Geneva, Switzerland, 2017.

20. US EPA. *Available Information on Assessment Exposure from Pesticides in Food;* U.S. Environmental Protection Agency Office of Pesticide Programs: Washington, DC, USA, 2000; pp. 1–16.
21. USEPA. Regional Screening Levels (RSLs)—Generic Tables. 2017. Available online: https://www.epa.gov/risk/regional-screening-levels-rsls-generic-tables-november-2017 (accessed on 17 September 2022).
22. Ma, R.; Wang, Y.; Sun, Z.; Zheng, C.; Ma, T.; Prommer, H. Geochemical evolution of groundwater in carbonate aquifers in Taiyuan, northern China. *Appl. Geochem.* **2011**, *26*, 884–897. [CrossRef]
23. Fan, J.J.; Tong, Y.Q.; Li, J.Y.; Wang, L.X.; Li, R.; Liu, Z.Y. Affecting Factors of high-fluorine water in our country and scheme to avoid fluorosis. *Saf. Environ. Eng.* **2008**, *15*, 14–16, (In Chinese with English abstract).
24. Narsimha, A.; Rajitha, S. Spatial distribution and seasonal variation in fluoride enrichment in groundwater and its associated human health risk assessment in Telangana State, South India. *Hum. Ecol. Risk* **2018**, *4*, 2119–2132. [CrossRef]
25. Singh, C.K.; Kumari, R.; Singh, N.; Mallick, J.; Mukherjee, S. Fluoride enrichment in aquifers of the Thar Desert: Controlling factors and its geochemical modelling. *Hydrol. Process* **2012**, *27*, 2462–2474. [CrossRef]
26. Lang, X.J.; Lin, W.J.; Liu, Z.M. Hydrochemical Characteristics of Geothermal Water in Guide Basin. *Earth Sci. J. China Univ. Geosci.* **2016**, *41*, 1723–1734.
27. Guo, W.C.; Shi, X.M. The development and utilization of Guide basin.s geothermal resources of Qinghai province. *Hydrogeol. Eng. Geol.* **2008**, *3*, 0079–0080, 0092. [CrossRef]
28. Xiao, J.; Jin, Z.; Zhang, F. Geochemical controls on fluoride concentrations in natural waters from the middle Loess Plateau, China. *J. Geochem. Explor.* **2015**, *159*, 252–261. [CrossRef]
29. Liu, D.; Xiao, C.; Liang, X. Distribution characteristics and risk assessment of fluorine in groundwater in Baicheng City. *People's Yangtze River* **2019**, *6*, 25–28.

water

MDPI

Article

The Impact of Molybdenum Mining on Cd Pollution along Wenyu Stream in Qinling Mountains, Northwest China

Huaqing Chen [1], Aning Zhao [1,*], Youning Xu [1], Jianghua Zhang [2] and Min Yang [3,*]

1 Xi'an Geological Survey Center of China Geological Survey, Xi'an 710054, China; chuaqing@mail.cgs.gov.cn (H.C.); xyouning@mail.cgs.gov.cn (Y.X.)
2 Field Observation Base for Environmental Geology of Typical Mines in Shaanxi Province, Ministry of Natural Resources, Xi'an 710054, China
3 School of Resources Engineering, Xi'an University of Architecture and Technology, Xi'an 710055, China
* Correspondence: zaning@mail.cgs.gov.cn (A.Z.); ymin@xauat.edu.cn (M.Y.)

Abstract: Mining has brought many environmental problems to the surrounding soil, water, and air, with toxic elements contaminating surface water, threatening ecological balance and human health. This study selected the Wenyu watershed downstream from a large molybdenum mine in the Qinling Mountains as the study area, aiming to explore the impact of molybdenum mining on surface water quality. The content characteristics of Cd, Pb, Cu, Cr and Hg in surface water, sediment, and rock samples were analyzed by field sampling and chemical testing. The results showed only obvious Cd pollution. The pollution status and ecological risk level of surface water and sediment samples in the Wenyu Stream watershed were evaluated using the single pollution index method, geo-accumulation index method, and Hakanson potential ecological risk assessment method. Finally, the sources of Cd pollution and the impact of mining on Cd distribution in the Wenyu Stream were comprehensively discussed. The research results showed that Cd content in the Wenyu Stream was significantly affected by mining activity and the coefficient of variation of Cd content reached 99.44%. Among 22 surface water samples, 21 samples met the Class II water standard, indicating a clean overall water quality of the Wenyu Stream, and only one sample exceeded the Class II water standard with a mild pollution level. All 15 sediment samples were polluted to varying degrees and the most severely polluted sample had reached a moderate to strong pollution level. Most of the samples were at a moderate pollution level. The potential ecological hazard indexes of Cd content were at medium to very strong risk level, indicating that the overall sediment in the main ditch of the Wenyu Stream was under a strong ecological risk level. The main sources of Cd pollution, including acid mine drainage, regional geological background, sediment release, and atmospheric dry and wet deposition, were discussed.

Keywords: molybdenum mining; cadmium pollution; surface water; sediment

check for updates

Citation: Chen, H.; Zhao, A.; Xu, Y.; Zhang, J.; Yang, M. The Impact of Molybdenum Mining on Cd Pollution along Wenyu Stream in Qinling Mountains, Northwest China. *Water* **2023**, *15*, 2779. https://doi.org/10.3390/w15152779

Academic Editors: Giovanni Esposito and Laura Bulgariu

Received: 28 June 2023
Revised: 28 July 2023
Accepted: 30 July 2023
Published: 31 July 2023

1. Introduction

Mining resources exploration has made great contributions to the development of the local economy and has brought a series of favorable conditions for local employment, taxation, etc. However, unreasonable and unscientific mining resource development has also caused a series of ecological environment deterioration problems, such as vegetation damage, landscape damage, groundwater level decline, water pollution, and heavy metal pollution in crops [1–4]. The problem of heavy metal pollution in surface water has always been the focus of many studies because it is related to the daily life of residents living near the river, such as drinking water quality, crop irrigation, and food safety. Predecessors have investigated the total amount of heavy metal elements in the sediments of Taiyu River and Taiyu Reservoir in Tongguan gold mining area and discussed the pollution degree of heavy metal elements in the river sediments concerning the gold mining activities [5].

Some studies also showed that the contents of Hg, Pb, Cd, As, Cr, Cu, and Zn in the sediments of the Xiyu river were adjacent to the Taiyu River and evaluated the pollution of the Xiyu River sediment affected by gold mining activities. The results indicated that Hg and Pb were the main pollution elements. Meanwhile, the elements with strong potential ecological hazards in the whole Xiyu River were revealed to be Hg and Pb through the evaluation of the potential ecological hazard index method [6]. Another study analyzed the content of heavy metal elements in the river water and sediment of typical rivers in the Xiaoqinling gold mining area and four rivers, with seven heavy metal elements in the mining area exceeding the national standard limit. The main occurrence form of heavy metal elements in rivers was the sedimentary state, and the adsorption and desorption of sediment were the main controlling factors for the changes of heavy metal elements in river sediment and river water along the river [7]. In southern China, the mining drainage water from Dabaoshan polymetallic mine in Guangdong Province has a long-term impact on the Hengshi River. Many studies have studied the spatial distribution of the concentration of six heavy metal elements (Mn, Cu, Zn, As, Cd, and Pb) in the suspended solids of the river. The results showed that the suspended solids in the Hengshi River were seriously polluted by heavy metals from Dabaoshan mining drainage water, and the concentration of heavy metals gradually decreased along the flow path, which was significantly positively correlated with the content of suspended solids [8]. Subsequently, 60 water samples were collected in the Dabaoshan mining area by 1 km square grid cells, and the distribution characteristics of 10 toxic elements, including Cr, Cd, Co, Ni, Cu, Zn, As, Sb, Hg, and Pb, were obtained through chemical testing. It was found that the most serious pollution was Cd, followed by Zn, Cu, Pb, Cr, Ni, and Hg. According to the analysis of the concentration center distribution of elements, the pollution sources are mainly pit soil, tailings, waste rock piles, and civilian mining sites [9]. Many previous studies on heavy metal pollution of rivers in mining areas in China have played a strong role in promoting the prevention and control of river pollution and achieved fruitful results. However, for the research on heavy metal pollution of rivers in the mining area, most predecessors used sediment to deduce the pollution of river water, while few studies have comprehensively evaluated water pollution by directly testing river water samples and studying the content characteristics of heavy metal elements in sediment, rock, and ore samples [10,11].

Cadmium (Cd) is a toxic heavy metal element which has the characteristics of high toxicity, difficult degradation, and easy residue and is one of the "five toxins" in the environment. Mining, beneficiation and metallurgy, sewage irrigation, and other activities will lead to the entry of Cd into the environment, which will not only affect the quality of water and soil, cause harm to the ecological environment, but also directly affect the quality and safety of agricultural products and threaten people's health through the food chain [12]. In recent years, heavy metal pollution incidents have occurred frequently, causing serious harm to people's production and life and local water environment ecology. Among them, the cadmium pollution incident in Longjiang, Guangxi province, in January 2012, had a significant social impact [13]. Since the 1920s, with the development of the global electrolysis industry, the annual production of Cd has significantly increased, and the environmental pollution problems caused by Cd have also emerged [14]. The most famous public nuisance event caused by Cd pollution is the Itai-itai disease event in the Jinzū River of Toyama Prefecture, Japan [15]. In addition, according to relevant research reports, the "cancer village" in Longling village, Hua County, Shaanxi Province, may also be related to Cd pollution [16]. Therefore, studying the pollution characteristics of representative harmful heavy metal element Cd is of great significance for river pollution prevention and control work.

Qinling Mountains is an important water conservation area for the South–North Water Transfer Project. In the journey of carrying out ecological civilization construction, its important position is increasingly prominent [17,18]. As a unique resource treasure trove and national-level ecological environment protection zone in recent years, Shaanxi Province has issued a series of ecological environment protection policies and regulations for the Qinling Mountains [19,20]. A previous study had discussed the overlying water and sediment in a reservoir downstream of a molybdenum mining area, but a comprehensive analysis of the pollution of surface water and sediment in streams or rivers has not yet been conducted [21]. In order to explore the impact of molybdenum mining activities on surface water quality, Wenyu Stream, within the influence range of a large molybdenum mine exploration in the Qinling Mountains, was selected as the study area. On the basis of analyzing the Cd, Pb, Cu, Cr and Hg contents in surface water, sediment, and surrounding rocks, only Cd exceeded the limit given by Soil Environmental Quality—Risk Control Standard for Soil Contamination of Agricultural Land (GB 15618-2018) and Environmental Quality Standard for Surface Water (GB 3838-2002) [22,23]. Then, the single pollution index method was used to evaluate the pollution status of Cd in surface water in Wenyu Stream. The geoaccumulation index and potential ecological risk evaluation were applied to study the pollution status and ecological risk degree of Cd in sediment and to explore the source of Cd in river water. This manuscript aims to provide a scientific basis for the sustainable development of the mining industry and the prevention and control of water and soil environmental pollution in the region.

2. Materials and Methods

2.1. Study Area

Wenyu Stream originates from the valley of the southeast slope of the Qinling Mountains, about 5 km northwest of Jindui town in the southeast of Shaanxi province. The stream is formed by several mountain streams with a total length of about 33 km from north to south. The mining and mineral processing area is 26 km away from the Luohe River downstream of the Wenyu Stream (Figure 1). The stream bed in the valley at the upper reaches of the stream is narrow, with a rapid and small flow. Only a small amount of sand and gravel are deposited at the bottom of the stream bed. Bedrocks are exposed in some sections, and farmland is rarely distributed on both banks. The terrain from the middle reaches of the stream to the downstream banks is relatively flat, the stream bed is becoming wider, the flow is slowing down, the flow is gradually increasing due to the continuous inflow of branch streams, and the sediment at the bottom of the stream bed is gradually increasing. The middle and upper reaches of Wenyu Stream are a large-scale molybdenum open-pit mining and beneficiation area. The stream is about 7 km long from north to south. Mining facilities, such as open-pit mining, waste piles, tailings pond, and concentrator, are distributed on the terraces on the east and west sides of the ditch. The study area is situated within the monsoon humid climate zone at the southern edge of the warm temperate zone, featuring distinct climatic patterns in mountainous regions. The mean annual temperature within this mining locale averages 11.5 °C. Precipitation levels typically range around 770 mm annually, with a significant portion of this precipitation concentrated between July and October, constituting approximately 60% of the total. Throughout the year, the prevailing wind direction is predominantly northwest-east, with an annual average wind speed of approximately 1.5 m/s [24]. The fault structure in the study area is the Yanmen sag fault, with strike of 70° to 90°, dip of 62° to 72° and width of 200 m to 300 m. The exposed strata are mainly volcanic rocks of the Middle Proterozoic Xionger group, including metamorphic andesitic porphyrite and tuffaceous slate. In the southeast of the study area, some quartzite of the Mesoproterozoic Daokou group is unconformably covered in the Xionger group. The intrusive rocks are mainly Yanshanian granite porphyry [25]. The main minerals in the large molybdenum mining area are molybdenite and pyrite, accompanied by sphalerite and chalcopyrite. The mine has been explored for over five decades since 1966.

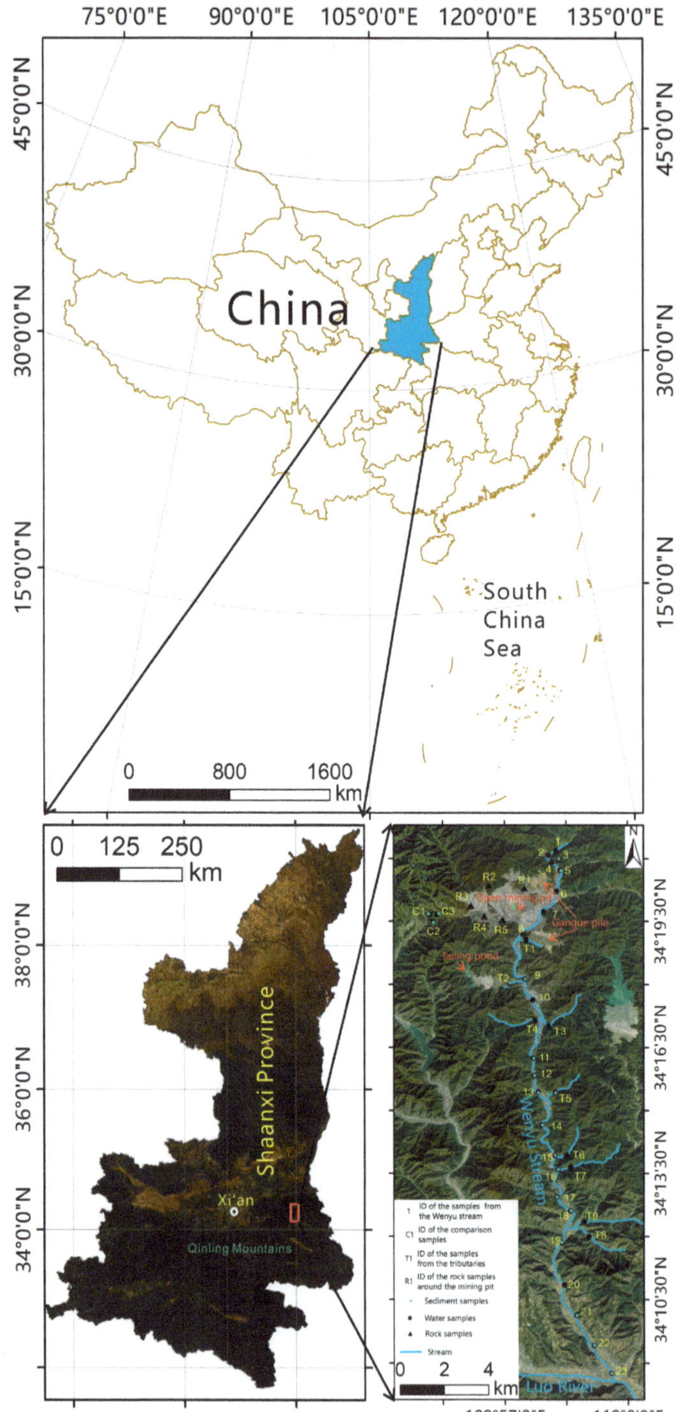

Figure 1. The map of the study area.

2.2. Methods

In this study, surface water and sediment samples were collected from north to south at an average density of one sample per kilometer in the Wenyu Stream Basin and within a process of about 26 km that may be affected by mining activities. Water samples and sediment samples upstream of the tributary inlet of Wenyu Stream were also collected. A total of 32 water samples and 17 sediment samples were collected in the field work from July to August in the year of 2022 (Figure 1). In order to compare and evaluate the impact of Molybdenum Mining on the enrichment of toxic elements in streams and the source of pollutants, three water and sediment samples were collected as the contrast samples at the ridge where there was no mining activity at the source of the Wenyu Stream (sampling points of C1, C2, and C3 in Figure 1) and five surrounding rock samples were collected around the open pit (Figure 1, sampling points of R1 to R5).

When collecting water samples, 500 mL clean polyethylene bottles were used to collect surface water samples, and small shovels were used to collect fine sediment at the bottom of the streambed. The sediment samples were put into cloth bags with plastic bags inside. The sediment samples were dried at room temperature, ground with a mortar, and sieved through a 200 mesh sieve for analysis and testing. Rock samples were processed to 200 mesh by coarse crushing, medium crushing, and shrinkage fine grinding for analysis and testing. A certain amount of sediment powder was weighed, an appropriate amount of aqua regia was added, and the mixture was heated on a heating plate to boil slightly for 2 h for digestion. After digestion, the solution was cooled, filtered with qualitative filter paper, and collected. The volume was fixed to a certain scale. A certain amount of rock powder was weighed and added into a tetrafluoroethylene crucible, distilled water was added to infiltrate, then an appropriate amount of perchloric acid was added, and it was heated on the electric heating plate to dissolve. The solution was removed when it was nearly evaporated to dryness, an appropriate amount of hydrofluoric acid and distilled water was added after cooling, and then evaporate to dryness. Finally, the crucible was removed, an appropriate amount of perchloric acid was added, and then it was heated until the white smoke of perchloric acid disappeared. After evaporation, an appropriate amount of hydrochloric acid and a small amount of distilled water were added, and then it was heated until the solution is transparent. The solution was cooled and then fixed to a certain scale. The analysis of Cd in water sample, sediment, and rock was completed by Xi'an Geological Survey Center of China Geological Survey. All testing steps were carried out in strict accordance with the Technical Code for China Geological Survey—Technical Requirements for Analysis of Ecological Geochemical Evaluation Samples (DD2005-03) [26]. The pH values of the water samples were measured in the laboratory with Rex phs-3c pH meter (Shanghai Inesa Scientific Instrumental Co., Ltd., Shanghai, China), the measurement resolution was 0.01 pH, and the accuracy was about ± 0.01 pH. The Cd in the water sample was determined by an inductively coupled plasma mass spectrometer (ICP-MS) (Thermo Fisher Scientific Inc., Waltham, MA, USA). Before determination, the water samples were acidified with 1% superior pure nitric acid to pH < 2. The contents of Cd in sediment samples were mainly determined by an induced coupled plasma atomic emission spectrometer (ICP-AES) (Thermo Fisher Scientific Inc., Waltham, MM, USA). Rock samples were determined by an atomic fluorescence spectrometer (AFS-2202E) (Beijing Haiguang Instrument Co., Ltd., Beijing, China).

The relative deviation limits of Cd, Pb, Cu, Cr, and Hg in the two standards of Soil Environmental Quality—Risk Control Standard for Soil Contamination of Agricultural Land (GB 15618-2018) and Environmental Quality Standard for Surface Water (GB 3838-2002) should be within 20%, and 10% of the duplicate samples of each batch of tests should be randomly selected for retest. In this study, 3 water samples and 2 sediment samples were selected as duplicate samples. The quality of sample test results shall be evaluated according to the relative deviation:

$$\mathrm{RD} = \frac{|A - B|}{(A + B)} \times 100\% \quad (1)$$

where RD is the relative deviation, A is the measured values, and B is the measured values of the duplicate samples.

The water quality in this study was evaluated using the single pollution index method. In accordance with the objectives of watershed classification for Wenyu Stream, which serves as a primary drinking water source protection area, the water quality of Wenyu Stream must meet the Class II surface water standard [27]. The Cd single pollution indexes of surface water and sediment were calculated by the following formula [28]:

$$P_{Ci} = (C_i - C_0)/C_0 \quad (2)$$

where P_{Ci} is the single pollution index of Cd element, C_i represents the measured concentration of Cd content (surface water in mg/L, sediment in mg/kg), C_0 is the limit value of Cd content in surface water quality standard Class II (mg/L) or the Cd content of the background from the contrast area (mg/kg) [23]. The single-exceedance factors can generally be divided into five levels to represent the range of pollution levels (Table 1).

Table 1. The relationship between single pollution index and the pollution degree level.

Pollution Level	No Pollution	Light Pollution	Moderate Pollution	Heavy Pollution	Extreme Pollution
The single pollution index (P_{Ci})	$P_{Ci} \leq 0$	$0 < P_{Ci} \leq 1$	$1 < P_{Ci} \leq 4$	$4 < P_{Ci} \leq 10$	$10 < P_{Ci}$

In order to comprehensively analyze the impact of pollutants on surface water/sediment, the Nemerow pollution index method was used to evaluate Cd, Pb, Cu, Cr, and Hg in surface water/sediment. Nemerow pollution index method reflects the comprehensive effect of various pollutants on surface water/sediment and highlights the impact of high-concentration pollutants on the environmental quality of surface water/sediment. The calculation formula is:

$$P_{Ni} = \sqrt{\frac{P_{imean}^2 + P_{imax}^2}{2}}$$

where P_{Ni} is the Nemerow pollution index of sample i, P_{imean} is the average value of all the single pollution indexes of sample i, P_{imax} is the max value of all the single pollution indexes of sample i. The Nemerow pollution index can generally be divided into five levels to represent the range of pollution levels [22,23] (Table 2).

Table 2. The relationship between the Nemerow pollution index and the pollution degree level.

Pollution Level	No Pollution	Light Pollution	Moderate Pollution	Heavy Pollution	Extreme Pollution
The Nemerow pollution index (P_{Ni})	$P_{Ni} \leq 1$	$1 < P_{Ni} \leq 2$	$2 < P_{Ni} \leq 3$	$3 < P_{Ni} \leq 5$	$5 < P_{Ni}$

The sediment is an important component of the riverbed structure, and the sediment not only serves as a repository for heavy metal pollutants but also as a potential secondary pollution source with potential impacts on water quality in the aquatic-sediment system [29]. The content of heavy metal elements in riverbed sediments is an important reference index for determining the quality of the water environment [30]. In this study, the geoaccumulation index method was used to evaluate 16 sediment samples collected in Wenyu Stream watershed. The geoaccumulation index method was proposed by Müller from the Institute of Sedimentology at Heidelberg University in 1969. It is a quantitative evaluation method for studying the degree of heavy metal pollution in sediments in water environments, and it is currently one of the most widely used methods for evaluating modern sediments with heavy metal pollution [31,32]. The specific calculation formula is provided below:

$$I_{geo} = \log_2\left(\frac{C_n}{kB_n}\right) \tag{3}$$

where B_n is the geochemical background, which is the average Cd content of three sediment samples from the contrast area (0.74 mg/kg), C_n is the measured content of pollutants in sediments (mg/kg), k is a constant generally taken to be 1.5 indicating that the background may vary due to differences in sedimentary characteristics and litho-geological features. The geoaccumulation indexes can generally be divided into seven levels [33] to indicate the range of pollution degrees (Table 3).

Table 3. Geoaccumulation index and pollution classification.

Pollution Level	No Pollution	No to Moderate Pollution	Moderate Pollution	Moderate to Severe Pollution	Severe Pollution	Severe to Extreme Pollution	Extreme Pollution
Index range	$I_{geo} \leq 0$	$0 < I_{geo} \leq 1$	$1 < I_{geo} \leq 2$	$2 < I_{geo} \leq 3$	$3 < I_{geo} \leq 4$	$4 < I_{geo} \leq 5$	$5 < I_{geo}$
Level	0	1	2	3	4	5	6

The ecological risk of Cd in river sediments was evaluated using the Potential Ecological Risk Index (PERI) method. PERI is a method developed by Swedish scholar Hakanson in 1980 for evaluating the pollution and ecological hazards of heavy metal elements in sediments and soils [34]. The advantage of this method is that it takes into account the concentration effect, ecological effect, and toxicity effect of heavy metal elements, which allows for both the calculation of an individual pollutant's potential ecological risk index ($E^i{}_r$), which measures the level of ecological risk caused by a single pollutant, as well as the calculation of the comprehensive ecological impact of multiple pollutants through a composite potential ecological risk index. In this study, we primarily used $E^i{}_r$ to evaluate the samples of riverbed sediment from the main channel of Wenyu Stream, and the formula used in the calculation is provided below:

$$E_r^i = T_r^i \times \left(\frac{E_s^i}{C_n^i}\right) \tag{4}$$

where $E^i{}_r$ is the potential ecological hazard index of a Cd element, $T^i{}_r$ is the toxicity response coefficient of Cd element, $E^i{}_s$ is the measured Cd content of sediment sample (mg/kg), and $C^i{}_n$ is the Cd background reference of sediment samples, in units of mg/kg. Based on previous research [35], the toxicity response coefficient $T^i{}_r$ for Cd in this study was set to 30, while the background reference $C^i{}_n$ was adopted as the average of the Cd content in three bottom sediment samples from the contrast area, which was 0.74 mg/kg. The relationship between the classification of single pollutant potential ecological hazard index and ecological risk level is shown in Table 4.

Table 4. The relationship between the potential ecological hazard index of a single pollutant and the degree of ecological risk.

Ecological Risk Degree	Slight	Medium	Strong	Very Strong	Extremely Strong
Index range	<40	$40 \leq E_r{}^i < 80$	$80 \leq E_r{}^i < 160$	$160 \leq E_r{}^i < 320$	$320 \leq E_r{}^i$

3. Results

3.1. The Test Quality of the Five Elements in Surface Water and Sediment Samples

In this study, three surface water samples and two sediment samples were randomly selected for duplicate sample testing, and the test results are shown in Table 5. The relative deviation of Cd, Pb, Cu, Cr, and Hg in surface water samples is between 0.00% and 7.69%, meeting the requirement of less than 20% relative deviation for parallel sample testing specified in the Environmental Quality Standard for Surface Water (GB 3838-2002) [23]. The relative deviation of Cd, Pb, Cu, Cr, and Hg in the sediment samples is between 2.05% and 5.43%, which also meets the requirement of parallel sample testing with a relative deviation of less than 20% specified in the Soil Environmental Quality Risk Control Standard for Soil Continuity of Agricultural Land (GB 15618-2018) [22].

Table 5. The test quality of the elements in surface water and sediment samples.

Elements			Cd	Cu	Pb	Cr	Hg
Water samples	15	Results (mg/L)	0.00200	0.01300	0.00100	0.00100	0.00005
		Duplicate results (mg/L)	0.00190	0.00140	0.00090	0.00110	0.00005
		RD (%)	2.56	3.70	5.26	4.76	0.00
	20	Results (mg/L)	0.00200	0.00900	0.00100	0.00100	0.00005
		Duplicate results (mg/L)	0.00210	0.01030	0.00110	0.00090	0.00005
		RD (%)	2.44	6.74	4.76	5.26	0.00
	T1	Results (mg/L)	0.00600	0.08700	0.00100	0.00100	0.00005
		Duplicate results (mg/L)	0.00700	0.08900	0.00110	0.00105	0.00005
		RD (%)	7.69	1.14	4.76	2.44	0.00
Sediment samples	3	Results (mg/kg)	3.78000	82.70000	620.00000	93.70000	1.28000
		Duplicate results (mg/kg)	3.95000	87.10000	646.00000	98.50000	1.17000
		RD (%)	2.20	2.95	2.05	2.50	4.49
	5	Results (mg/kg)	1.76000	50.50000	221.00000	66.70000	0.13800
		Duplicate results (mg/kg)	1.89000	56.30000	236.00000	63.20000	0.13100
		RD (%)	3.56	5.43	3.28	2.69	2.60

3.2. The Cd Exceedance in the Surface Water of the Wenyu Stream

The statistical results of Cd content in 35 water samples from the Wenyu Stream and contrast area are shown in Figure 1. Figure 2a showed that the Cd contents of 14 samples from total 23 samples in the main ditch of the Wenyu Stream were significantly higher than those in the contrast areas which were not influenced by mining activities. Moreover, the average Cd content in the main ditch of Wenyu Stream was twice as high as that in the contrast area, with the highest value being 10 times higher. The results indicated that the cumulative effect caused by Cd contamination in Wenyu Stream was significant due to mining activities. Among the nine samples in the tributary ditches of the Wenyu Stream, six samples did not exceed the contrast areas named T4, T5, T6, T7, T8, and T9. Samples from T1, T2, and T3 were two, four, and six times higher in Cd content than the contrast areas, respectively. Additionally, from a statistical perspective, the coefficient of variation for Cd content reaches 99.44%, and the Cd content presents an uneven distribution from upstream to downstream indicating that Cd pollution in the Wenyu Stream was significantly influenced by mining activities (Table 6). The intensive human activities in the study area mainly include mining, transportation, ore processing, and storage of surrounding rock wastes and tailings.

According to the Cd limit of Class II water from the Environmental Quality Standards for Surface Water Standard (GB3838-2002), 22 of the 23 samples in the main ditch of the Wenyu Stream met the standards of Class I and Class II water accounting for 95.65%, indicating the overall water quality of Wenyu Stream was clear. The Class I water samples accounted for 52.17%, including samples 1, 2, 3, 4, 5, 6, 7, 8, 18, 21, 22, and 23 in Figure 1b. The Class II water samples accounted for 43.48%, including samples 9, 11, 12, 13, 14, 15, 16, 17, 19, and 20 in Figure 2b. The sample of Class V water accounted for 6.25%, including sample 10 in Figure 2b. The water quality of six water samples from the nine tributaries of Wenyu Stream were Class I water, which were samples T4, T5, T6, T7, T8, and T9, respectively. Sampling point T1 was Class II water and sampling points T2 and T3 were Class V water.

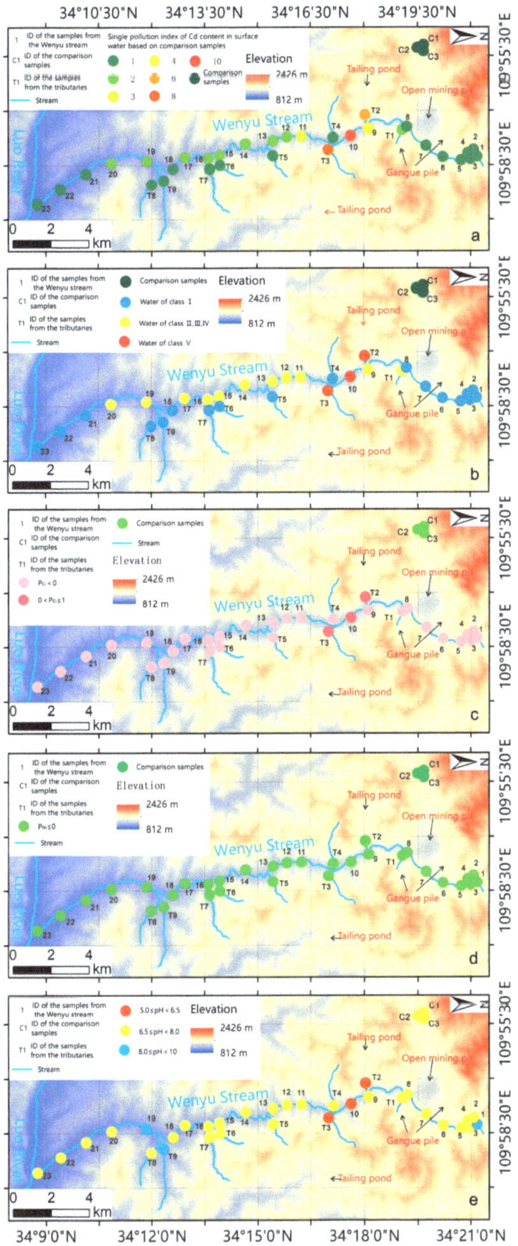

Figure 2. The evaluation maps of the surface water in Wenyu Stream. (**a**). The single pollution indexes of Cd content in surface water based on comparison samples. (**b**). The water quality classes divided according to the Environmental Quality Standards for Surface Water Standard (GB3838-2002). (**c**). The single pollution indexes of Cd content in surface water based on Class II water limit issued by the Environmental Quality Standards for Surface Water Standard (GB3838-2002). (**d**). The Nemerow pollution indexes of Cd, Pb, Cu, Cr, and Hg contents in surface water based on Class II water limit issued by the Environmental Quality Standards for Surface Water Standard (GB3838-2002). (**e**). The water pH of the surface water samples.

Table 6. Statistical characteristics of Cd, Pb, Cu, Cr, and Hg content in surface water of the Wenyu Stream.

Elements	Mode (mg/L)	Mean (mg/L)	Median (mg/L)	Range (mg/L)	Range/Mode	Range/Average	Standard Deviation (mg/L)	Coefficient of Variation (%)	Limits (mg/kg) [23]	Average Single Pollution Index
Cd	0.001	0.001	0.001	0.001–0.01	9.000	4.299	0.002	99.440	0.005	0.960
Pb	0.001	0.001	0.001	0.001–0.001	0.000	0.000	0.000	0.000	0.010	0.000
Cu	0.001	0.021	0.008	0.001–0.280	86.000	4.151	0.019	93.800	1.000	−0.984
Cr	0.001	0.001	0.001	0.001–0.001	0.000	0.000	0.000	0.000	0.050	−0.980
Hg	0.00005	0.00005	0.00005	0.00005–0.00005	0.000	0.000	0.000	0.000	0.00005	0.000

As shown in Figure 2c, the Cd single pollution index of sample 10 in the main ditch of the Wenyu Stream was 1, attributed to mild pollution, and the other 22 samples were not beyond the limit. Among the nine water samples in the tributary ditches of Wenyu Stream, seven samples from T1, T4, T5, T6, T7, T8, and T9 were unpolluted by Cd content. Samples T2 and T3 were attributed to mild pollution. The Nemerow pollution indexes of the surface water samples shown in Figure 2d were all below 1. Concerning the position of the over-limit sample 10 in Figure 2b,e, this point was located near the mineral processing production area, approximately 2.0 km away from the mining area. Additionally, a large tailing pond is located on the west bank, about 1 km upstream. The T2 sample at this location was Class V water. Therefore, the abnormal Cd content of sample 10 may be related to the drainage water from the tailings pond.

3.3. *The Cd Accumulation Degree of Stream Sediment*

From the statistical results of Cd contents in sediment samples (Table 7), the range of Cd contents was 7.06, with significant variations, and the coefficient of variation was 43.93%. Moreover, the Cd concentration in each sample from upstream to downstream exhibits an asymmetrical distribution, indicating a significant disturbance by human activities. The results of the geoaccumulation index analysis showed that all 15 sediment samples from the main ditch of Wenyu Stream exhibit varying degrees of pollution, with point numbers 15 and 16 having the highest levels of pollution, reaching moderate to severe levels (Figure 3a). The least pollution level included point numbers 5, 13, and 22, ranging from not polluted to moderate pollution. Most points fall within moderate levels of pollution. The T2 and T5 sediment samples in two tributary ditches of Wenyu Stream also reached moderate levels of pollution (Figure 3c). The Cd content in sediment from the entire watershed, both individually and averagely, was significantly higher than that in the contrast area. This reflects the impact of external heavy metal element inputs on Wenyu Stream, leading to the accumulation and enrichment effects of Cd in the river water. As shown in Figure 3b, only tree Nemerow pollution indexes of the sediment samples were between 1 and 2, indicating light pollution. Previous studies analyzed and evaluated the heavy metal pollution of sediments in the middle and upper reaches of the Beijiang River in northern Guangdong Province by testing eight elements of Cu, Pb, Zn, Cd, Ni, Cr, As, and Hg using the geoaccumulation index. The results showed that the pollution of toxic elements in the main stream of the Beijiang River was more serious than that in the tributaries due to long-term accumulation, and the main sources of pollution were the existence of multiple mining and smelting enterprises near the river [36]. In addition, the contents of toxic elements in tailings and river sediments downstream of six tailings ponds in southern China were compared and analyzed. The study found that the contents of Cu, Pb, Zn, Cd, Ni, Cr, As, Hg, and Sb in river sediments downstream of some tailings ponds were higher than those in tailings, indicating that the accumulation effect of elements could cause the content of toxic elements in sediments to be higher than that in pollution sources [37]. These results demonstrate that mining activities are significant contributors to the accumulation and exceedance of heavy metals in river sediments near mining areas, and studying sediment pollution and enrichment characteristics is crucial for reflecting river water pollution conditions.

Table 7. Statistical characteristics of Cd, Pb, Cu, Cr, and Hg content in sediment samples from the Wenyu Stream.

Elements	Mean (mg/kg)	Median (mg/kg)	Range (mg/kg)	Range/Average	Standard Deviation	Coefficient of Variation(%)	Limits [22] (mg/kg)	Average Single Pollution Index
Cd	3.71	3.41	1.760–7.660	1.90	1.63	43.930	4.000	1.559
Pb	178.100	138.000	23.100–564.000	3.037	141.579	79.494	1000.000	−0.771
Cu	57.988	49.200	16.600–151.000	2.318	36.226	62.471	200.000	−0.673
Cr	63.759	64.600	29.600–98.400	1.097	19.464	30.528	1300.000	−0.949
Hg	0.144	0.094	0.023–0.776	5.211	0.119	82.563	6.000	−0.955

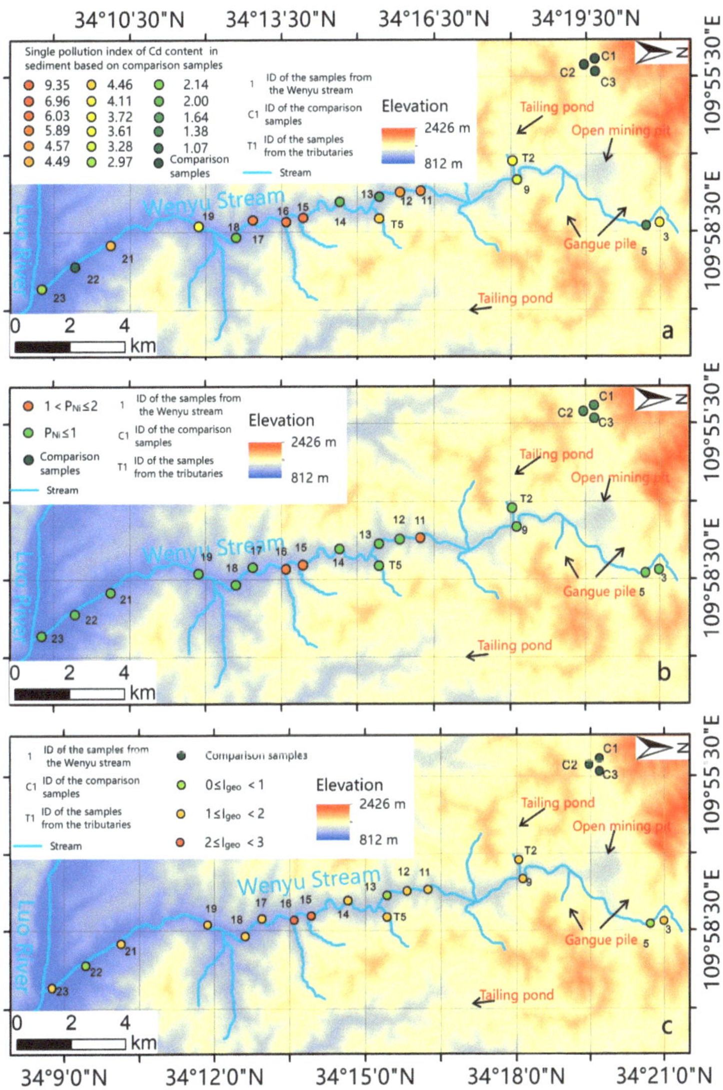

Figure 3. The evaluation maps of the sediment in the Wenyu Stream. (**a**). The single pollution indexes of Cd content in sediment based on comparison samples. (**b**). The Nemerow pollution indexes of Cd, Pb, Cu, Cr, and Hg contents in sediment based on the Soil Environmental Quality–Risk Control Standard for Soil Contamination of Agricultural Land (GB 15618-2018). (**c**). The geoaccumulation indexes of Cd content in sediment.).

3.4. Potential Ecological Risk of Cd Content in Sediment

According to the results of the potential ecological risk index, the potential ecological risk index ($E^i{}_r$) for Cd in each sample were obtained. As shown in Figure 4, the $E^i{}_r$ values for Cd in 15 sediment samples in the main ditch of the Wenyu Stream range from moderate to very high-risk levels, indicating that the overall level of sediment pollution in the main ditch of the Wenyu Stream reached a high level of ecological risk. According to the classification principle in Table 4, samples with strong ecological risk levels accounted for 40.00%, strong ecological risk levels accounted for 46.67%, and samples with moderate ecological risk levels accounted for 13.33%. The Cd potential risks in T2 and T5 samples from two tributaries of the Wenyu Stream were strong and very strong. Therefore, it can be inferred that the contamination of Cd in sediment in the Wenyu Stream has seriously threatened the ecological security of the study area. On the one hand, the accumulation effect caused by mining activities was significant because the content of Cd in the sediment of the Wenyu Stream was high and significantly higher than the contrast area. On the other hand, it is also closely related to the high toxicity response coefficient of Cd element. Several common Cd-containing compounds in nature, such as CdS and CdO, have certain toxicity and are one of the main elements harmful to human health [38,39]. The results of this study are consistent with previous soil heavy metal pollution survey results conducted in the Lalin River Basin in northeastern China [40].

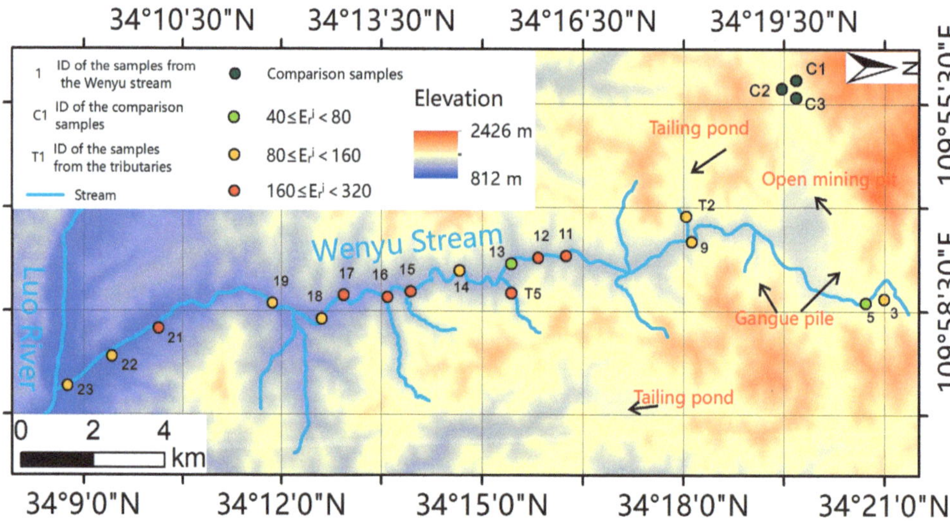

Figure 4. The potential ecological risk maps of the sediment in the Wenyu Stream.

It is worth noting that the average value of the sediment in the contrast area was selected as the background reference ($C^i{}_n$) in the evaluation of the potential ecological risk index of Cd in this study. This indicator implicitly ignored the influence of natural geological background factors on sediment Cd content in the study area and only considered the impact of mining activities, which conformed to the overall exploration ideas of this article. If both effects need to be considered simultaneously, the element indicator for background reference $C^i{}_n$ could be selected as the highest value of heavy metal elements in sediment before global industrialization [41–43]. Furthermore, if it is determined that riverbed sediment mainly comes from upstream and soil erosion on both sides of the river, then the background value of heavy metal elements in soils in the province where the research area is located can be used [44,45]. Different backgrounds have a significant impact on calculating potential ecological hazard indices. Using the background value of heavy metal elements in the research area as a reference value can better reflect the degree of pollution caused by human activities [46,47], which is the basis for selecting the nearby control area sediment average as the background reference value for $C^i{}_n$ in this study.

3.5. The Cd Content of Surrounding Rock

As shown in Table 8, the average Cd content in five rock samples collected from the surrounding area of the large open-pit molybdenum mine is 4.10 times the Cd background in Shaanxi Province and 4.90 times the Cd background in the Guanzhong Basin, indicating a high natural geological background for Cd within a certain range of the molybdenum mine upstream of the Wenyu Stream. Previous studies on the geochemical region of southern Shaanxi Province have also shown that chalcophile elements, tungsten-molybdenum group elements, such as Mo, Cr, V, Cu, Zn, Cd, and Ti, usually occur in high concentrations and have enriched features. In particular, Cd is highly concentrated in some areas, with local concentrations even exceeding five times the background average, forming an excess of microelements [47], which is similar to the findings of this study.

Table 8. The Cd contents in surrounding rock samples and soil Cd background in Shaanxi Province and the Guanzhong Basin.

ID	Cd Content (mg/kg)	Mean (mg/kg)	The Cd Background of Shaanxi Province (mg/kg) [47]	The Cd Background of Soils of Guanzhong Basin (mg/kg) [48]
R1	0.20			
R2	0.24			
R3	0.36	0.58	0.14	0.118
R4	0.11			
R5	2.00			

4. Discussion

This study evaluated the surface water and sediment Cd content in the Wenyu Stream, which is downstream of a large molybdenum mine in the Qinling Mountains. After determining the main pollution problems, the potential sources of pollution were investigated and analyzed. The results can provide targeted countermeasures for subsequent treatment and restoration. Currently, two levels of source apportionment were discussed in previous studies: The first level involves qualitative identification of the main sources of pollutants in environmental media, known as source identification. The second level involves quantitative calculation of the contribution of various sources of pollution, known as source apportionment [49]. Many researchers referred to both levels as source apportionment [50]. This study focused on source identification for pollution in the Wenyu Stream to provide technical references for pollution control at the source of the mining area.

4.1. Acid Mine Drainage Source of Cadmium in the Wenyu Stream

As shown in Figure 2, both T2 and T3 samples located in the drainage ditches of the tailings pond leachate discharge canal and the mineral processing drainage canal, respectively, have Cd content exceeding Class II water limit and reaching Class V water. Sample 10, located between them, was the most heavily polluted point in the main ditch of the Wenyu Stream. In addition, pH testing results of water samples show that the pH values of T2 and T3 points are 6.00 and 6.13, respectively. This indicates that both the tailings pond upstream of T2 and the mineral processing plant upstream of T3 discharge Cd-contaminated acid drainage water into the main ditch of the Wenyu Stream. According to previous studies, acid mine drainage mainly forms in three ways [51]: ① In the process of ore processing, if acid reagent is used for mineral processing, the wastewater discharged is acid water, ② During mine production, a large number of waste rocks and tailings containing sulfide minerals are stacked in the open air, which are easily oxidized to form metal ions and sulfate ions. In rainy weather, they will be dissolved in water and slowly enriched to form acidic wastewater through leaching, ③ During the mining of the deposit, a large amount of groundwater seeps into the working face, which is in long-term contact with rocks and minerals containing sulfur and heavy metals, and partially dissolves to form acid mine water. Obviously, the acidic wastewater at T2 and T3 points in this study conforms to the first two production methods mentioned above. Combined with previous research results, the unprocessed metallurgical wastewater of Dabaoshan multimetal mine in northern Guangdong Province was directly discharged into Hengshi River, causing serious harm to the environment along the the Hengshi River Basin [52]. The acid and polluted wastewater discharged by Dexing Copper Mine in Jiangxi Province contaminated the Le'an River and caused harm to animals, plants, and farmland along the riverbanks [53]. The results of previous research have shown that acidic wastewater generated by mining activities and its illegal discharge are another important cause of Cd pollution in the Wenyu Stream [54].

4.2. The Surrounding Rock Source of Cd in the Wenyu Stream

According to the test results of rock samples, the Cd content distribution is extremely uneven, with the highest value exceeding 18 times the lowest value. The rock sample with the highest Cd content is a pit surrounding rock rich in pyrite. In addition, Cd is often distributed in sulfide ores, such as sphalerite and others associated with large-scale molybdenum mines [55]. This phenomenon is related to the properties of Cd's copper affinity and sulfur affinity as well as its similar geochemical behavior to Zn [56,57]. As a trace heavy metal element in rocks, the average abundance of Cd is 0.2 mg/kg in the crust [42]. Under strong oxidizing conditions, Cd can form oxide minerals, such as CdO and $CdCO_3$, and can also be oxidized into $CdSO_4$ in aqueous solutions [42]. In weak oxidizing environments, the mineral sphalerite containing Cd can be rapidly oxidized and dissolved, producing cadmium sulfide (CdS) [58]. The large-scale molybdenum mine upstream of the Wenyu Stream has been operating for over 50 years. Long-term mining, mineral processing, transportation, and other mining activities have changed the original occurrence environment of Cd, which is released from minerals and rocks, migrated and transformed in the supergene geological environment, and finally accumulated and enriched, causing river pollution.

4.3. Sediment Sources of Cd in the Wenyu Stream

River pollution caused by heavy metal adsorption and release in sediment-water systems is one of the hot issues in research. Most of the heavy metal pollutants will be adsorbed by suspended particles after entering the river watershed, and the suspended particles will be accelerated to sink by gravity in the water body, when particles accelerate to sink to a state where gravity and resistance are equal, they will sink at a stable settling speed [59]. Generally, the flow of river water body is relatively stable, that is, the handling effect of hydrodynamic force is gentle. However, the Wenyu Stream is formed by the convergence of mountain streams, and the frequent agitation of the river makes the load of sediment particles exceed its carrying capacity. At the same time, sediment particles are prone to generate intermolecular forces with heavy metal pollutant molecules [60–62]. For example, cadmium ions will generate electrostatic adsorption with coarse and fine sediment particles, which will combine through chemical bonds and accelerate the settlement of pollutants [63–65]. The flow velocity at the bottom of the river water body is slow, and the sediment at the bottom is less affected by hydraulic transport. Less heavy metal pollutants in the sediment re-enter the water body due to river impact and transport. Therefore, the existing form of heavy metal pollutants after sedimentation is greatly affected by the water environment.

The pH is the main factor affecting the adsorption and desorption of heavy metals in the water environment. H^+ ions will desorb metal ions from the sediment and re-enter the water environment through ion competition [65]. Previous studies on the release kinetics of heavy metals in Guangzhou urban water polluted sediments under acidic conditions can be divided into two stages. The first stage is the rapid desorption of heavy metals on the sediment surface after the deposition of particulate suspended solids, and the second stage is the slow diffusion of heavy metals from the internal pores of the sediment to the external solution [42]. The first stage is the release process after sedimentation of sediment particles mentioned above. In the second stage, there is a concentration difference between the surface interstitial water of sediment and the overlying water (river water) [59]. As the free water in the sediment void, interstitial water connects the sediment with the overlying water. Through ion exchange, molecular diffusion, and biological disturbance, interstitial water diffuses and migrates to the river with interstitial water as the intermediate medium, thus affecting the water quality of the river. A total of eight toxic elements, including Cu, Pb, Zn, Cd, Ni, Cr, as and Hg, have been tested and analyzed in 25 sediment samples from the main stream of the Zhongbei River and its tributaries in the northern mountainous area of Guangdong Province. They used the geoaccumulation index method to analyze and evaluate the heavy metal pollution in the sediment of the area. The study found that the concentration of heavy metals in the main stream was more serious than that in the branch ditch due to the large amount of pollutants in the main stream. The pollution of this degree was due to the existence of multiple mining yards and smelters in the area. Other scholars' research found that the content of heavy metals in river sediments downstream of tailings pond is higher than that in tailings, indicating that the cumulative effect of heavy metals [41] will lead to the content of heavy metals in sediments higher than that in pollution sources. In the study on the release law of elements in the river sediment of Harbin reach of the Songhua River in Northeast China, it was found that different temperatures, pH values, water disturbance levels, and salt concentrations will have different degrees of influence on the release of heavy metals [43].

After the development of molybdenum ore in the study area, Cd, Zn, and other heavy metal ions were released after entering the downstream Wenyu Stream watershed. Comparing the times of exceeding the background value of Cd content in river water in Figure 2a with the times of exceeding the background value of Cd content in sediment in Figure 3a, it was found that the times of exceeding the standard of Cd content in river sediment samples are 4.18 times that of the Wenyu Stream water samples. Meanwhile, comparing the Nemerow indexes between the surface water in Figure 2d and the sediment in Figure 3b, the Nemerow indexes of pollutants in sediment were larger, indicating the sediment was in the stage of pollutant enrichment. The Nemerow index of pollutants in sediment is larger, indicating that the sediment is in the stage of pollutant enrichment. After the release of heavy metal ions through different ways in the development into the river, on the one hand, with the continuous migration, transformation, and accumulation of river water flow from the downstream of tailings pond to the upstream of the river, the pH, and water quality structure of the water body are gradually changed, and the function of the river water body is reduced. On the other hand, when it infiltrates into the groundwater, the suspended solids become sediments and are transferred to the river sediment, which changes the water quality to a certain extent. Exchangeable heavy metal ions may also be released into the river water and have an impact again. Previous studies on the treatment of river sediment pollution in Guangdong Province have shown that the treatment of heavy metal elements in river sediment can be carried out by physical remediation methods with the help of engineering technology, such as artificial dredging, chemical remediation methods with chemical remediation agents applied to the sediment, or biological remediation methods with the use of surrounding microorganisms and plant communities [44]. In order to deal with the relationship between China's rapid economic development and environmental and ecological protection in mining areas, relevant functional staff must pay more attention to the treatment of river sediment pollution, understand the types of heavy metal pollutants in sediment and treatment methods, so as to effectively control and repair the ecological health of rivers.

4.4. The Atmospheric Dust Fall Sources of Cd in the Wenyu Stream

In addition, the effects of atmospheric dry and wet deposition on the input of Cd elements into rivers also occupy a significant proportion. On the one hand, mining activities in densely populated areas generate a large amount of dust through heavy machinery operation. On the other hand, the surface rocks and minerals exposed to air undergo geological weathering processes that produce dust particles containing heavy metal ions and sulfate ions. These particles are eventually transported directly into rivers via wind or leached into surface runoff during rainfall, ultimately entering rivers. Predecessors have observed atmospheric dust fall in another mining area 20 km east of the study area and found that the cumulative multiple of the atmospheric dust content in the mining area relative to the background value of soil Cd in the area has reached 1.67, which indicates that the dust from mining activities is also a way of causing heavy metal pollution in the mining area [66–68]. Based on Figure 2, it can be inferred that water samples 8, 9, 10, and 11, within a 1–5 km radius of the upper reaches of the Wenyu Stream, where open-pit mines and dumpsites are prone to generating high levels of dust, may have Cd concentrations exceeding background values by factors of 0, 3, 9, and 2, respectively, while downstream sites have the highest concentration ratios exceeding background values by up to 1. According to relevant studies, the enrichment of Cr, Co, Mn, Ni, As, Cd, Cu, Pb, Zn, Tl, and rare earth elements in sediments of Beijing's Beiyun River is similar to that of metal elements in atmospheric particulate matter. It can be inferred that heavy metal elements in atmospheric particulate matter have a significant contribution rate to heavy metal elements in the sediments of the Beiyun River, mainly through rainfall into rivers [69], which is similar to the viewpoint presented in this study. Another study has found that the spatial distribution of air dust pollution is consistent with the distribution of pollution sources [70]. Meanwhile, three aspects of atmospheric dry-and-wet deposition, human activities, and parent rock weathering in the Gongnaisi River Basin in Xinjiang Uygur Autonomous Region have impacts on the enrichment of heavy metal elements in the river [71,72]. These studies indicate that the effects of atmospheric dry and wet deposition generated by mining activities may be another important cause of riverine Cd contamination. In future research, we will carry out atmospheric dust fall observation in the mining area to provide data support for the comprehensive discussion of mine environmental problems in the study area.

5. Conclusions

In conclusion, this study aimed to explore the effects of mineral exploitation on river water quality in the Wenyu Stream downstream of a large molybdenum mining area in Qinling Mountains. The results showed that Cd pollution was a significant problem in the Wenyu Stream due to mining activities.

The analysis of surface water, sediment, rock, and ore samples by ground sampling and chemical testing revealed that 14 out of 23 sampling points had significantly higher Cd contents than those in the contrast area, which was not affected by mining activities. The variation coefficient of Cd content reached 99.44%, indicating that the Cd contents in the Wenyu Stream are highly variable and significantly influenced by mining activities. Only one point exceeded the Class II water quality limit, and the pollution exceeding multiple was 1 with slight pollution. The results of geoaccumulation index analysis showed that the 15 sediment samples from the main ditch of the Wenyu Stream had different degrees of pollution: The most seriously polluted samples 15 and 16 reached the moderate to strong pollution level. Samples 5, 13, and 22 with the lightest pollution were no pollution to moderate pollution. Most of the samples were moderately polluted. The potential ecological risk indexes of Cd in 15 sediment samples from the main ditch of the Wenyu Stream were in the range of medium risk to very strong risk, indicating the sediment in the main ditch of the Wenyu Stream is suffering strong ecological risk.

Acid mine drainage, original geological background, sediment release, and atmospheric dust fall were identified as the primary sources of Cd pollution in the Wenyu

Stream watershed. Acid mine drainage is a common form of pollution caused by mining activities, which can lead to the release of heavy metals into rivers and streams. Original geological background also plays a crucial role in determining the levels of heavy metal concentrations in soils and water bodies. Sediment release is another important source of heavy metal pollution, as it can transport pollutants from mines to nearby waterways. Finally, atmospheric dust fall can also contribute to heavy metal pollution through its interaction with rain and wind.

To address these issues, it is essential to implement effective monitoring and management strategies for heavy metal pollution in rivers and streams caused by mineral exploitation. This includes reducing emissions from mines, improving waste disposal practices, and promoting sustainable development in mining areas. For example, mining companies could adopt more environmentally friendly technologies, such as hydrometallurgy and leaching techniques, to reduce their environmental impact. Additionally, governments could impose stricter regulations and penalties on polluting industries to encourage them to adopt cleaner production methods.

Furthermore, public education and awareness-raising campaigns are crucial for promoting sustainable development in mining areas. By educating local communities about the impacts of heavy metal pollution on aquatic ecosystems and human health, we can encourage people to take action to protect their environment. This could include supporting initiatives, such as rainwater harvesting and wastewater treatment plants, to reduce the amount of pollutants released into rivers and streams.

In conclusion, this study highlights the importance of protecting wetlands and riparian areas from heavy metal pollution caused by mineral exploitation. By understanding the sources and effects of pollution, we can develop effective strategies to mitigate its impact on aquatic ecosystems and promote sustainable development in mining areas. It is essential that we work together to find solutions to this pressing environmental issue before it is too late.

Author Contributions: All authors participated in the development of this study. Conceptualization, Y.X. and H.C.; methodology, A.Z.; software, M.Y.; validation, J.Z.; formal analysis, H.C.; investigation, J.Z.; writing—original draft preparation, M.Y.; writing—review and editing, M.Y.; project administration, Y.X.; funding acquisition, A.Z. All authors have read and agreed to the published version of the manuscript.

Funding: This research was funded by the National Natural Science Foundation of China, grant number 42272342, the China Geological Survey Foundation, grant number DD20230457, the Social Science Foundation of Shaanxi, grant number 2021D068 and the Natural Science Basic Research Program of Shaanxi, program number 2021JM-350.

Data Availability Statement: Raw data are available upon reasonable request addressed to the corresponding author on reasonable request.

Acknowledgments: We are thankful to the Key Laboratory for Geohazards in Loess Areas, the Chinese Ministry of Natural Resources, and the Xi'an Geological Survey Center of China Geological Survey for their contributions to this research. The authors would like to thank the reviewers and editors for their very helpful and constructive reviews of this manuscript.

Conflicts of Interest: The authors declare no conflict of interest.

References

1. Xu, Y. Investigation and research on the mine geological environment: Present status and outlook. *Geol. Bull. China* **2008**, *27*, 1235–1244.
2. Li, J.; Zhang, J. Discussion on the work of mine geo-environmental investigation of China. *Hydrogeol. Eng. Geol.* **2018**, *45*, 169–172.
3. Wu, Q. Study of classification of geologic environmental problems in mines in China. *Hydrogeol. Eng. Geol.* **2003**, *5*, 107–112.
4. Tepe, A.Y. Toxic Metals: Trace Metals–Chromium, Nickel, Copper, and Aluminum. *Encycl. Food Saf.* **2014**, *2*, 356–362.
5. Xu, Y.; Zhang, J. Contents of heavy metals in bottom sediments of the Taiyu River in the Tongguan gold mining area, Shaaxi, China, and contamination assessments. *Geol. Bull. China* **2008**, *27*, 1263–1671.

6. Zhang, J.; Zhao, A.; Chen, H.; Xu, Y.; He, F. Evaluation of potential ecological risk of heavy- metal pollution in bottom mud of the Xiyu River in the Xiaoqinling gold mining area, China. *Geol. Bull. China* **2008**, *27*, 1286–1291.
7. Liu, R.; Xu, Y.; Li, X.; Zhang, J.; Chen, H.; He, F.; Qiao, G.; Ke, H.; Shi, Y. The distribution and coupling relationship of heavy metal content in water and sediments in the gold mining area. *Geol. Bull. China* **2014**, *33*, 1220–1230.
8. Huang, J.; Qin, M.; Ma, W.; Yu, J.; Peng, X.; Yang, L. Characteristics analysis and ecological risk assessment of heavy metals contamination in suspended solids in a river affected by acid mine drainage. *Environ. Chem.* **2016**, *35*, 2315–2326.
9. Chen, Q.; Zhang, X.; Hu, M. Assessment of aquatic pollution of heavy metals in a copper- iron area. *Environ. Sci. Technol.* **2006**, *29*, 64–71.
10. Kodat, M.; Tepe, Y. A holistic approach to the assessment of heavy metal levels and associated risks in the coastal sediment of Giresun, southeast Black Sea. *Heliyon* **2023**, *9*, e16424.
11. Aydın, H.; Tepe, Y.; Ustaoğlu, F. A holistic approach to the eco-geochemical risk assessment of trace elements in the estuarine sediments of the Southeastern Black Sea. *Mar. Pollut. Bull.* **2023**, *189*, 114732. [CrossRef] [PubMed]
12. Song, Y.; Xu, H.; Ren, L.; Gong, P.; Zhou, Q. Eco-toxicological Effects of Heavy Metals on the Inhibition of Seed Germination and Root Elongation of Chinese Cabbages in Soils. *Environ. Sci.* **2002**, *23*, 103–107.
13. Wang, J.; Ma, Q.; Zhao, X.; Zhong, S.; Xu, Z. Influence of emergent cadmium pollution on fish species and health risk assessment in Longjiang River in Guangxi Autonomous Region. *Ecol. Environ. Sci.* **2019**, *28*, 974–982.
14. Sun, D. Research Progress on Phytoremediation of Cd contaminated soil. *Environ. Prot. Circ. Econ.* **2014**, *34*, 53–55.
15. Avenell, S. Japan's long environmental sixties and the birth of a green leviathan. *Jpn. Stud.* **2012**, *32*, 423–444. [CrossRef]
16. Jiang, J. The relationship of environmental geochemistry to human health and agricultural production. *Geophys. Geochem. Exploration.* **2004**, *28*, 330–332.
17. Zhang, S.; Cui, Y. Protection and monitoring of the headwaters area in the Qinling mountains for south to north water transfer project. *Yangtze River* **2007**, *38*, 40–42.
18. Yang, M.; Zhao, A.; Ke, H.; Chen, H. Geo-Environmental Factors'Influence on the Prevalence and Distribution of Dental Fluorosis: Evidence from Dali County, Northwest China. *Sustainability* **2023**, *15*, 1871. [CrossRef]
19. Fang, Z.; Xu, W.; Zhang, J.; Xiao, Y.; Zhang, L. Designing protected area systems in the Qinling Mountains based on biodiversity and ecosystem service evaluation. *Acta Ecol. Sin.* **2017**, *37*, 5334–5341.
20. Wen, Y. Construction of Qinling Ecological Demonstration Area. *J. Anhui Agri. Sci.* **2011**, *39*, 14278–14280.
21. Song, Z.; Song, G.; Tang, W.; Yan, D.; Zhao, Y.; Zhu, Y.; Wang, J.; Ma, Y. Molybdenum contamination dispersion from mining site to a reservoir. *Ecotoxicol. Environ. Saf.* **2021**, *208*, 111631. [CrossRef] [PubMed]
22. *GB 15618-2018*; Soil Environmental Quality–Risk Control Standard for Soil Contamination of Agricultural Land. China Environmental Science Press Co., Ltd.: Beijing, China, 2019.
23. *GB 3838-2002*; Environmental Quality Standards for Surface Water. China Environmental Science Press Co., Ltd.: Beijing, China, 2002.
24. Hu, J.; Qiu, S.; Luo, N.; Qing, G.; Huang, C. Multi-Dimensional Spatial and Temporal Variations of Ecosystem Service Values in the Li River Basin, 1990–2020. *Remote Sens.* **2023**, *15*, 2996. [CrossRef]
25. Ren, T.; Zheng, J. Geological characteristics of the deep orebody in the north open pit of Jinduicheng molybdenum deposit, Huaxian County, Shaanxi Province. *World Nonferrous Met.* **2019**, *10*, 252–254.
26. *DD2005-03*; Technical Code for China Geological Survey-Technical Requirements for Analysis of Ecological Geochemical Evaluation Samples (DD2005-03). China Geological Survey: Beijing, China, 2005.
27. Luonan County Strengthens Pollution Control and Helps Rural Revitalization. Available online: https://www.luonan.gov.cn/xwzx/zxbd/112508.htm (accessed on 7 December 2021).
28. *NY/T 395-2012*; Technical Rules for Monitoring of Environmental Quality of Farmland Soil. China Agriculture Press Co., Ltd.: Beijing, China, 2012.
29. Niu, H.; Wu, Q.; Chen, X. Distribution characteristics and correlations of heavy metals in the surface sediments in Guangzhou section of the Pearl River. *Ecol. Environ.* **2006**, *15*, 954–959.
30. Li, L.; Zeng, X.; Li, G.; Mei, X. Heavy metal pollution of Wenyu River sediment and its risk assessment. *Acta Sci. Circumstantiate* **2007**, *27*, 289–297.
31. Müller, G. Index of geoaccumulation in sediments of the Rhine River. *Geojournal* **1969**, *2*, 108–118.
32. Farkas, A.; Erratico, C.; Viganò, L. Assessment of the environmental significance of heavy metal pollution in surficial sediments of the River Po. *Chemosphere* **2007**, *68*, 761–768. [CrossRef]
33. Müller, G. Schwermetalle in den sedimenten des Rheins-Vernderungen seit 1971. *Umschav* **1979**, *79*, 778–783.
34. Hakanson, L. An ecological risk index aquatic pollutioncontrol: A sedimentological approach. *Water Res.* **1980**, *14*, 975–1001. [CrossRef]
35. Xu, Z.; Ni, S.; Tuo, X.; Zhang, C. Calculation of Heavy Metals' Toxicity Coefficient in theEvaluation of Potential Ecological Risk Index. *Environ. Sci. Technol.* **2008**, *31*, 112–115.
36. Xu, Z.; Yang, X.; Wen, Y.; Chen, G.; Fang, J. Evaluation of the Heavy Metals Contamination and Its Potential Ecological Risk of the Sediments in Beijiang River's Upper and Middle Reaches. *Environ. Sci.* **2009**, *30*, 3262–3268.
37. Gao, Y.; Xu, Y.; Zhang, J. Evaluation of Cd pollution of a molybdenum ore area in Dongchuan River basin of the Qinling Mountain. *Geol. Bull. China* **2018**, *37*, 2241–2250.

38. Abbasi, S.; Keshavarzi, B.; Moore, F.; Shojaei, N.; Sorooshian, A.; Soltani, N.; Delshab, H. Geochemistry and environmental effects of potentially toxic elements, polycyclic aromatic hydrocarbons and microplastics in coastal sediments of the Persian Gulf. *Environ. Earth Sci.* **2019**, *78*, 492. [CrossRef]
39. Zhang, X.; Xin, C.; Li, C. Geochemical Characteristics of Heavy Metal's Contamination and Its Surface Geochemical Mechanism in Baiyin City, Gansu Province. *Geol. Sci. Technol. Inf.* **2010**, *29*, 124–131.
40. Li, J.; Chen, H.; Teng, Y.; Dong, J. Contamination characteristics and source apportionment of soil heavy metals in Lalin River basin. *Trans. Chin. Soc. Agric. Eng.* **2016**, *32*, 226–233.
41. Qiao, S.; Jiang, J.; Xiang, W.; Tang, J. Distribution of heavy metals in sediments in lakes in Wuhan with assessment on their potential ecological risk. *Resour. Environ. Yangtze Basin* **2005**, *3*, 353–357.
42. Deng, H.; Li, L.; Kim, J.; Ling, F.; Beckingham, L.E.; Wammer, K.H. Bridging environmental geochemistry and hydrology. *J. Hydrol.* **2022**, *613*, 128448. [CrossRef]
43. Gong, X.; Chen, C.; Zhou, W.; Jian, M.; Zhang, Z. Assessment on Heavy Metal Pollution in the Sediment of Poyang Lake. *Environ. Sci.* **2006**, *27*, 732–736.
44. Bai, W.; Quan, X.; Tan, H.; Zhao, Q.; Zhao, G.; Zhu, Q. Pollution of heavy metals in sediments from Changshou lake in Chongqing and Its potential ecological risk assessment. *Earth Environ.* **2011**, *39*, 382–387.
45. Zhu, W.; Huang, T.; Chai, B.; Zhang, Y.; Lu, J. The impact of the redox conditions on the speciation of the heavy metals in the sediment of the Fenhe reservoir and their assessments of the potential ecological risk. *J. Arid Land Resour. Environ.* **2009**, *23*, 34–40.
46. Wang, L.; Zhou, X.; Zheng, B.; Fu, Q. Sediments eco-environmental quality assessment in the Changjiang Estuary and its adjacent waters. *Acta Ecol. Sin.* **2008**, *28*, 2191–2198.
47. Jia, Z.; Liang, T.; Lin, J. Study on heavy metal contamination and potential ecological risk in Hong Kong Rivers. *Acta Sci. Nat. Univ. Pekin.* **1997**, *33*, 485–492.
48. Pan, A.; He, Y.; Ma, R. Zoning of environmental geochemistry in Shaanxi Province. *Adv. Earth Sci.* **2004**, *19*, 439–443.
49. Xiao, R.; Wang, S.; Li, R.; Wang, J.; Zhang, Z. Soil heavy metal contamination and health risks associated with artisanal gold mining in Tongguan, Shaanxi, China. *Ecotoxicol. Environ. Saf.* **2017**, *141*, 17–24. [CrossRef]
50. Simeonov, V.; Einax, J.; Tsakovski, S.; Kraft, J. Multivariate statistical assessment of polluted soils. *Cent. Eur. J. Chem.* **2005**, *3*, 1–9. [CrossRef]
51. Zhang, C.; Luo, Y.; Wu, L. Methods for source apportionment of soil pollutants and their advances in application to aoil environmental research. *Soils* **2007**, *39*, 190–195.
52. Rezaie, B.; Anderson, A. Sustainable resolutions for environmental threat of the acid mine drainage. *Sci. Total Environ.* **2020**, *717*, 137211. [CrossRef]
53. Qin, J.; Jiang, X.; Yan, Z.; Zhao, H.; Zhao, P.; Yao, Y.; Chen, X. Heavy metal content and microbial characteristics of soil plant system in Dabaoshan mining area, Guangdong Province. *PLoS ONE* **2023**, *18*, e0285425. [CrossRef] [PubMed]
54. Yuan, J.; Ding, Z.; Bi, Y.; Li, J.; Wen, S.; Bai, S. Resource Utilization of Acid Mine Drainage (AMD): A Review. *Water* **2022**, *14*, 2385. [CrossRef]
55. Ighalo, J.O.; Kurniawan, S.B.; Iwuozor, K.O.; Aniagor, C.; Ajala, O.J.; Oba, S.; Iwuchukwu, U.F.; Ahmadi, S.; Igwegbe, C.A. A review of treatment technologies for the mitigation of the toxic environmental effects of acid mine drainage (AMD). *Process Saf. Environ. Prot.* **2022**, *157*, 37–58. [CrossRef]
56. Xu, L.; Bi, X.; Zhang, X.; Huang, M.; Liu, G. Mantle contribution to the generation of the giant Jinduicheng porphyry Mo deposit, Central China: New insights from combined in-situ element and isotope compositions of zircon and apatite. *Chem. Geol.* **2023**, *616*, 121238. [CrossRef]
57. Ye, X.; Pan, Z.; Li, C.; Liu, T.; Xia, B. The present situation and prospects of geochemical researches on cadmium. *Acta Petrol. Mineral.* **2005**, *24*, 339–348.
58. Wang, J.; Jiang, Y.; Sun, J.; She, J.; Yin, M.; Fang, F.; Xiao, T.; Song, G.; Liu, J. Geochemical transfer of cadmium in river sediments near a lead-zinc smelter. *Ecotoxicol. Environ. Saf.* **2020**, *196*, 110529. [CrossRef] [PubMed]
59. Shi, M. Shanxi Section of WeiHe River Sediment Releasing Regularity and Its Influence on Water Quality Research. Master's Thesis, Xi'an University of Technology, Xi'an, China, 2016.
60. Kubier, A.; Wilkin, R.T.; Pichler, T. Cadmium in soils and groundwater: A review. *Appl. Geochem.* **2019**, *108*, 104388. [CrossRef] [PubMed]
61. Kumar, S.; Islam, A.R.M.T.; Islam, H.M.T.; Hasanuzzaman, M.; Ongoma, V.; Khan, R.; Malick, J. Water resources pollution associated with risks of heavy metals from Vatukoula Goldmine region, Fiji. *J. Environ. Manag.* **2021**, *293*, 112868. [CrossRef]
62. Wang, J.; Sun, Y.; Qian, J.; Wu, J.; Pan, T. Simulated study on phosphorus release of Chao Lake sediment. *Acta Sci. Circumstantiae* **2002**, *6*, 738–742.
63. Fang, X.; Peng, B.; Wang, X.; Song, Z.; Zhou, D.; Wang, Q.; Qin, Z.; Tan, C. Distribution, contamination and source identification of heavy metals in bed sediments from the lower reaches of the Xiangjiang River in Hunan province, China. *Sci. Total Environ.* **2019**, *689*, 557–570. [CrossRef]
64. Zhang, Y.; Zhang, H.; Zhang, Z.; Liu, C.; Sun, C.; Zhang, W.; Marhaba, T.F. pH effect on heavy metal release from a polluted sediment. *J. Chem.* **2018**, *2018*, 7597640. [CrossRef]
65. Qiao, L.; Fan, B.; Zhang, Y. Reasearch composition and formation mechanism of rivee sediment. *Environ. Sci. Manag.* **2022**, *47*, 88–92.

66. Yan, H.; Zhang, H.; Shi, Y.; Zhou, P.; Li, H.; Wu, D.; Liu, L. Simulation on release of heavy metals Cd and Pb in sediments. *Trans. Nonferrous Met. Soc. China* **2021**, *31*, 277–287. [CrossRef]
67. Zhang, J.; Liu, Y.; Wen, M.; Zheng, C.; Chai, S.; Huang, L.; Liu, P. Distribution Characteristics and Ecological Risk Assessment of Nitrogen, Phosphorus, and Some Heavy Metals in the Sediments of Yueliang Lake inWestern Jilin Province, Northeast China. *Water* **2022**, *14*, 3306. [CrossRef]
68. Xu, Y.; Zhang, J.; Chen, S.; Ke, H. Distribution of contents of heavy metals in soil profile contaminated in different ways in Xiaoqinling gold mining area. *J. Agro-Environ. Sci.* **2008**, *27*, 200–206.
69. Yuan, H.; Yin, H.; Yang, Z.; Yu, J.; Liu, E.; Li, Q.; Tai, Z.; Cai, Y. Diffusion kinetic process of heavy metals in lacustrine sediment assessed under different redox conditions by DGT and DIFS model. *Sci. Total Environ.* **2020**, *741*, 140418. [CrossRef] [PubMed]
70. Yang, Z.; Nie, Y.; Hu, C. Migration and transformation of heavy metals at solid/water interface in Beiyunhe River. *Chin. J. Environ. Eng.* **2012**, *6*, 3455–3459.
71. Ye, L.; Huang, M.; Zhong, B.; Wang, X.; Tu, Q.; Sun, H.; Wang, C.; Wu, L.; Chang, M. Wet and dry deposition fluxes of heavy metals in Pearl River Delta Region (China): Characteristics, ecological risk assessment, and source apportionment. *J. Environ. Sci.* **2018**, *70*, 106–123. [CrossRef]
72. Sun, Z.; Mou, X.; Tong, C.; Wang, C.; Xie, Z.; Song, H.; Sun, W.; Lv, Y. Spatial variations and bioaccumulation of heavy metals in intertidal zone of the Yellow River estuary, China. *Catena* **2015**, *126*, 43–52. [CrossRef]

water

MDPI

Article

Dechlorination of Hexachlorobenzene by Ni/Fe Bimetallic Nanoparticles and the Influence of Co-Existing Heavy Metal Ions

Yuanying Huang [1,2,*], Siwen Liu [1,2], Qian Wang [3], Guoxin Huang [4,*], Xueqi Zhang [1] and Yang Liu [1]

1 National Research Center for Geoanalysis, Beijing 100037, China
2 Key Laboratory of Ministry of Natural Resources for Eco-Geochemistry, Beijing 100037, China
3 Qingdao Environmental Engineering Assessment Centre, Qingdao 266000, China
4 Chinese Academy of Environmental Planning, Beijing 100043, China
* Correspondence: yuanyinghuang304@163.com (Y.H.); huanggx@caep.org.cn (G.H.)

Abstract: Hexachlorobenzene (HCB) is one of the most persistent environmental pollutants of global concern. Ni/Fe nanoparticles, with their small particle size, large surface area, and high reactivity, are a promising candidate for HCB degradation. In this work, we investigated the kinetics and products of the dechlorination of HCB by Ni/Fe nanoparticles and how the presence of heavy metal ions Cd(II) and Zn(II) influences the reaction. It is found that 400 μg/L HCB can be rapidly removed by 7.5 g/L Ni/Fe nanoparticles and the removal percentage reaches 99% in 48 h. The removal is facilitated by adsorption and sequential dechlorination of HCB, producing PCB, 1,2,3,4-TeCB, and 1,2,3-TCB as the main products, with 1,2,3,5/1,2,4,5-TeCB, 1,2,4-TCB, and 1,2-DCB as the minor products. The addition of heavy metal ions Cd(II) and Zn(II) does not significantly affect the removal rate of HCB but hinders the adsorption and degradation of the byproducts through competitive adsorption. Additionally, the concentration of both Cd(II) and Zn(II) decreases rapidly and achieves over 98% removal in 4 h. Our study reveals that Ni/Fe nanoparticles can remove HCB and heavy metals Cd(II) and Zn(II) concurrently, with the extent of HCB dechlorination reduced compared to that without heavy metal. These findings may inform the application of Ni/Fe nanoparticles in the treatment of water bodies and soil contaminated by both halogenated aromatics and heavy metal.

Keywords: nanoparticle; Ni/Fe; hexachlorobenzene; dechlorination; heavy metal

Citation: Huang, Y.; Liu, S.; Wang, Q.; Huang, G.; Zhang, X.; Liu, Y. Dechlorination of Hexachlorobenzene by Ni/Fe Bimetallic Nanoparticles and the Influence of Co-Existing Heavy Metal Ions. *Water* **2024**, *16*, 2855. https://doi.org/10.3390/w16192855

Academic Editor: Christos S. Akratos

Received: 20 August 2024
Revised: 29 September 2024
Accepted: 30 September 2024
Published: 8 October 2024

1. Introduction

Hexachlorobenzene (HCB), a typical chlorinated aromatic compound and an important chemical intermediate, had been widely used in fungicides, military pyrotechnic smokes, and synthetic rubber peptizing agents in the past [1,2]. It is a persistent, bioaccumulative, and toxic compound that may damage human organs and cause a lot of diseases including dermatitis, severe numbness in the cardiovascular and nervous systems, and carcinogenesis [3–6]. Although most countries have banned the production and usage of HCB since the beginning of the 21st century, HCB is still prevalent in the environment due to its extensive historical use, ongoing unintentional emission, and long half-life. The current global burden of HCB on the environment was estimated to be 10,000–26,000 tons [4]. It has been detected in most parts of the ecosystem, including water, soil, air, animals, plants, and even anthropologically in human milk [7]. Thus, appropriate remediation technologies are in urgent need for HCB-contaminated water and soil.

In recent years, zero-valent Fe nanoparticles have drawn much attention as a promising reagent for the chemical degradation of chlorinated pollutants, because of their ease of synthesis, large surface area, and effectiveness in the treatment of a variety of organic pollutants and heavy metals [8]. Nevertheless, several drawbacks of the material have significantly limited its application. Firstly, Fe nanoparticles oxidize easily upon contact with air or in aqueous solution, forming a layer of iron oxide or hydroxide on the surface of the nanoparticles, significantly reducing their reactivity. Additionally, the reduction of 1,2,4,5-tetrachlorobenzene by Fe nanoparticles was shown to be slow and incomplete, with a half-life of 36 h and detection of only a small subset of the theoretical byproducts [9]. To counter these problems, an effective method is to deposit a second, less reactive metal on the surface of iron to form bimetallic nanoparticles. The second metal acts as a catalyst to enhance the electron-donating ability of Fe as well as protects the surface of the nanoparticles from being oxidized. Many metals have been shown to effectively catalyze dechlorination reactions when coupled with iron, such as Pt, Cu, Pd, and Ag [10–14]. However, only Pd/Fe and Cu/Fe nanoparticles have been applied to the treatment of HCB [10,14]. Nickel is the ninth most produced metal in the world and a typical hydrogenolysis catalyst with the ability to catalyze the hydrodechlorination of a variety of chloroaromatics such as 2,4-dichlorophenol, triclosan, and Aroclor 1242 [15–18]. Therefore, Ni/Fe nanoparticles hold great promise to serve as a commercially viable and effective reagent for HCB dechlorination.

Contaminated water bodies and soil usually contain a mixture of organic and metallic pollutants [19]. Examples are farmlands in post-mining areas and leachate from landfills, where the main heavy metal pollutants identified are Cd, Ni, Cu, As, Hg, Pb, and Zn [20]. Thus, it is necessary to consider the influence of co-existing heavy metal ions when developing remediation technologies for HCB. Most optimally, the heavy metal ions may be removed concurrently with the organic pollutants. Although there are a few studies on the effects of various ions on the degradation of chlorinated compounds by metallic nanoparticles, the influence of most heavy metal ions on the dechlorination of HCB is not well understood [16,21].

In this study, we synthesize Ni/Fe nanoparticles via chemical deposition and determine their surface characteristics with BET, TEM, and SEM. We then investigate the reaction kinetics and mechanism of HCB dechlorination by Ni/Fe nanoparticles and study the influence of heavy metals Cd(II) and Zn(II) on the removal and degradation of HCB. The change in heavy metal concentration is also monitored to determine whether they can be simultaneously removed by the nanoparticles.

2. Materials and Experiments

2.1. Chemicals

The standard materials of chlorinated benzenes were purchased from CNW Technologies GmbH (Dusseldorf, Germany). Deoxygenated–deionized water was used for solution preparation. Other reagents were from commercial companies of China, including $FeCl_3$, $NaBH_4$, Cd $(NO_3)_2$, $ZnSO_4$, NaOH (analytical pure), thick HCl (superior pure), hexane (chromatographically pure), and anhydrous ethanol (chromatographically pure). All reagents were above analytical grade.

2.2. Equipment

The instruments used in this work include the following: (1) H-8100 transmission electron microscope (Hitachi, Tokyo, Japan), (2) Quanta FEG 450 thermal field emission environmental scanning electron microscope (FEI, Hillsboro, OR, USA), (3) Autosorb-1-type ratio surface and porosity analyzer (Quantachrome Instruments, Boynton Beach, FL, USA), (4) SHA-B-type water bath oscillator (Ronghua Instrument, Jintan, Jiangsu, China), (5) DHG-9070A-type drying box (Shanghai Yiheng Technology Co., Ltd., Shanghai, China), (6) C-MAG HS4S25 magnetic stirrer (IKA, Staufen, Germany) and (7) AP-9908S vacuum pump (Tianjin Altsnes Instrument Co., Ltd., Tianjin, China).

Gas chromatography was conducted on a GC 2010 gas chromatograph (Shimadzu, Kyoto, Japan) equipped with a DB-5 capillary column from Agilent Technologies, Santa Clara, CA, USA (30 m × 0.25 mm × 0.25 μm film thickness, (5%-phenyl)-methylpolysiloxane non-polar column). High-purity helium (purity 99.999%) was used in the analysis with a flow rate of 1 mL/min. The injector temperatures were set at 250 °C. The GC oven temperature was programmed as follows: the temperature was held at 60 °C for 1 min, then increased at 5 °C/min to 200 °C, which was held for 3 min; it was then increased at 30 °C/min to 250 °C and held for 2 min. The detector temperature was 300 °C.

2.3. Experiment Design

2.3.1. Synthesis and Characterization of Ni/Fe Nanoparticles

All glassware was soaked with 1:1 HNO_3 for 48 h, rinsed with tap water and deionized water, and air-dried. In a nitrogen flow, 1.6 mol/L $NaBH_4$ aqueous solution was slowly added into an equal volume of 1.0 mol/L $FeCl_3 \cdot 6H_2O$ (containing V/V = 30% anhydrous ethanol) solution. After all $NaBH_4$ was added, the reaction was stirred for 5 min to afford iron nanoparticles via the following reaction:

$$4Fe^{3+} + 3BH_4^{-} + 9H_2O \rightarrow 4Fe\downarrow + 3H_2BO_3^{-} + 12H^{+} + 6H_2 \quad (1)$$

The generated black solid particles were transferred to a sand core funnel with a built-in 0.2 μm filter membrane and were filtered and washed with deoxygenated deionized water and then with anhydrous ethanol 2-to-3 times. The black particles were then quickly transferred to the reaction bottle and dried at 50 °C overnight. After cooling, the bottle was sealed carefully to protect the nano-Fe particles from exposure to air.

Nanoscale Ni/Fe particles were prepared by the deposition of $NiCl_2$ ethanol solution onto nanoscale Fe particles. As a typical procedure, 0.2 g of nanoscale Fe was added to 5 mL of 32 mM $NiCl_2$ in a 12 mL vial and the solution was stirred for several minutes. As Ni(II) was reduced, a thin layer of Ni was deposited on the surface of Fe, which gave nano-Ni/Fe bimetallic particles. After being washed for several times with deoxygenated deionized water to neutrality, the freshly prepared nanoparticles were immediately used in the degradation experiments.

The synthesized Fe and Ni/Fe nanoparticles were characterized by the BET method to find their specific surface areas. The experiment was performed by the Analytics and Chemistry Laboratory of China University of Geosciences (Beijing, China). The particles were also imaged by transmission electron microscopy (TEM) and scanning electron microscopy (SEM) to determine their nanostructures.

2.3.2. Degradation Experiments

To investigate the nano-Ni/Fe degradation of HCB in a heavy-metal-free condition and in the presence of single or double heavy metals, we set up four systems of simulated polluted water: a heavy-metal-free system (HCB), a system with added Cd(II) (HCB + Cd), a system with added Zn(II) (HCB + Zn), and that with both Cd(II) and Zn(II) (HCB + Cd + Zn). In each system, the initial concentration of HCB was 400 μg/L with 0.1% acetone. For the heavy-metal-containing samples, analytical pure Cd $(NO_3)_2/ZnSO_4$ was added during preparation to obtain Cd^{2+}/Zn^{2+} concentrations of 5 mg/L.

The experiments were conducted in 12 mL reaction bottles. To a 5 mL sample solution, an excess of nano-Ni/Fe particles (7.5 g/L) was added. The solution pH was not adjusted. The oxygen in the headspace was immediately purged with N_2 for seconds. The reaction was conducted in a shaking table at 120 rpm in the dark (25 ± 2 °C). Each run was performed in duplicate. At selected time intervals, an aliquot of the liquid phase was drawn from the reaction vial and the organic products were extracted with hexane. The extract was analyzed by gas chromatography and the experimental results took the average of three parallel samples.

3. Results and Discussion

3.1. Micro-Appearance Analysis

Figure 1 shows the TEM images of Fe nanoparticles prepared in water, v/v 30% ethanol, and v/v 50% ethanol. The particles formed a characteristic chain-like structure, a typical behavior due to magnetic attraction between the metal particles. The diameter of the particles was in the range of 20–100 nm. It was found that the addition of ethanol changed the size and morphology of the nanoparticles under otherwise identical experimental conditions. As shown in the TEM images, the particle size tends to be larger and more severely clustered without ethanol. The addition of ethanol caused the particles to be more dispersed, but there was no significant difference upon increasing the ethanol concentration from 30% to 50%. Thus, to maximize the adsorption capacity of the nanoparticles while saving cost, 30% ethanol was used as the solvent for the $FeCl_3 \cdot 6H_2O$ solution during the synthesis of the nanoparticles.

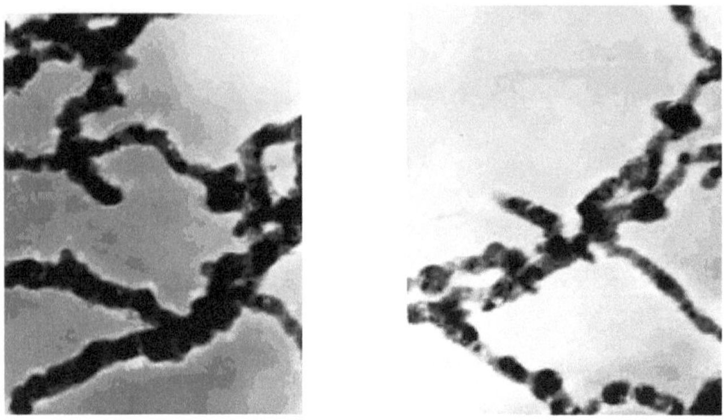

(a) (b) (c)

Figure 1. Transmission electron microscopy (TEM) images of nano-Fe particles (**a**) in water, (**b**) in 30% ethanol, (**c**) in 50% ethanol (Magnification: 80,000×).

SEM images of Fe and Ni/Fe nanoparticles (Figure 2) showed that they had similar diameters, in the range of 20–60 nm, which was in line with the results obtained by Zhang et al. (20–100 nm) and by Naser et al. (40–50 nm) [15,22]. The nanoparticles form chain-like assemblies, where the Fe nanoparticles appeared as smooth "fiber", while the Ni/Fe nanoparticles (2%) appeared slightly more "granular". The formation of the chain-like structure was thought to be driven by magnetic attraction between the nanoparticles due to the ferromagnetic properties of Ni and Fe metal [23]. The surface area of the synthesized Fe and Ni/Fe nanoparticles were characterized by the BET method and was found to be 25.41 m^2/g for the Fe nanoparticles and 27.70 m^2/g for the Ni/Fe nanoparticles. The similarity in surface area indicated that they may have a similar adsorption capacity. It is worth noting that the size and morphology of the nanoparticles are largely dependent on the synthesis conditions, such as the solvent used, the type and concentration of the ferric salt and the reductant, etc., which in turn affected the reactivity of the obtained nanoparticles [23–25].

3.2. Dechlorination of HCB by Nanoscale Ni/Fe

Figure 3 shows the degradation of HCB to chlorobenzenes in water over time. The concentration of HCB dropped rapidly in the first 4 h (removal percentage: 79%) and then gradually decreased, reaching 99% removal in 48 h. Judging from the mass balance, the initial drop in concentration was due to adsorption onto the nanoparticles, followed by a period of slower removal dominated by dechlorination [12]. The main products of HCB dechlorination were pentachlorobenzene (PCB), tetrachlorobenzenes (TeCBs), and trichlorobenzenes (TCBs), with a trace amount of 1,2-dichlorobenzene (DCB). Among the three isomers of TeCB, 1,2,3,4-TeCB was the main product, and 1,2,3,5- and 1,2,4,5-TeCB were only produced in small amounts. For TCB, 1,2,3-TCB was the main product, with very little 1,2,4-TCB detected in 72 h of reaction. The formation of the n-chlorobenzene series suggests a mechanism of sequential dechlorination where HCB was degraded from n-Cl-to (n-1)-Cl-benzenes. In the first 4 h, PCB was the main product, whose concentration first increased and then decreased, peaking at 12 h. TeCBs exhibited a similar trend of variation, but peaked at 24 h, before which their concentration was lower than that of PCB, indicating that they were formed further down the reaction series.

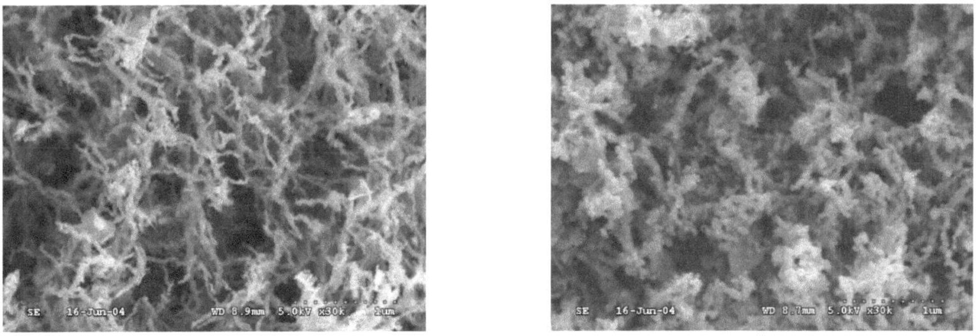

(a) (b)

Figure 2. Scanning electron microscope (SEM) images of (**a**) Fe nanoparticles and (**b**) Ni/Fe nanoparticles (2%).

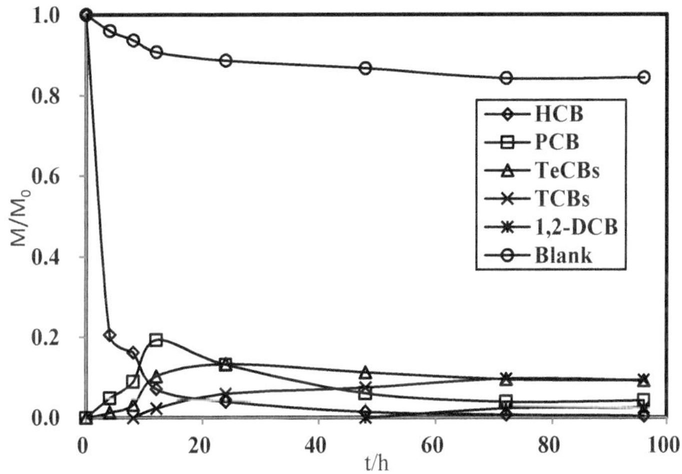

Figure 3. Action time of HCB dechlorination by Ni/Fe.

The reductive dechlorination of chlorinated organic compound by Ni/Fe nanoparticles proceeds through catalytic hydrogenation, where Fe acts as the reductant and loses electrons to become Fe^{2+} [26]. H^+ in water gains electrons and becomes activated [H], which is adsorbed onto the catalytic Ni surface. Meanwhile, the C-Cl bond in the chlorinated hydrocarbon (RCl) adsorbed on the surface of the nanoparticles breaks, with the Cl atom being replaced by an H atom, converting RCl into RH and Cl^-. Based on byproduct analysis, a reaction pathway is proposed for the degradation of HCB by Ni/Fe nanoparticles (Figure 4). HCB loses one chlorine atom to form PCB, which undergoes further dechlorination to form 1,2,3,4-, 1,2,3,5-, and 1,2,4,5-TeCB, as well as 1,2,3-TCB and 1,2,4-TCB. The final product was 1,2-DCB. The highly selective ortho-hydrogenation is thought to be primarily due to the steric hindrance of the chlorine atoms [9].

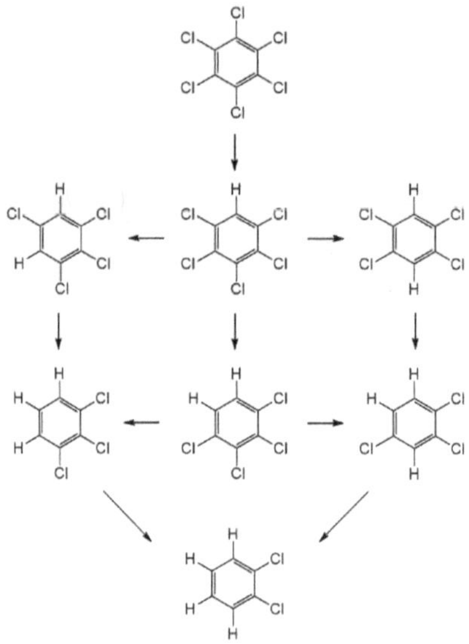

Figure 4. Pathway of degradation of hexachlorobenzene by Ni/Fe nanoparticles.

Several other studies have investigated the rate and byproducts of HCB removal by Fe-based bimetallic nanoparticles. Zhu et al. showed that 4 g/L Cu/Fe nanoparticles in 200 μg/L HCB solution with pH = 4 buffer facilitated a 85% removal percentage in 12 h [14]. Shih et al. demonstrated that with 2 g Pd/Fe nanoparticles in 15 mL HCB solution (the initial concentration was not provided), the removal percentage of HCB reached 70% in 24 h and then plateaued [10]. In our study, 93% removal of HCB was achieved in 12 h with 7.5 g/L Ni/Fe nanoparticles in 400 μg/L HCB solution, which was faster than the rates reported in both studies. In terms of product distribution, the main products obtained by Zhu et al. and in the present study after 96 h of reaction were TeCB and TCB, whereas Shih et al. reported that DCB was the main product after 100 h of reaction, suggesting that more complete degradation was achieved with Pd/Fe nanoparticles. The differences may be attributed to different initial concentrations of reactants, reaction conditions, and catalytic abilities of the coating metals.

Theoretical calculations suggested that the activation energies of removing chlorine atoms become greater as the number of Cl substituents decreases [27]. Therefore, the degradation of DCB to mono-substituted CB and benzene may be difficult in the current reaction conditions. Nevertheless, the less-substituted chlorobenzenes exhibit lower toxicity and are more easily removed by microorganisms, so chemical degradation by nanoparticles may alleviate the toxic stress on the ecosystem and assist the self-repair of the environment.

3.3. *Dechlorination of HCB in the Presence of Heavy Metals*

3.3.1. HCB Dechlorination by Ni/Fe Nanoparticles in the Presence of a Single Heavy Metal

Figure 5 illustrates the concentrations of HCB and its degradation products at different time intervals for three reaction systems: (1) with no heavy metal, (2) with Cd(II), and (3) with Zn(II). As is apparent in Figure 5a, HCB was removed rapidly by Ni/Fe nanoparticles in all reaction systems, and the addition of heavy metal did not significantly affect the removal rate. Figure 5b shows that the concentration of PCB first increased then decreased in all reaction systems, peaking at 12 h. The addition of Cd(II) elevated the PCB concentration throughout the experiment, which may be attributed to competitive adsorption between PCB and Cd(II) [15]. On the contrary, the addition of Zn(II) showed almost no effect on the concentration of PCB.

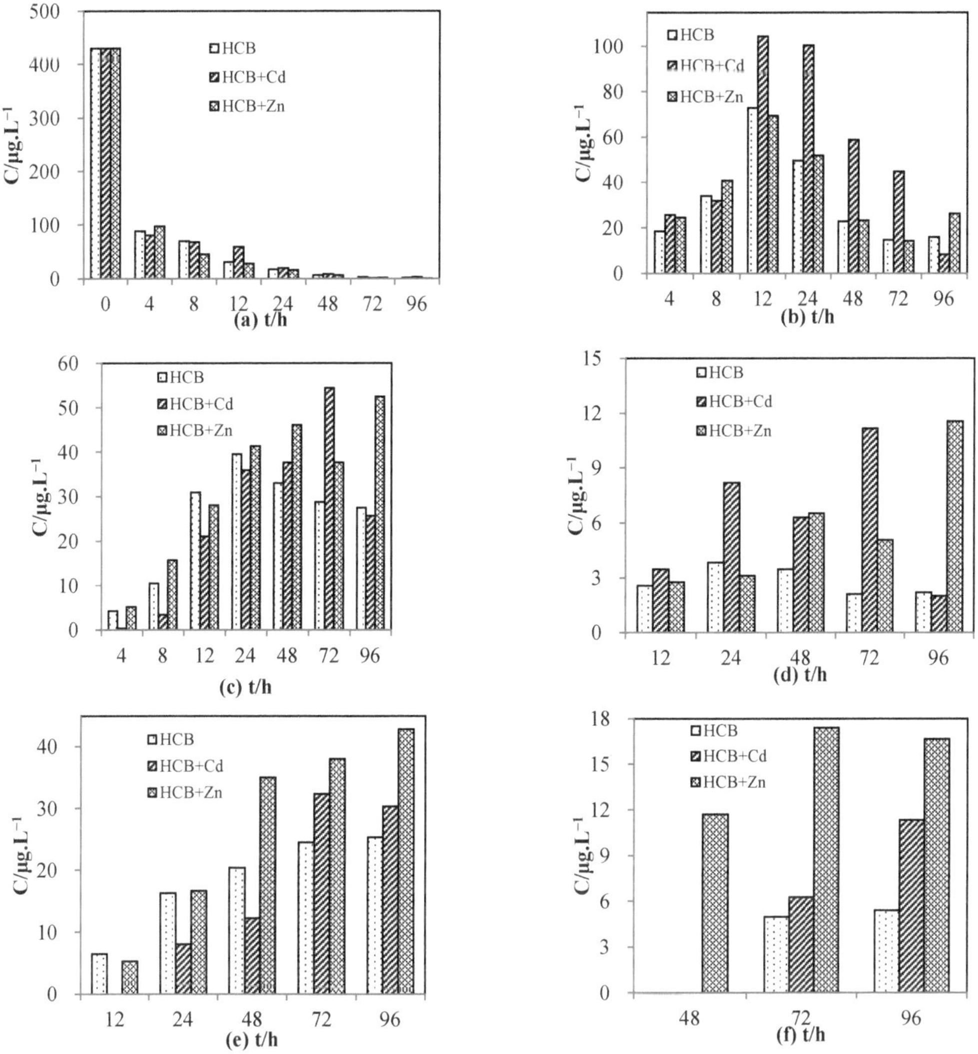

Figure 5. Change in the concentrations of hexachlorobenzene and its degradation products over time. (**a**) hexachlorobenzene, (**b**) pentachlorobenzene, (**c**) 1,2,3,4-tetrachlorobenzene, (**d**) 1,2,3,5/1,2,4,5-tetrachlorobenzene, (**e**) 1,2,3-trichlorobenzene, (**f**) 1,2-dichlorobenzene.

Figure 5c shows that the concentration of 1,2,3,4-TeCB first increased and then decreased, peaking at 24 h for the system with HCB only and 72 h and 48 h for the systems with added Cd(II) and Zn(II), respectively. The delayed peaks indicated that the addition of heavy metal slowed down the degradation. Furthermore, the concentration of 1,2,3,4-TeCB was lower in the beginning of the reaction in the system with Cd(II), but rose to a high level as the reaction proceeded. It reached a maximum of 54.42 μg/L in the system with Cd(II) and 52.46 μg/L in the system with Zn(II), which were higher than the maximum of the heavy-metal-free system (35.89 μg/L). This phenomenon suggested that with added heavy metals, the generation of TeCB was slowed down at the beginning of the reaction. As the reaction went on, the removal of TeCB via adsorption and dechlorination became the dominant process, which was also slowed down by the addition of heavy metal, leading to an increased concentration. 1,2,3,5/1,2,4,5-TeCB was produced in a small amount and its variation was more irregular (Figure 5d). However, the delay of the maximum and the increased concentration in the later part of the reaction can still be observed.

Figure 5e shows that the concentration of 1,2,3-TCB was initially lower with added Cd(II) and Zn(II). It then increased, surpassing that without added heavy metal and staying at a high level. The concentration plateaued at 72 h for the system with added Cd(II), whereas it kept increasing until the end of the experiment with added Zn(II). The observation that the maximum of TCB was delayed or even unobtainable in the time frame of the reaction indicated that its removal was inhibited by the addition of heavy metals.

Figure 5f shows that 1,2-DCB was produced at a low level, with only a trace amount detected at 48 h for the system with added Zn(II) and 72 h for the other two systems. The concentration of DCB increased with added Cd(II) and Zn(II), suggesting that the heavy metals affected its adsorption efficiency.

The changing trends of the byproducts agree with previously published results on the catalytic dechlorination of 2,4-dichlorophenol over Ni/Fe nanoparticles, where addition of an inhibitory substance (humic acid) delayed the peak of the byproduct o-chlorophenol and lowered the initial concentration of phenol, the final product [15]. On the other hand, the lack of change in the removal rate of HCB may be explained by its non-polarity. It is hypothesized that HCB may occupy different active sites on the surface of the nanoparticles from charged ions and polar organic molecules, therefore experiencing less adsorptive competition. This is supported by previous studies where the presence of Na^{+} and Mg^{2+} was shown to have no impact on HCB's degradation rate [21].

3.3.2. HCB Dechlorination by Ni/Fe Nanoparticles in the Presence of Multiple Heavy Metals

Figure 6 shows the concentration of HCB and its degradation products in a heavy-metal-free system and in a system with both Cd(II) and Zn(II). As shown in Figure 6a, the concentration of HCB was lower with Cd(II) + Zn(II), and the removal percentage in 4 h increased from 79% to 88%. It seems that the co-existence of Cd(II) and Zn(II) did not affect the removal rate of HCB but promoted it slightly.

Figure 6b shows that the concentration of PCB peaked at 48 h with Cd(II) + Zn(II), and stayed at a high level from 12 h to 48 h. The delayed peak and plateauing were not observed in the systems with a single heavy metal, suggesting that the reaction was further slowed down by the addition of both heavy metals. The variation in 1,2,3,4-TeCB in the system with Cd(II) + Zn(II) resembled that with a single heavy metal, except that the concentration kept increasing until the end of the experiment instead of peaking at a certain point like in the system with Cd(II) (Figure 6c). This implies that TeCB's degradation was further inhibited by the addition of both heavy metals and therefore could not reach a maximum in the time frame of the experiment. Moreover, as can be seen from Figure 6c,d, the maximum concentration of 1,2,3,4-TeCB and 1,2,4,5/1,2,3,5-TeCB was higher than that of the systems with a single heavy metal, suggesting that adsorptive competition was more pronounced in the presence of both heavy metals.

The concentration of 1,2,3-TCB with the addition of both heavy metals was the lowest among all four systems in the beginning. It then kept rising and became the second highest, only slightly lower than the system with Zn(II) (Figure 6e). A very low amount of 1,2,4-TCB was also detected during the experiment. For 1,2-DCB, a small amount was produced and its variation with the addition of two heavy metals was similar to that with Cd(II) (Figure 6f).

Overall, the addition of two heavy metals caused notable changes in the variation in the chlorobenzenes. Specifically, the removal of HCB was slightly accelerated, while the adsorption and dechlorination reactions involving PCB, TeCB, and TCB were further hindered compared to the systems with a single heavy metal. The variation in DCB did not differ much from that of the system with Cd(II). The observed phenomena show that the effects of multiple influencing ions are not merely additive but complex, varying for different organic molecules. One of the possible causes is that different adsorbates have distinct preferences for binding sites. Additionally, the adsorbed species are known to alter the surface characteristics of the nanoparticles and modify their adsorption capacity.

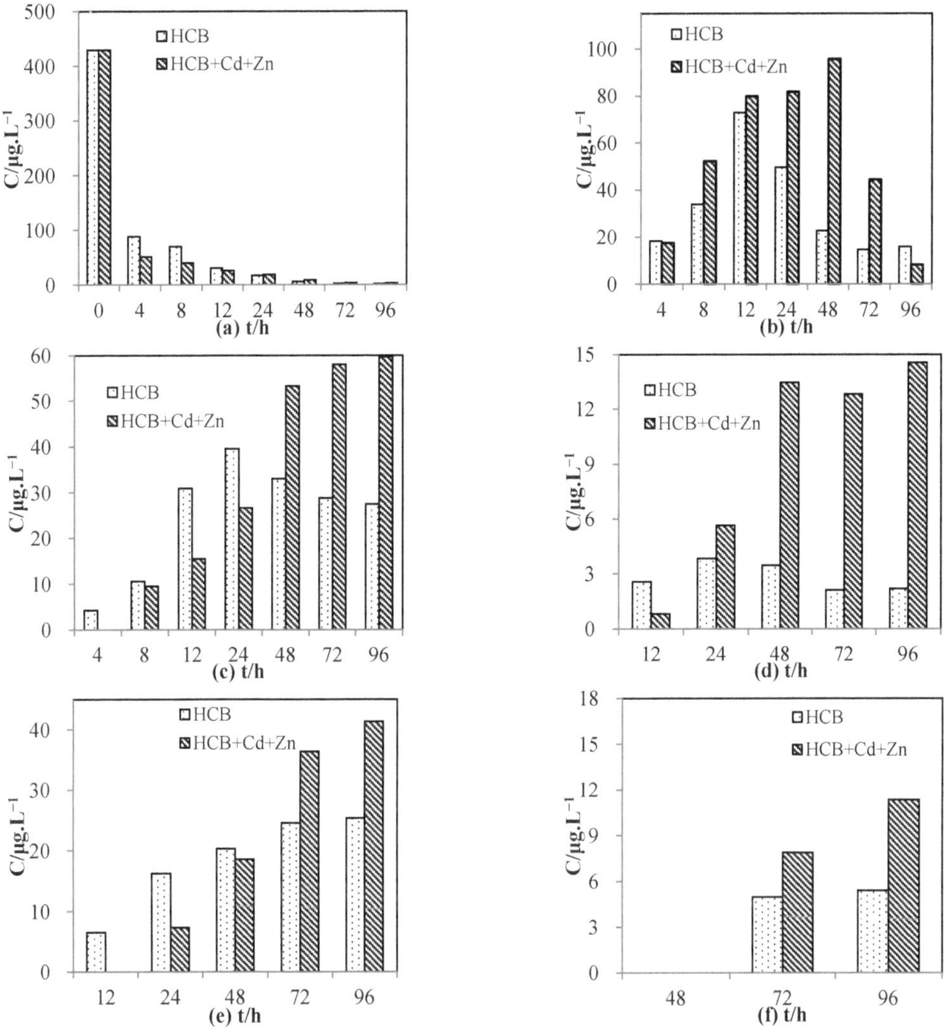

Figure 6. Change in the concentrations of hexachlorobenzene and its products over the reaction time. (**a**) hexachlorobenzene, (**b**) pentachlorobenzene, (**c**) 1,2,3,4-tetrachlorobenene, (**d**) 1,2,3,5/1,2,4,5-tetrachlorobenzene, (**e**) 1,2,3-trichlorobenzene, (**f**) 1,2-dichlorobenzene.

3.3.3. Change in Heavy Metal Concentration

Figure 7 shows the normalized concentration of Cd(II) and Zn(II) in the reaction system containing a single type of heavy metal and the reaction system with both heavy metals.

In all cases, the concentration of the heavy metals decreased rapidly, reaching a removal percentage of over 98% in 4 h, and remained at a low level. The removal of Zn(II) was slightly better than that of Cd(II). Since the standard reduction potentials of Zn(II) and Cd(II) are more negative or very close to that of iron, the main removal mechanism is believed to be adsorption or surface complex formation instead of metal displacement [28]. The finding that the heavy metals can be effectively removed by Ni/Fe nanoparticles suggests that competition between the heavy metals and CBs for adsorption sites may be the main cause for the reduced rate of dechlorination.

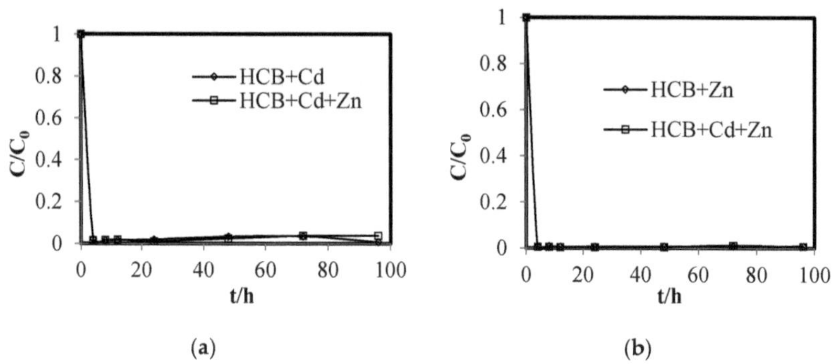

Figure 7. Normalized concentrations of (**a**) Cd(II) and (**b**) Zn(II) in the reaction systems.

Our results demonstrate that HCB can be quickly removed by Ni/Fe nanoparticles via adsorption and sequential dechlorination, producing PCB, 1,2,3,4-TeCB, and 1,2,3-TCB as the main products, with 1,2,3,5/1,2,4,5-TeCB, 1,2,4-TCB, and 1,2-DCB detected in trace amounts. The effect of heavy metal ions on the reaction rate and product composition was examined by analyzing the variation in concentration for all products, where a low initial concentration, increased maximum, and the delay or absence of peaks were identified as signals of inhibited adsorption and degradation. It can be seen from the experiment results that the addition of heavy metals has little effect on HCB's removal but affects the adsorption and degradation of the byproducts. The presence of both Cd(II) and Zn(II) further reduces the extent of the sequential dechlorination process. On the other hand, the removal of heavy metal ions was not affected, probably because the amount of HCB added was low and thus had no noticeable effect on the adsorption of metal ions to the nanoparticles. Future investigation was required to clarify how the change in reaction conditions, such as starting HCB and heavy metal concentration, pH, temperature, etc., affects the reaction.

4. Conclusions

Our study has investigated the kinetics and products of HCB dechlorination by Ni/Fe nanoparticles, and the influence of added heavy metal ions Cd(II) and Zn(II) on the reaction. It is found that HCB can be rapidly removed by Ni/Fe nanoparticles via adsorption and sequential dechlorination. The main products are pentachlorobenzene, 1,2,3,4-tetrachlorobenzene, and 1,2,3-trichlorobenzene, with 1,2,3,5/1,2,4,5-tetrachlorobenzene, 1,2,4-trichlorobenzene, and 1,2-dichlorobenzene as the minor products. The addition of Cd(II) and Zn(II) has minimal effect on the removal rate of HCB, but inhibits the adsorption and degradation of the products, as can be seen in the changes in their concentration–time profiles. The inhibitory effect may be attributed to competitive adsorption between the heavy metals and the chlorobenzenes. The presence of both heavy metals generally caused greater inhibition than a single type of heavy metal. On the other hand, the heavy metals were effectively removed by Ni/Fe nanoparticles in all reaction systems, with the removal of Zn(II) slightly better than that of Cd(II). The results show that Ni/Fe nanoparticles can remove HCB and heavy metal ions concurrently, with the degradation of hexachlorobenzene made slower by the presence of heavy metals. These findings are poised to guide further applications of bimetallic nanoparticles in the treatment of water and soil with mixed pollutants.

Author Contributions: Conceptualization, Y.H., S.L. and Q.W.; Data curation, Y.H. and Q.W.; Formal analysis, Y.H., Q.W. and G.H.; Funding acquisition, S.L.; Investigation, Y.H., Q.W. and G.H.; Methodology, Y.H. and G.H.; Resources, S.L.; Supervision, S.L.; Visualization, Q.W.; Writing—original draft, Y.H.; Writing—review and editing, S.L., Q.W., G.H., X.Z. and Y.L. All authors have read and agreed to the published version of the manuscript.

Funding: The research was funded by the National Key Research and Development Program of China (No. 2023YFC3709104) and the Research Fund of China Geological Survey (DD20230754).

Data Availability Statement: Data is contained within the article.

Acknowledgments: The authors would like to thank Jingyan Hao, Lei Yang, and Longlong Zhang for their kind help during this study.

Conflicts of Interest: The authors declare that they have no known competing financial interests or personal relationships that could have appeared to influence the work reported in this paper.

References

1. Böhm, L.; Grančič, P.; Scholtzová, E.; Heyde, B.J.; Düring, R.-A.; Siemens, J.; Gerzabek, M.H.; Tunega, D. Adsorption of the Hydrophobic Organic Pollutant Hexachlorobenzene to Phyllosilicate Minerals. *Environ. Sci. Pollut. Res.* **2022**, *30*, 36824–36837. [CrossRef] [PubMed]
2. Pala, N.; Jiménez, B.; Roscales, J.L.; Bertolino, M.; Baroni, D.; Figuerola, B.; Avila, C.; Corsolini, S. First Evidence of Legacy Chlorinated POPs Bioaccumulation in Antarctic Sponges from the Ross Sea and the South Shetland Islands. *Environ. Pollut.* **2023**, *329*, 121661. [CrossRef] [PubMed]
3. Bailey, R.E. Global Hexachlorobenzene Emissions. *Chemosphere* **2001**, *43*, 167–182. [CrossRef]
4. Barber, J.L.; Sweetman, A.J.; Van Wijk, D.; Jones, K.C. Hexachlorobenzene in the Global Environment: Emissions, Levels, Distribution, Trends and Processes. *Sci. Total Environ.* **2005**, *349*, 1–44. [CrossRef]
5. Fan, J.; Liu, C.; Guo, D.; Han, L.; Zhang, C.; Jin, H. Biochar and Biochar-Polylactic Acid Composite Enhance Biodegradation of Hexachlorobenzene in Soil by Altering Microbial Community. *Appl. Soil Ecol.* **2022**, *177*, 104521. [CrossRef]
6. Zhang, X.; Zhang, X.; Zhang, Z.-F.; Yang, P.-F.; Li, Y.-F.; Cai, M.; Kallenborn, R. Pesticides in the Atmosphere and Seawater in a Transect Study from the Western Pacific to the Southern Ocean: The Importance of Continental Discharges and Air-Seawater Exchange. *Water Res.* **2022**, *217*, 118439. [CrossRef]
7. Qiu, X.; Zhu, T.; Li, J.; Pan, H.; Li, Q.; Miao, G.; Gong, J. Organochlorine Pesticides in the Air around the Taihu Lake, China. *Environ. Sci. Technol.* **2004**, *38*, 1368–1374. [CrossRef] [PubMed]
8. Liu, W.-J.; Qian, T.-T.; Jiang, H. Bimetallic Fe Nanoparticles: Recent Advances in Synthesis and Application in Catalytic Elimination of Environmental Pollutants. *Chem. Eng. J.* **2014**, *236*, 448–463. [CrossRef]
9. Li, J.; Wang, F.; Yang, X.; Gu, C.; Zhang, Y.; Jiang, X. Degradation of Highly-Chlorinated Benzenes by Nanoscale Iron and Palladized Iron in Aqueous System. *Environ. Sci.* **2011**, *32*, 692–698.
10. Shih, Y.; Chen, Y.-C.; Chen, M.; Tai, Y.; Tso, C.-P. Dechlorination of Hexachlorobenzene by Using Nanoscale Fe and Nanoscale Pd/Fe Bimetallic Particles. *Colloids Surf. A Physicochem. Eng. Asp.* **2009**, *332*, 84–89. [CrossRef]
11. Nie, X.; Liu, J.; Yue, D.; Zeng, X.; Nie, Y. Dechlorination of Hexachlorobenzene Using Lead–Iron Bimetallic Particles. *Chemosphere* **2013**, *90*, 2403–2407. [CrossRef] [PubMed]
12. Xu, Y.; Zhang, W. Subcolloidal Fe/Ag Particles for Reductive Dehalogenation of Chlorinated Benzenes. *Ind. Eng. Chem. Res.* **2000**, *39*, 2238–2244. [CrossRef]
13. Nie, X.; Liu, J.; Zeng, X.; Yue, D. Rapid Degradation of Hexachlorobenzene by Micron Ag/Fe Bimetal Particles. *J. Environ. Sci.* **2013**, *25*, 473–478. [CrossRef]
14. Zhu, N.; Luan, H.; Yuan, S.; Chen, J.; Wu, X.; Wang, L. Effective Dechlorination of HCB by Nanoscale Cu/Fe Particles. *J. Hazard. Mater.* **2010**, *176*, 1101–1105. [CrossRef]
15. Zhang, Z.; Cissoko, N.; Wo, J.; Xu, X. Factors Influencing the Dechlorination of 2,4-Dichlorophenol by Ni–Fe Nanoparticles in the Presence of Humic Acid. *J. Hazard. Mater.* **2009**, *165*, 78–86. [CrossRef]
16. Lin, Y.; Jin, X.; Owens, G.; Chen, Z. Simultaneous Removal of Mixed Contaminants Triclosan and Copper by Green Synthesized Bimetallic Iron/Nickel Nanoparticles. *Sci. Total Environ.* **2019**, *695*, 133878. [CrossRef]
17. Zhang, Z.; Hu, S.; Baig, S.A.; Tang, J.; Xu, X. Catalytic Dechlorination of Aroclor 1242 by Ni/Fe Bimetallic Nanoparticles. *J. Colloid Interface Sci.* **2012**, *385*, 160–165. [CrossRef] [PubMed]
18. USGS. *Mineral Commodity Summaries 2023*; Mineral Commodity Summaries; USGS: Reston, VA, USA, 2023; p. 210.
19. Zhang, Y.; Liu, J.; Zhou, Y.; Gong, T.; Wang, J.; Ge, Y. Enhanced Phytoremediation of Mixed Heavy Metal (Mercury)–Organic Pollutants (Trichloroethylene) with Transgenic Alfalfa Co-Expressing Glutathione S-Transferase and Human P450 2E1. *J. Hazard. Mater.* **2013**, *260*, 1100–1107. [CrossRef]
20. Zhao, W.; Gu, C.; Ying, H.; Feng, X.; Zhu, M.; Wang, M.; Tan, W.; Wang, X. Fraction Distribution of Heavy Metals and Its Relationship with Iron in Polluted Farmland Soils around Distinct Mining Areas. *Appl. Geochem.* **2021**, *130*, 104969. [CrossRef]

21. Su, Y.; Hsu, C.-Y.; Shih, Y. Effects of Various Ions on the Dechlorination Kinetics of Hexachlorobenzene by Nanoscale Zero-Valent Iron. *Chemosphere* **2012**, *88*, 1346–1352. [CrossRef]
22. Naser, R.; Shahwan, T. Comparative Assessment of the Decolorization of Aqueous Bromophenol Blue Using Fe Nanoparticles and Fe-Ni Bimetallic Nanoparticles. *Desalination Water Treat.* **2019**, *159*, 346–355. [CrossRef]
23. Zhang, L.; Manthiram, A. Chains Composed of Nanosize Metal Particles and Identifying the Factors Driving Their Formation. *Appl. Phys. Lett.* **1997**, *70*, 2469–2471. [CrossRef]
24. Shen, J.; Li, Z.; Yan, Q.; Chen, Y. Reactions of Bivalent Metal Ions with Borohydride in Aqueous Solution for the Preparation of Ultrafine Amorphous Alloy Particles. *J. Phys. Chem.* **1993**, *97*, 8504–8511. [CrossRef]
25. Mitov, M.; Popov, A.; Dragieva, I. Nanoparticles Produced by Borohydride Reduction as Precursors for Metal Hydride Electrodes. *J. Appl. Electrochem.* **1999**, *29*, 59–63. [CrossRef]
26. Schrick, B.; Blough, J.L.; Jones, A.D.; Mallouk, T.E. Hydrodechlorination of Trichloroethylene to Hydrocarbons Using Bimetallic Nickel-Iron Nanoparticles. *Chem. Mater.* **2002**, *14*, 5140–5147. [CrossRef]
27. Garbou, A.M.; Liu, M.; Zou, S.; Yestrebsky, C.L. Degradation Kinetics of Hexachlorobenzene over Zero-Valent Magnesium/Graphite in Protic Solvent System and Modeling of Degradation Pathways Using Density Functional Theory. *Chemosphere* **2019**, *222*, 195–204. [CrossRef]
28. Li, X.; Zhang, W. Sequestration of Metal Cations with Zerovalent Iron Nanoparticless A Study with High Resolution X-ray Photoelectron Spectroscopy (HR-XPS). *J. Phys. Chem. C* **2007**, *111*, 6939–6946. [CrossRef]

water

MDPI

Article

Extension of Iber for Simulating Non–Newtonian Shallow Flows: Mine-Tailings Spill Propagation Modelling

Marcos Sanz-Ramos [1,*], Ernest Bladé [1], Martí Sánchez-Juny [1] and Tomasz Dysarz [2]

[1] Flumen Research Institute, Polytechnique University of Catalonia (UPC)—International Centre of Numerical Method in Engineering (CIMNE), 08034 Barcelona, Spain; ernest.blade@upc.edu (E.B.); marti.sanchez@upc.edu (M.S.-J.)
[2] Department of Hydraulic and Sanitary Engineering, Faculty of Environmental Engineering and Mechanical Engineering, Poznań University of Life Sciences, 60-649 Ponzan, Poland; tomasz.dysarz@up.poznan.pl
* Correspondence: marcos.sanz-ramos@upc.edu

Abstract: Mine tailings are commonly stored in off-stream reservoirs and are usually composed of water with high concentrations of fine particles (microns). The rupture of a mine-tailings pond promotes, depending on the characteristics of the stored material, the fluidization and release of hyper-concentrated flows that typically behave as non–Newtonian fluids. The simulation of non–Newtonian fluid dynamics using numerical modelling tools is based on the solution of mass and momentum conservation equations, particularizing the shear stress terms by means of a rheological model that accounts for the properties of the fluid. This document presents the extension of Iber, a two-dimensional hydrodynamic numerical tool, for the simulation of non–Newtonian shallow flows, especially those related to mine tailings. The performance of the numerical tool was tested throughout benchmarks and real study cases. The results agreed with the analytical and theoretical solutions in the benchmark tests; additionally, the numerical tool also revealed itself to be adequate for simulating the dynamic and static phases under real conditions. The outputs of this numerical tool provide valuable information, allowing researchers to assess flood hazard and risk in mine-tailings spill propagation scenarios.

Keywords: numerical modelling; Iber; viscous-plastic flows; gypsum tailings; pyroclastic/pyritic tailings

Citation: Sanz-Ramos, M.; Bladé, E.; Sánchez-Juny, M.; Dysarz, T. Extension of Iber for Simulating Non–Newtonian Shallow Flows: Mine-Tailings Spill Propagation Modelling. *Water* **2024**, *16*, 2039. https://doi.org/10.3390/w16142039

Academic Editors: Ruiping Liu and Youning Xu

Received: 17 June 2024
Revised: 8 July 2024
Accepted: 16 July 2024
Published: 18 July 2024

1. Introduction

Mine-tailings disasters represent one of the most significant environmental and humanitarian challenges associated with the mining industry [1–3]. Tailings, the waste materials produced from ore processing, contain a mixture of finely ground rock, water, and residual amounts of chemicals used in the extraction process. When containment structures fail or are overwhelmed by natural or anthropic forces, huge amounts of these tailings can be released, causing severe environmental contamination and posing serious risks to human health and livelihoods [2,4–6].

The complexity and scale of mine-tailings disasters present daunting challenges for emergency responders, environmental regulators, and affected communities alike. Effective response and recovery efforts require a deep understanding of the behaviour of tailings materials, including their transport mechanisms and dispersion patterns, as well as the potential for further environmental degradation over time. Moreover, addressing the root causes of these disasters demands comprehensive risk assessment, improved engineering practices, and robust regulatory frameworks to prevent future occurrences [7–9].

In recent years, heightened awareness of the risks associated with mine-tailings storage has spurred efforts to enhance disaster preparedness and mitigate the impacts of potential spills. Advances in technology and modelling tools have enabled increasingly accurate predictions of tailings behaviour and the development of strategies to minimize risks to both human populations and the environment [10]. In response to the increasing awareness of these risks, there has been a growing interest in the development of numerical tools to predict the propagation and dispersion of mine-tailings spills. Thus, mine-tailings spill propagation modelling is a crucial component of assessing and mitigating the environmental impact of mining activities [11].

These models aim to simulate the movement of tailings from the source of the spill and subsequently through the environment, considering factors such as topography, hydraulics, sediment transport, and the physical and chemical characteristics of the tailings materials. One of the key challenges in mine-tailings spill propagation modelling is accurately representing the complex behaviour of the fluid once released into the environment. The rheological properties of tailings, commonly assimilated to non-Newtonian flows, can vary widely, depending on factors such as particle size distribution, mineralogy, and water content [12–14]. Additionally, the terrain over which the spill travels can greatly influence its behaviour, with steep slopes, channels, and vegetation all affecting the flow dynamics [15–18].

The simulation of the dynamics of non–Newtonian flows, such as mine tailings, can be performed by means of numerical modelling tools that are based on the solution of mass and momentum conservation equations [19–22]. To that end, the shear stress terms might be particularised by a rheological model that accounts for the properties of the fluid [19,23–25]. Additional numerical treatments might be accomplished with the aim of accurately reproducing both the dynamic and the static behaviour of the fluid over steep slopes and irregular geometries [16,26–31].

The purpose of the research presented here is the evaluation of the new features of Iber, a freely distributed depth-averaged two-dimensional hydrodynamic numerical tool (www.iberaula.com (accessed on 15 June 2024)), recently extended for modelling of non–Newtonian shallow flows. The code integrates several rheological models to consider different flow behaviours and a specific numerical scheme that determines the stopping of the fluid according to the rheological properties. The accuracy and efficiency of the simulations realised according to the implemented methods are carefully assessed in benchmarks and real case studies, showing a good agreement with the observations and the absence of numerical instabilities.

2. Materials and Methods

2.1. Numerical Simulation Tool: IberNNF

Physically-based numerical tools for the simulation of non–Newtonian flows utilise mass and momentum balance equations implemented for shallow flow conditions [16,18,32–36]. These equations, when applied to water, are known as shallow water equations (SWE). They are derived from the Navier–Stokes equations through a temporal averaging to filter out turbulent fluctuations (Reynolds Averaged Navier–Stokes, or RANS), and a depth averaging to convert the three-dimensional equations into two-dimensional ones [30,37].

The SWE, applied in a two-dimensional domain (2D-SWE), form a hyperbolic nonlinear system of three partial differential equations: one corresponding to mass conservation and the other two to momentum conservation in the x and y directions of space:

$$\frac{\partial h}{\partial t} + \frac{\partial q_x}{\partial x} + \frac{\partial q_y}{\partial y} = 0 \tag{1}$$

$$\frac{\partial q_x}{\partial t} + \frac{\partial}{\partial x}\left(\frac{q_x^2}{h} + g\frac{h^2}{2}\right) + \frac{\partial}{\partial y}\left(\frac{q_x q_y}{h}\right) = ghS_{o,x} - \frac{\tau_x}{\rho} \tag{2}$$

$$\frac{\partial q_x}{\partial t} + \frac{\partial}{\partial x}\left(\frac{q_x q_y}{h}\right) + \frac{\partial}{\partial y}\left(\frac{q_y^2}{h} + g\frac{h^2}{2}\right) = ghS_{o,y} - \frac{\tau_y}{\rho} \tag{3}$$

where h is the flow depth, q_x and q_y are the two components of depth-averaged specific discharge, g is the projected acceleration due to gravity, $S_{o,x}$ and $S_{o,y}$ are the components of the bottom slope, and τ_x and τ_y are the two components of the shear stress associated with the rheological model.

Once discretised on a computational mesh composed by elements, the first equation of the previous system shows the variation of the flow depth over time ($\partial h/\partial t$) in an element due to volume inflows and outflows ($\partial q_x/\partial x$, $\partial q_y/\partial y$) through its boundaries. The second and third equations correspond to the variations of the discharges ($\partial q_x/\partial t$, $\partial q_y/\partial t$) as functions of the forces acting on the fluids. Four terms can be distinguished (only shown for the x-component): inertia $\partial(q_x^2/h)/\partial x$ and $\partial(q_x q_y/h)/\partial y$, pressure $\partial\left(gh^2/2\right)/\partial x$, gravity through the bottom slope ($S_{0,x}$), and shear stress (τ_x).

Iber is a freely distributed, two-dimensional, depth-averaged hydrodynamic tool [38,39] currently able to simulate different environmental shallow-water flows and transport processes [16,40–52]. The previous systems of equations are solved through the Roe scheme [53], a first-order Godunov-type upwind scheme for convective fluxes and the geometric slope source term.

Iber has been recently enhanced to simulate non–Newtonian flows (IberNNF), such as dense snow avalanches [15,16]. To that end, a single-phase fluid and the shear stress grouping proposed by Julien and León [23] are assumed, besides the utilisation of an ad-hoc numerical scheme that ensures the balance among the different terms of SWE [16]. Thus, the rheological model of any particular fluid can be integrated through the shear stresses term as the addition of five components: the dispersive contribution (τ_d), the turbulent contribution (τ_t), the viscous contribution (τ_v), the Mohr–Coulomb contribution (τ_{mc}), and the cohesive contribution (τ_c).

In this work, with the aim of characterizing the static and dynamic behaviour of non–Newtonian shallow flows such as those found in mine-tailings spill propagation, two friction models have been implemented in the code. The following rheological models are expressed as friction slope (S_f), being $\tau = \rho g h S_f$.

Since the proposal of the Bingham rheological model [54], several approaches have been introduced to deal with the difficulties associated with directly obtaining the shear stress proportional to the flow velocity [55]. Some authors [56,57], assuming an incompressible and homogeneous flow, derived the following expression for the viscous (τ_v) and the Mohr–Coulomb (τ_{mc}) contributions:

$$S_f = \frac{1}{\rho g h}\left(\frac{3}{2}\tau_y + 3\frac{\mu_B v}{h}\right) \tag{4}$$

where τ_y is the yield stress, ρ is the fluid density, h is the flow depth, μ_B is the fluid viscosity, v is the flow velocity, and g is the gravitational acceleration.

Voellmy [58] presented a rheological model that considers the turbulent friction (τ_t) and the Mohr–Coulomb (τ_{mc}) terms as follows:

$$S_f = \mu + \frac{v^2}{\xi h} \tag{5}$$

where ξ is the turbulent friction coefficient, and μ is the Coulomb friction coefficient.

2.2. Test Cases

Different test cases were utilised to analyse the performance of this new extension of the numerical tool Iber for the simulation of non–Newtonian shallow flows, particularly with reference to mine-tailings propagation. The fluid was characterised by selecting a particular rheological model (Voellmy or simplified Bingham) that can represent its behaviour during the dynamic and static phases.

The first test case, originally presented by Bryant [59], consists of an idealised dam-break problem for viscous–plastic fluids that attempts to represent the release of a stored fluid and the subsequent propagation process. The experiment defines an initial fluid height of 30.5 m and length of 305 m which is immediately released over a flat terrain. Hungr [20] presented, on the bases of this experiment, an analytical solution for a plastic fluid (ρ = 1835 kg/m^3 and total steady-state shear strength of 2390 Pa), a particular case of a simplified Bingham rheological model, and the numerical results obtained with the 1D numerical tool DAN. This experiment attempts to explore the capability of a fluid to stop and rest, given a non-horizontal free surface, such may occur with mud flows, debris flows, etc.

The second test case consists of a constant inclined plane (0.23 m/m) with a non–horizontal initial flow surface, aiming to represent the release and stop of a pyroclastic avalanche, a particular non–Newtonian flow that can be triggered by rainfall after a volcanic eruption. A detailed description of the numerical test can be found in de' Michieli Vitturi et al. [60], who simulated the avalanche dynamics with the Voellmy rheological model (μ = 0.3 and ξ = 300 m/s^2). According to the rheological properties, the fluid should stop on the inclined terrain surface. This test extends the previous one to the context of steep slopes, in which non–Newtonian unconfined flows can stop in the presence of a non–horizontal surface.

The third test case is the mine-tailings propagation caused by the failure of a gypsum pond which occurred in 1966, in East Texas (USA). Jeyapalan et al. [61] revisited the hydraulics of the pond failure, reporting the relevant observed data: the fluid was spread asymmetrically, and inside the impoundment the fluid was mobilized to points around 110 m perpendicular to the dyke and 140 m parallel to the dyke. This caused the mobilisation of around of 80,000–130,000 m^3 of gypsum tailings, which flowed some 300 m beyond the dyke in 60–120 s. From a previous analysis, the fluid was characterized as Bingham type with a flow density of 1400 kg/m^3, a yield stress of 1000 Pa, and a viscosity of 50 Pa·s.

Finally, the mine-tailing pond failure of Los Frailes (Aznalcóllar, Spain), which occurred on 25 April 1998, was utilised to test the numerical tool under real conditions. When the failure occurred, after a very fast breach formation that affected the two lagoons [62–64], the stored fluids (pyroclastic and pyritic tailings) were spilled and probably mixed. Their properties changed during the spill [65], and thus, the flow probably propagated as a unique hyper-concentrated fluid [66,67]. Despite the uncertainties in the hydraulics of the event [66], the fluid was assumed to behave as a Bingham plastic, with the aim of testing the performance of the numerical tool for simulating non–Newtonian shallow flows. Additionally, the spilled hydrograph proposed by Consultec Ingenieros [68] was used as inflow at the breakpoint. A bulk density of 3100 kg/m^3 was assumed, being the parameters of the rheological varied from 0 to 50 Pa for the yield stress (τ_y) and from 0 to 2000 Pa·s for the fluid viscosity (μ_B). The test was oriented as a sensitivity analysis instead of as a reconstruction of the flood event (Table 1).

Table 1. Summary of the simulated scenarios in the studied case of the mine-tailing pond failure of Los Frailes.

Parameter	Range	Increments
Yield stress, τ_y [Pa]	0–50	5
Viscosity, μ_B [Pa·s]	0–2000	5

3. Results

3.1. Idealised Dam-Break Problem for Viscous–Plastic Fluids

The calculation domain of 2000 m length was discretized with one-dimensional elements of 1 m length. According to Hungr [20], the rheological model selected was the simplified Bingham, but one only considering a total yield stress contribution of 2390 Pa (i.e., $\rho g h S_f$ = 2390 Pa). Results were compared with an approximated energy solution presented in Hungr [20], with a runout distance of 1896 m, and those of a DAN model, a one-dimensional numerical model developed ad hoc by the same author.

Figure 1a (green line) shows the fluid profile at the end of the simulation, which stopped at 130 s with a runout distance of 1875 m. The results show a good agreement with the detention time and runout distance of the analytical solution (Figure 1a, red line), in addition to the results of the DAN model (Figure 1a, blue dots). A non-horizontal free surface was obtained with the fluid at rest.

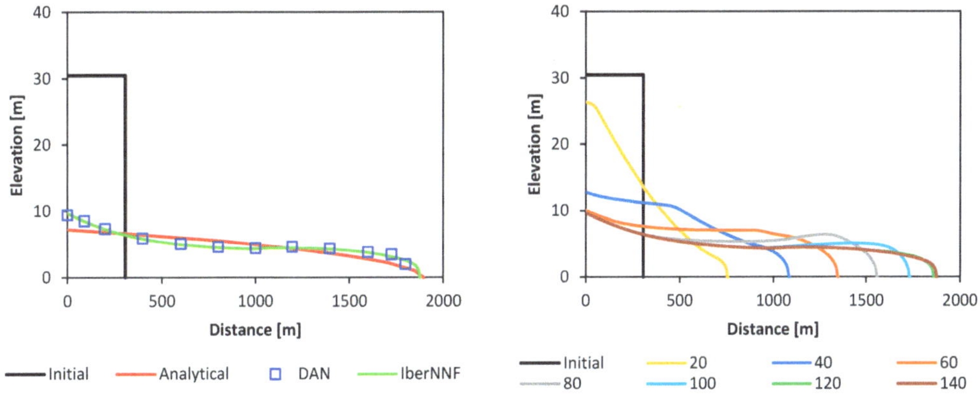

Figure 1. Dam-break of an idealised mine-tailings dam. (**a**) Final runout profile of the presented code (green line), compared to the analytical solution (red line) and the numerical solution of the DAN model (blue dots), as presented by Hungr [20]. (**b**) Runout profile evolution obtained with the presently described code, for each 20 s.

The evolution of the simulated fluid profile is plotted in Figure 1b for each 20 s, showing the particular behaviour of the fluid according to the rheological model used. This results in a non–horizontal final shape with fluid depths at the leading-edge higher than intermediate positions during the dynamic phase. The resulting wave-front velocities were 22.6, 16.5, 13.1, 10.6, 8.6, 6.5, and 1.4 m/s respectively, decreasing gradually.

This test shows the good performance of the presented numerical model, especially in those configurations where non-velocity terms of the rheological model are neglected. Hence, the fluid stops according to the yield stress contribution, which depends on the flow depth at each location and time step.

3.2. Fluid Detention on Sloping Terrain

A particular test over steep slopes is presented by de' Michieli Vitturi et al. [60], aiming to show the performance of IMEX_SfloW2D, a numerical tool developed to simulate pyroclastic avalanches in volcanic regions. A non-stable condition for an initial fluid over an inclined plane was imposed and then immediately released. Such initial condition was implemented by raster data, being the fluid-rheology Voellmy type. The domain was discretised with one-dimensional elements of 1 m in length.

Figure 2 compares the results of both numerical models. Circles correspond to IMEX_SfloW2D, while coloured lines represent the results obtained with the presented numerical model. In both cases the fluid stopped over the inclined plane, showing a quite similar solution. Since the free surface of the left side of the initial volume has a slope less than the Coulomb friction coefficient (μ), the avalanche did not move upstream; otherwise, the avalanche should move to reach an equilibrium state according to the fluid characteristics.

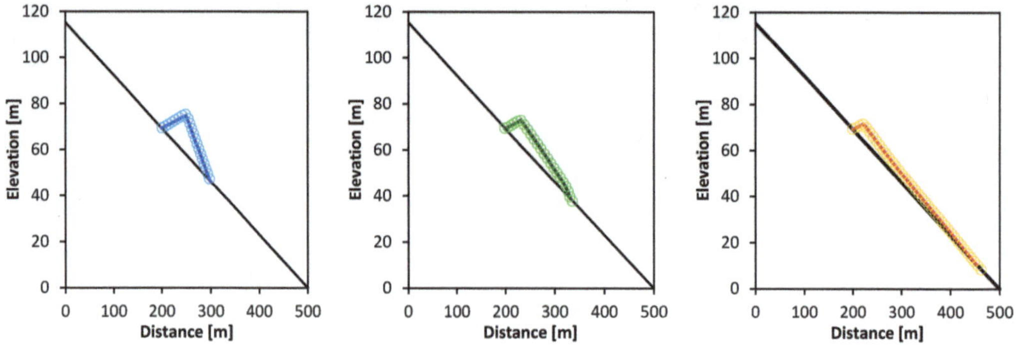

t = 0 s   t = 5 s   t = 40 s

Figure 2. Idealised test case of a pyroclastic avalanche proposed by de' Michieli Vitturi et al. [60]. Numerical results for a Voellmy-like fluid with $\mu = 0.3$ and $\xi = 300$ m/s^2: original data (circles) and simulated results (lines). The black line represents the topography.

3.3. *Idealised Gypsum Spill (East Texas, USA)*

The impoundment dimensions of the gypsum pond presented by Jeyapalan et al. [61] were idealized to 280 m in length and 110 m in width, with the breach spanning 120 m in width and 20 m in thickness. The initial condition was assumed to be a constant height of 11 m, and the fluid was released instantaneously. The computational domain, spanning 510 m in length and 400 m in width, was discretized into triangular-shaped elements of 2 m mean side length. The total number of elements was 105,936. The fluid was assumed to be Bingham type ($\tau_y = 1000$ Pa and $\mu_B = 50$ Pa·s).

Figure 3 depicts the spill propagation each 30 s. The fluid spread asymmetrically due to the idealised geometry of the impoundment. The maximum simulated runout was 265 m, the wave front being stopped at 50 s, while the detention of the fluid was completely produced at 102 s. The spread volume from the impoundment was 170,620 m^3, while the remainder was 176,804 m^3, with a maximum height of 7 m.

t = 0 s   t = 30 s

Figure 3. *Cont.*

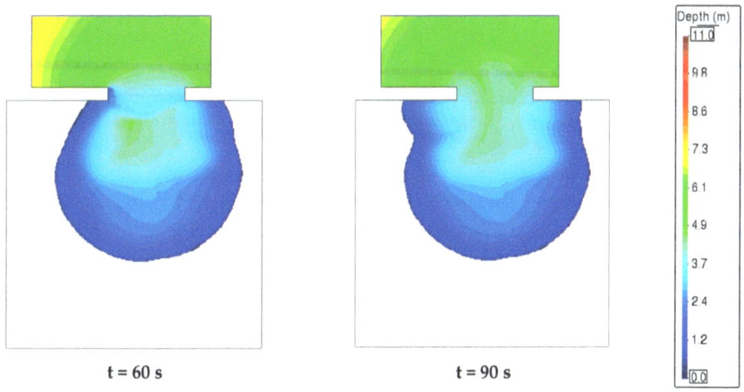

Figure 3. Results of the simulation considering a Bingham-like fluid in an idealised geometry of the gypsum pond failure that occurred in 1966 in East Texas (USA). Depth evolution each 30 s.

3.4. *Pond Failure of Los Frailes (Spain)*

The simulation of the spill was limited to the upper part of the more than 85 km of the total observed flood, extending from the point of failure (Figure 4a, red star) to El Guijo gauge station (Figure 4a, blue circle), located 7.1 km downstream of that point. The modelled area spans approximately 1000 ha over a distance of 9 km and encompasses a predominantly flat terrain comprising alluvial terraces, including the convergence point of the Guadiamar and Agrio rivers (Figure 4a). An unstructured mesh composed of triangular elements with a mean side length of 10 m was utilised to discretise the calculation domain, generating 219,330 elements (Figure 4b).

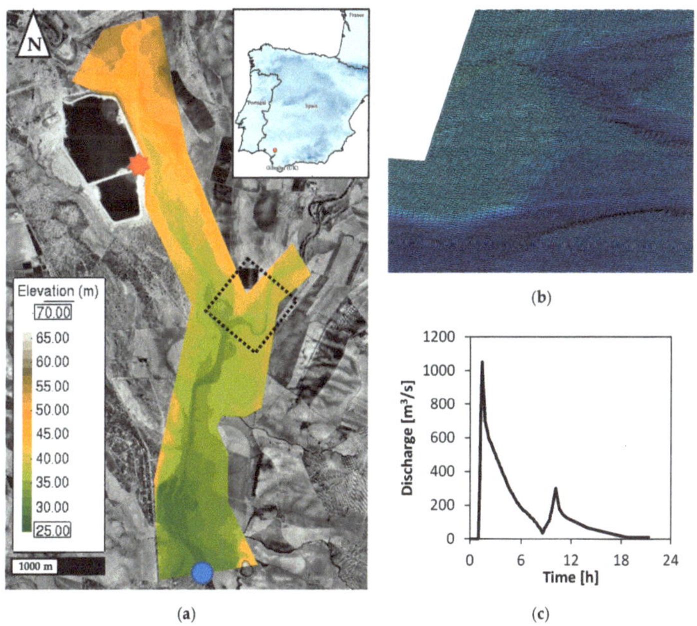

Figure 4. (**a**) General location of the Los Frailes mine pit, topography of the terrain, breakpoint (red star) and gauge station (blue circle). (**b**) Mesh discretisation at the Agrio River and Guadiamar River junction (dotted rectangle in (**a**)). (**c**) Estimated hydrograph generated after the pond break.

Post-failure topographical data, consisting of a Digital Terrain Model (DTM) with a 5 × 5 m pixel size, were utilized to update the elevations of the nodes due to the unavailability of an original dataset acquired previous to the event. It is worth noticing that modifications to the riverbed and banks occurred subsequent to the spill event as part of river restoration efforts [69–71], potentially leading to differences in the topographical features.

The simulation process involved varying the Bingham parameters to analyse the performance of the numerical model. The parameters τ_y and μ_B were systematically varied within ranges of 0 to 50 Pa and 0 to 2000 Pa·s (see Table 1). An initial analysis was made with the aim of understanding the flow behaviour by independently varying the yield stress and Bingham viscosity parameters. Subsequently, a combination of both parameters was tested together.The inlet boundary condition was defined by the hydrograph (Figure 4c) of the spill originated at the junction of the two lagoons, as represented by the red star in Figure 4a. This hydrograph was derived through an iterative process of inverse convolution from the recorded data at the gauge station [68]. This process was first carried out by means of the numerical tool HEC-1, and then with the numerical tool HEC-RAS to obtain the discharge–volume relations, and, finally, the "Modified Puls" routing method was used to obtain the hydrograph at the breaking point. A critical flow regime was imposed as an outlet condition at the El Guijo gauge station.

In the first analysis, the yield stress was again revealed to have a significant role in fluid retention. As expected, an increase in τ_y resulted in a higher volume of retained spill. Figure 5 illustrates this effect, showing that for low values of τ_y, the flow extension increases near the discharge point, but with smaller flow depths (Figure 5a). Conversely, higher τ_y values lead to a smaller extension near the breakpoint, but greater extension downstream, with higher flow depths observed in both areas (Figure 5d).

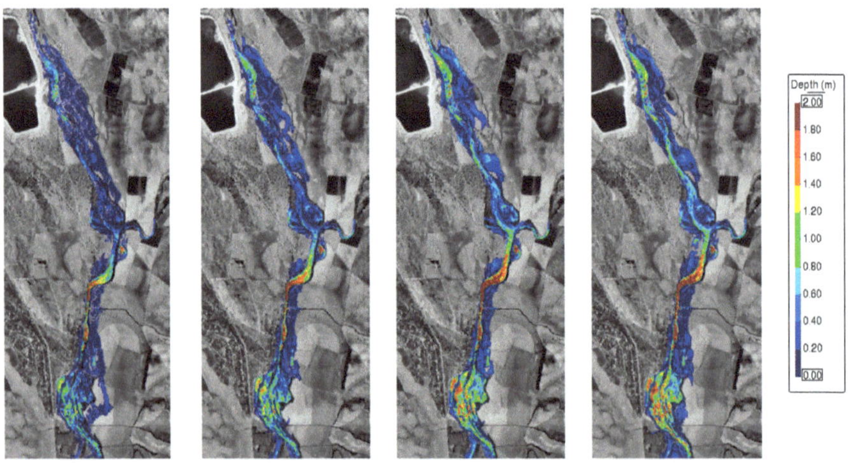

(a) (b) (c) (d)

Figure 5. Retained flow at the end of the simulation, considering null viscosity (μ_B = 0 Pa·s) and different values of the yield stress: (**a**) τ_y = 5 Pa; (**b**) τ_y = 15 Pa; (**c**) τ_y = 25 Pa; and (**d**) τ_y = 50 Pa.

On the other hand, Bingham viscosity affects flow behaviour in terms of flow propagation velocity, albeit without retention capacity, exhibiting a Manning-like behaviour in water flows. Figure 6a illustrates simulated hydrographs at the outlet boundary for various values of μ_B (25, 50, 200, 1000, 1500, and 2000 Pa·s), with τ_y = 0 Pa. An attenuation of the hydrograph is observed as Bingham viscosity increases, with the second peak becoming almost negligible for values greater than 200 Pa·s. For μ_B values less than 50 Pa·s, the simulated results closely match the observed ones (Figure 6a). The arrival time of the flood at the gauge station is well-captured for μ_B = 25 Pa·s, although there is a slight underestimation of the peak discharge. The numerical model also reproduces the second peak well for low values of μ_B.

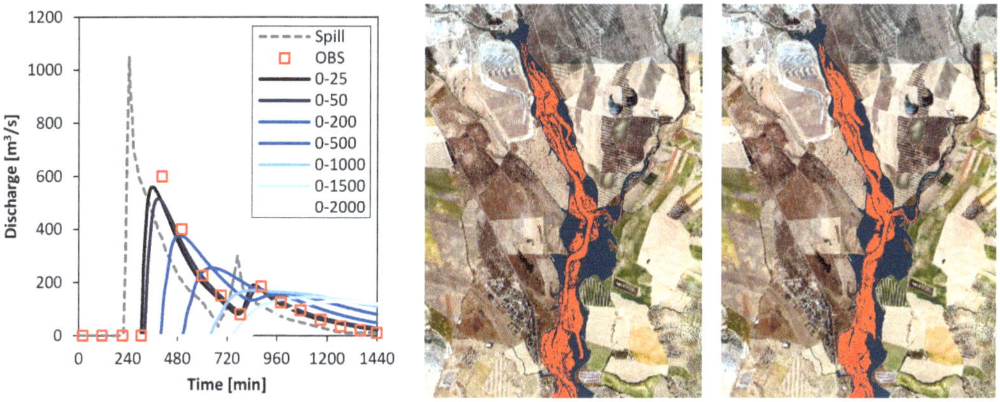

(a) (b) (c)

Figure 6. (**a**) Estimated hydrograph (red squares) and results of the different μ_B values tested (purple lines). Simulated flood extension (red polygon) and observed flood recorded by Landsat TM on 30 April: (**b**) τ_y = 25 Pa and μ_B = 5 Pa·s; (**c**) τ_y = 25 Pa and μ_B = 15 Pa·s.

Figure 6b,c illustrate the best-fit result of combining τ_y and μ_B values considering the previous analysis. The simulated flood extension is compared with the satellite image taken in the study area five days after the disaster, on 30 April (dark-blue polygon), which was then used to define amendments in contaminated soils [13]. These figures show the extension of the computed flood (red polygon) overlaid on the registered flood (dark-blue polygon) for two different combinations of the Bingham parameters: τ_y = 25 Pa and μ_B = 5 Pa·s (Figure 6b); and τ_y = 25 Pa and μ_B = 15 Pa·s (Figure 6c).

Despite both parameter combinations having the same τ_y value, significant differences in flood extension are observed due to the reduction of velocity for μ_B = 15 Pa. When the velocity is low because the resistance forces are high, there is greater retention of the flood. This effect is particularly noticeable downstream of the river's junction, where a more continuous extension of the computed flood is observed.

4. Discussion

4.1. On the Numerical Approach

In general, in the numerical modelling of non–Newtonian fluids using 2D shallowwater equation (2D-SWE)-based models, the flow is simulated as a continuum. In this approach, the individual movement of particles that could occur in nature cannot be simulated. However, most fluidified mine tailings behave as a continuum, and simulation tools based on 2D-SWE can be used here to describe their dynamics as long as the different terms of 2D-SWE are well balanced [16,27,72]. Nevertheless, the intrinsic assumptions for obtaining the system of 2D-SWE, together with the uncertainty/variability in field observations [73], may lead to uncertainties in the flood extent and in the internal movements of the particles, but this occurs for all numerical models, since they are simplifications of reality.

In numerical flow modelling, it is necessary to establish a wet–dry depth limit for an accurate characterization of wave fronts while preserving mass, which is a threshold used to consider whether there is flow in a mesh element or not. This is a relevant parameter for water flow, especially for flooding in flat areas [74–76], and for hydrological processes modelling [40,47,77], but it also applies to non–Newtonian fluids such as those occurring in mine-tailings spill propagation modelling. Very large wet–dry limits, greater than a few centimetres, can significantly alter flow propagation, especially as to the flow front [38,40]. The wet–dry limit must be properly defined, considering, in general, the geometric dimension of the problem, mesh size, expected flow depth, and, particularly, fluid properties. Additionally, selecting the appropriate numerical scheme to handle wet–dry fronts, especially the drying method, is crucial for preserving mass conservation [28,29,31,53].

There are few references on wet–dry limit treatment for non–Newtonian shallow flow models based on the solution of the 2D-SWE. For this type of flow, the individual and/or aggregate particle size is generally smaller than a few millimetres; since the fluid is simulated as a continuum, wet–dry limit values from 0.001 to 0.01 m can be sufficient to properly define the dynamics and extent of the flood. A value of 0.01 m was chosen for all simulations, but further investigation is necessary to fully understand the role of this parameter in the simulation of non–Newtonian shallow flows.

4.2. On the Rheological Models

Rheological models to describe both the dynamic and static phase of non–Newtonian shallow flows exist for a wide field of applications. In particular, in contexts related to environmental flows, and, especially, shallow flows, several rheological models have been developed to describe the relationship between the shear stress and the shear rate [78].

From the simplest potential law to the full—and complex—Bingham model, several rheological models exist in the literature, the development of each one being oriented to achieve a particular reproduction of a fluid behaviour. The aim of the present work is not to test as many rheological models as possible—or all that exist. Although this would allow a broader simulation of the behaviour of non-Newtonian shallow fluids, some rheological models are not appropriate to fluids that have a non–horizontal free surface at rest.

Rheological models that only integrate velocity-dependent terms do not allow the detention of the fluid, such as occurs with water flows in unconfined geometries. This would be the case of Manning-based rheological models.

The Manning rheological model, an empirical equation widely utilised in hydraulics and hydrology, applies to uniform flow in open channels and is a function of the channel velocity, flow area, and channel slope:

$$S_f = \frac{n^2 v^2}{h^{4/3}} \tag{6}$$

where n is the Manning coefficient, v is the flow velocity, and h is the flow depth. It is related to turbulent friction (τ_t), which is utilised by several authors in simulating hyper-concentrated flows [79–82]. The unique value for calibration is the Manning coefficient (n).

In work similar to the Manning rheological model, while considering constant sediment concentration and uniform flow, Macedonio and Pareschi [83] proposed the following relation of the shear stress: $\tau = \tau_y + \mu_1 (dv/dz)^\alpha$, where τ_y is the yield stress, μ_1 is a proportionality coefficient, and α is the flow behaviour index.

When $\alpha = 2$, a dilatant flow behaviour is expected:

$$S_f = \frac{n^2 v^2}{h^3} \tag{7}$$

The same authors also presented the application of the Manning equation to viscous flows by particularizing the parameter $\alpha = 1$. This allows for the representation of viscous flows:

$$S_f = \frac{n^2 v}{h^2} \tag{8}$$

A similar behaviour is expected in those models that also integrate non–velocity-dependent terms, such as the Voellmy and simplified Bingham, when the velocity-dependent terms are neglected. O'Brien [84] derived an expression for the representation of the shear stress of mudflows, being a quadratic equation that integrates the Mohr–Coulomb term, the viscous term, and the turbulent term as follows:

$$S_f = \frac{\tau_y}{\rho g h} + \frac{K \mu_B v}{8 \rho g h^2} + \frac{n^2 v^2}{h^{4/3}} \quad (9)$$

where τ_y is the yield stress, ρ is the fluid density, g is the gravitational acceleration, h is the flow depth, K is a resistance parameter, μ_B is the flow viscosity, v is the flow velocity, and n is the Manning coefficient. With this approach, the fluid continues flowing when the yield stress contribution is omitted.

Bartelt et al. [85] developed a resistance term related to the cohesion, a physical property of the fluid. This rheological model is non–velocity-dependent and is expressed as follows:

$$S_f = \frac{1}{\rho g h}\left(C_B\,(1-\mu)\left(1 - e^{-\frac{\rho g h}{C_B}}\right)\right) \quad (10)$$

where ρ is the fluid density, g is the gravitational acceleration, h is the flow depth, C_B is the cohesion, and μ is the Coulomb friction coefficient. Although this model is functional and allows for the stopping of the fluid within unconfined and irregular slopes, it is commonly used together with the Voellmy model.

4.3. On the Performance of Iber in Simulating Mine-Tailings Spill Propagation

The test case of a sloping terrain presented in de' Michieli Vitturi et al. [60] was originally developed for the simulation of pyroclastic avalanches, such as those produced in post-eruptive events in volcano regions. However, some mining activities generate similar materials, which, in contact with water, have similar behaviours when released after the break of a mine pond [62,63,66,86–92]. Apart from the particularities of the test, the extension of Iber for simulating non–Newtonian shallow flows was proved to be valid, even for steep slopes. The fluid can stop on steep terrains thanks to the specific numerical scheme utilised in which non–velocity-dependent terms of the rheological model are used to counterbalance the pressure forces [16]. This can be useful in several situations, such as for mine ponds located in mountain regions [2,93,94]. The use of a well-balanced numerical scheme avoids oscillations leading to numerical instabilities; otherwise, the imbalance can make them incapable of simulating even quiescent states over both simple and complex geometries [27–31,95,96].

The simulation of a gypsum spill caused by the failure of a pond in East Texas in 1966 [61] was utilised to show the performance of the numerical model over two-dimensional domains. Some differences were observed due to important assumptions and simplifications applied not only in the domain discretisation, but also in the topographical data. The simulated runout distance was smaller than the reported one, which might have been caused by the assumption of a completely horizontal flat terrain and, particularly, the rheological model utilised. But other factors should not be discarded.

In this regard, the yield stress term in Equation (4) is multiplied by a factor of 1.5; this means the contribution of a 50% higher yield stress term than found in similar numerical models that use a Bingham-based rheological model without this factor [84,97,98].

Aiming to show the influence of this factor in the simulations, the same case was re-calculated, omitting this factor, which showed a better performance in term of runout. Figure 7 compares the flow behaviour considering the increased factor of 1.5 in the yield stress terms (upper row), according to the proposal of [56,57], and without considering this factor (lower row). In this last case, the runout distance reached 300 m, in agreement with the observations [61]. Despite the differences in the resistance terms, the leading edge of fluid stopped at 50 s in both cases, keeping a similar shape as to the flood extent. In this sense, in both cases the released volume was higher than the estimations, but was within the same order of magnitude as the values presented by Jeyapalan et al. [61]. Despite the fact that the freezing time agrees with the observations, the uncertainties in the breach formation (breaking time and shape evolution of the breach) and the possible fluidification of the stored fluid [65] might favour a lack of representation of the fluid behaviour during the dynamic and static phase [66].

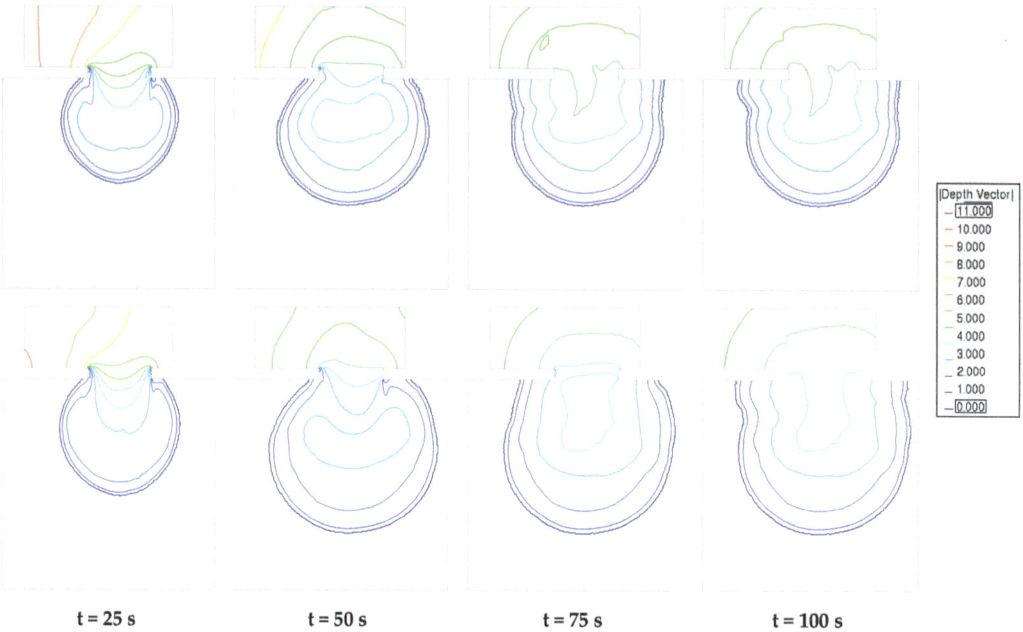

Figure 7. Idealised gypsum spill. Contour line depth evolution when using the simplified Bingham rheological model with an increased factor of 1.5 in the yield stress, according to [56,57] (**upper row**), and without the increased factor in the yield stress term (**lower row**).

Finally, despite the fact that the Los Frailes mine pond failure was not utilised to reproduce the event, the presented results highlight some discrepancies with the observations. The topographical data utilized was the most recent Digital Terrain Model (DTM) provided by the Instituto Geográfico Nacional. However, significant morphological changes occurred after the accident due to the removal operations of the deposited muds, and the river restoration program known as "El Corredor Verde del Guadiamar" [70].

Data recorded at El Guijo gauge station suggested the presence of two peaks, but its poor quality necessitated an ad hoc hydraulic study to reconstruct the spill's hydrograph [66]. Nevertheless, this hydrograph was then utilized as an inlet condition at the failure point, potentially generating differences in the spill reconstruction. A portion of the spilled fluid, predominantly composed of pyrite, remained and ceased flowing, indicating a mudflow-like behaviour. A Bingham-type fluid was assumed, but a hyper-concentrated behaviour (sediment-laden flow) cannot be discarded, given the observations made during the spill [66,67]. In any case, the performance of the extension of Iber to simulate real scenarios of non–Newtonian shallow flows, such as mine-tailing spills, was proved to be suitable.

4.4. *On the Code Optimization and Computing Time for GPU-Computing*

Iber is a numerical tool explicit in time; hence, the CFL stability criterion applies over the computational time step [99]. This condition establishes a relation between the water depth, the flow velocity, the wave celerity, the size of the element, and the maximum permissible computational time step. The extension of Iber for the simulation of non–Newtonian shallow flows uses the computational time step proposed by Cea and

Bladé [40], with a value of CFL = 0.45 being used in all test cases for numerical stability. Iber automatically calculates the calculation time steps for solving the equations.

Thus, high flow depths and velocities, in addition to small-sized elements, notably increase the computational time. The current version of IberNNF is CPU-based, which is partially parallelised using OpenMP. This means that, at most, only "speed-ups" corresponding to the number of CPU cores can be obtained. When dealing with the simulations of mine-tailings propagation case studies, the problem domain or/and the resolution needed to assess the fluid dynamics could notably increase the computation time.

Future work should focus on the application of general-purpose computing on graphics processing unit (GPGPU) techniques [100,101] to parallelise the code for GPU computing, such as seen in the water-based hydrodynamic, habitat, and sediment-transport modules of Iber, called R-Iber [50,102]. With that, together with a code optimization, the computing process will reach speed-ups above 100-times that of the sequential version. With the aim of testing the current applications, the model is freely distributed as a part of the software suite through www.iberaula.com (accessed on 15 June 2024).

4.5. On the Associated Risk of Mining Activities

Mining activities, while essential for the extraction of valuable minerals and resources, pose several associated risks that can impact the environment, human health, and local communities. To that end, guidelines for the design, construction, monitoring, and closure of tailings dams are mandatory to manage and prevent hazardous situations. Following these recommendations can greatly reduce the risks of failures and dangerous conditions for both people and the environment [2,94].

Nevertheless, more than 150 major tailings-dam failures have been reported since 1961, with a wide variety of materials being released [93]. This means 2.5 failures per year, on average, 82 of them having occurred in the 2000–2024 period. This reveals a clear necessity for additional—and in-depth—management, monitoring, prevention, and planning strategies to reduce the associated flood risks of mining activities [2,103].

The development of numerical simulation tools can help in addressing several engineering problems related to mining activities and can also help in reducing the risk of a failure and the subsequent occurrence of situations dangerous for people and the environment, due to the release and propagation of uncontrolled spills. Understanding the characteristics of tailings ponds, including their construction and filling processes, as well as the type of fluid stored (and whether it maintains Newtonian properties), is essential [66] for properly addressing not only the reconstruction of a past event but also the evaluation of the risk of future failure or malfunction scenarios in mine-tailings dams that might cause flood events.

5. Conclusions

Numerical modelling of non–Newtonian fluids using 2D shallow water equations (2D-SWE) simulates the flow as a continuum. Despite the fact that it might not capture individual particle movements accurately, this approach is quite useful for the simulation of the propagation of a mine-tailings spill after a dam failure.

To that end, specific rheological models are used to describe both static and dynamic behaviours. However, not all rheological models are suitable for fluids with non–horizontal free surfaces at rest. Only those that contain non–velocity-dependent terms, such as Bingham or Voellmy models, among others, can retain the flow in unconfined geometries and step slopes.

This work presents the extension of the hydraulic numerical tool Iber for the simulation of non–Newtonian flows, in particular, mine tailings. This extension draws on the developments carried out for the simulation of snow avalanches, but the friction terms have been particularized to rheological models capable of representing the dynamic and static behaviours in the propagation of mine tailings.

The different case studies utilised to test the code proved that the numerical tool is capable of simulating this type of fluid under different situations, such as the obtaining of a non–horizontal free surface for a fluid at rest, and the stopping of the fluid in an area with a slope, according to the properties of the fluid, as well as demonstrating its usefulness in real cases. The code performed suitably, providing reliable results without numerical instabilities, even when using a wet–dry limit of 0.01 m, which provided a good description of the flooded area.

The development or enhancement of numerical tools for evaluating flood hazards and risks associated with mining dam failure is crucial. These tools should not only assess the potential impact of such failures, but also propose effective mitigation measures and contribute to the development of guidelines for constructing safer structures. These tools must accurately predict the extent and severity of flooding that could result from a mining dam failure. They should provide insights into vulnerable areas and describe potential damage to infrastructure and communities and the likelihood of loss of life. By doing so, they enable stakeholders to make informed decisions about emergency response planning, evacuation routes, and the implementation of protective measures, in addition to contributing to the development of guidelines for constructing safer mining structures in the future.

Author Contributions: Conceptualization, M.S.-R. and E.B.; methodology, M.S.-R., M.S.-J. and E.B.; software, M.S.-R. and E.B.; validation, M.S.-J. and T.D.; resources, M.S.-R.; data curation, M.S.-R. and E.B.; writing—original draft preparation, M.S.-R.; writing—review and editing, E.B., M.S.-J. and T.D.; supervision, E.B. All authors have read and agreed to the published version of the manuscript.

Funding: This research received no external funding.

Data Availability Statement: The data presented in this study are available on request from the corresponding author.

Acknowledgments: The authors thank Beningno Bayán, retired engineer of Confederación Hidrográfica del Guadalquivir and member of the 'Plan Doñana 2005' scientific commission, for his collaboration in helping to understand the physics of the spill, and his generous work of information-searching, particularly as to the case study 'Pond Failure of Los Frailes'.

Conflicts of Interest: The authors declare no conflict of interest.

References

1. Hudson-Edwards, K.A.; Macklin, M.G.; Jamieson, H.E.; Brewer, P.A.; Coulthard, T.J.; Howard, A.J.; Turner, J.N. The impact of tailings dam spills and clean-up operations on sediment and water quality in river systems: Rios Agrio–Guadiamar, Aznalcollar, Spain. *Appl. Geochem.* **2003**, *18*, 221–239. [CrossRef]
2. Kheirkhah Gildeh, H.; Halliday, A.; Arenas, A.; Zhang, H. Tailings Dam Breach Analysis: A Review of Methods, Practices, and Uncertainties. *Mine Water Environ.* **2021**, *40*, 128–150. [CrossRef]
3. Owen, J.R.; Kemp, D.; Lèbre, É.; Svobodova, K.; Pérez Murillo, G. Catastrophic tailings dam failures and disaster risk disclosure. *Int. J. Disaster Risk Reduct.* **2020**, *42*, 103361. [CrossRef]
4. Roche, C.; Thygesen, K.; Baker, E. *Mine Tailings Storage: Safety Is No Accident*; United Nations Environment Programme: Nairobi, Kenya; GRID-Arendal: Arendal, Norway, 2017; ISBN 978-82-7701-170-7.
5. Ngole-Jeme, V.M.; Fantke, P. Ecological and human health risks associated with abandoned gold mine tailings contaminated soil. *PLoS ONE* **2017**, *12*, e0172517. [CrossRef] [PubMed]
6. Martínez-Gomariz, E.; Barbero, C.; Sanchez-Juny, M.; Forero-Ortiz, E.; Sanz-Ramos, M. Dams or ponds classification based on a new criterion to assess potential flood damage to roads in case of failure. *Nat. Hazards* **2023**, *117*, 625–653. [CrossRef]
7. Hudson-Edwards, K.A.; Jamieson, H.E.; Lottermoser, B.G. Mine wastes: Past, present, future. *Elements* **2011**, *7*, 375–380. [CrossRef]
8. Kossoff, D.; Dubbin, W.E.; Alfredsson, M.; Edwards, S.J.; Macklin, M.G.; Hudson-Edwards, K.A. Mine tailings dams: Characteristics, failure, environmental impacts, and remediation. *Appl. Geochem.* **2014**, *51*, 229–245. [CrossRef]
9. Bowker, L.N.; Chambers, D.M. The Risk, Public Liability and Economics of Tailings Storage Facility Failures. *Earthwork Act* **2015**, *24*, 56.
10. Mudd, G.M. The Environmental sustainability of mining in Australia: Key mega-trends and looming constraints. *Resour. Policy* **2010**, *35*, 98–115. [CrossRef]
11. Dong, L.; Deng, S.; Wang, F. Some developments and new insights for environmental sustainability and disaster control of tailings dam. *J. Clean. Prod.* **2020**, *269*, 122270. [CrossRef]

12. López-Pamo, E.; Barettino, D.; Antón-Pacheco, C.; Ortiz, G.; Arránz, J.C.; Gumiel, J.C.; Martínez-Pledel, B.; Aparicio, M.; Montouto, O. The extent of the Aznalcollar pyritic sludge spill and its effects on soils. *Sci. Total Environ.* **1999**, *242*, 57–88. [CrossRef] [PubMed]
13. Madejón, P.; Domínguez, M.; Madejón, E.; Cabrera, F.; Marañón, T.; Murillo, J. Soil-plant relationships and contamination by trace elements: A review of twenty years of experimentation and monitoring after the Aznalcóllar (SW Spain) mine accident. *Sci. Total Environ.* **2018**, *625*, 50–63. [CrossRef] [PubMed]
14. Tornero, V.; Arias, A.M.; Blasco, J. Following the Aznalcóllar toxic spill. *Environ. Res.* **2011**, *111*, 1033–1036. [CrossRef] [PubMed]
15. Sanz-Ramos, M.; Andrade, C.A.; Oller, P.; Furdada, G.; Bladé, E.; Martínez-Gomariz, E. Reconstructing the Snow Avalanche of Coll de Pal 2018 (SE Pyrenees). *GeoHazards* **2021**, *2*, 196–211. [CrossRef]
16. Sanz-Ramos, M.; Bladé, E.; Oller, P.; Furdada, G. Numerical modelling of dense snow avalanches with a well-balanced scheme based on the 2D shallow water equations. *J. Glaciol.* **2023**, 1–17. [CrossRef]
17. Zugliani, D.; Rosatti, G. TRENT2D❄: An accurate numerical approach to the simulation of two-dimensional dense snow avalanches in global coordinate systems. *Cold Reg. Sci. Technol.* **2021**, *190*, 103343. [CrossRef]
18. Medina, V.; Hürlimann, M.; Bateman, A. Application of FLATModel, a 2D finite volume code, to debris flows in the northeastern part of the Iberian Peninsula. *Landslides* **2008**, *5*, 127–142. [CrossRef]
19. Xia, C.; Tian, H. A Quasi-Single-Phase Model for Debris Flows Incorporating Non–Newtonian Fluid Behavior. *Water* **2022**, *14*, 1369. [CrossRef]
20. Hungr, O. A model for the runout analysis of rapid flow slides, debris flows, and avalanches. *Can. Geotech. J.* **1995**, *32*, 610–623. [CrossRef]
21. Wu, T.; Vuong, T.; Wang, C.; Chu, C.; Lin, C. Model development for simulating mudslide and the case study of the failure of the gypsum tailings dam in East Texas in 1966. *Nat. Hazards Earth Syst. Sci. Discuss.* **2020**. *preprint*. [CrossRef]
22. Ding, W.; Wang, G.; Yang, Q.; Xu, Y.; Gao, Y.; Chen, X.; Xu, S.; Han, L.; Yang, X. Risk Assessment and Control of Geological Hazards in Towns of Complex Mountainous Areas Based on Remote Sensing and Geological Survey. *Water* **2023**, *15*, 3170. [CrossRef]
23. Julien, P.Y.; Leon, C. Mudfloods, mudflows and debris flows, classification in rheology and structural design. In Proceedings of the International Workshop on the Debris Flow Disaster, Caracas, Venezuela, 14–16 December 1999.
24. Hu, J.; Li, K.; Su, W.; Zhao, X. Numerical Simulation of Drilling Fluid Flow in Centrifugal Pumps. *Water* **2023**, *15*, 992. [CrossRef]
25. Melo, M.; Eleutério, J. Probabilistic Analysis of Floods from Tailings Dam Failures: A Method to Analyze the Impact of Rheological Parameters on the HEC-RAS Bingham and Herschel-Bulkley Models. *Water* **2023**, *15*, 2866. [CrossRef]
26. Bladé, E.; Gómez-Valentín, M.; Dolz, J.; Aragón-Hernández, J.L.; Corestein, G.; Sánchez-Juny, M. Integration of 1D and 2D finite volume schemes for computations of water flow in natural channels. *Adv. Water Resour.* **2012**, *42*, 17–29. [CrossRef]
27. Bladé, E.; Gómez, M.; Sánchez-Juny, M.; Dolz, J. Source term treatment of SWEs using the surface gradient upwind method. *J. Hydraul. Res.* **2012**, *50*, 447–448. [CrossRef]
28. Bermúdez, A.; Dervieux, A.; Desideri, J.-A.; Vázquez, M.E. Upwind schemes for the two-dimensional shallow water equations with variable depth using unstructured meshes. *Comput. Methods Appl. Mech. Eng.* **1998**, *155*, 49–72. [CrossRef]
29. LeVeque, R.L. *Finite Volume Methods for Hyperbolic Problems*; Cambridge University Press: Cambridge, MA, USA, 2002; Volume 31.
30. Toro, E.F. *Riemann Solvers and Numerical Methods for Fluid Dynamics*; Springer: Berlin/Heidelberg, Germany, 2009; Volume 40, ISBN 978-3-540-25202-3.
31. Vázquez-Cendón, M.E. Improved treatment of source terms in upwind schemes for the shallow water equations in channels with irregular geometry. *J. Comput. Phys.* **1999**, *148*, 497–526. [CrossRef]
32. Chevrel, M.O.; Labroquère, J.; Harris, A.J.L.; Rowland, S.K. PyFLOWGO: An open-source platform for simulation of channelized lava thermo-rheological properties. *Comput. Geosci.* **2018**, *111*, 167–180. [CrossRef]
33. Christen, M.; Kowalski, J.; Bartelt, P. RAMMS: Numerical simulation of dense snow avalanches in three-dimensional terrain. *Cold Reg. Sci. Technol.* **2010**, *63*, 1–14. [CrossRef]
34. Eglit, M.; Yakubenko, A.; Zayko, J. A Review of Russian Snow Avalanche Models—From Analytical Solutions to Novel 3D Models. *Geosciences* **2020**, *10*, 77. [CrossRef]
35. Pirulli, M.; Sorbino, G. Assessing potential debris flow runout: A comparison of two simulation models. *Nat. Hazards Earth Syst. Sci.* **2008**, *8*, 961–971. [CrossRef]
36. Stefania, S.; Zugliani, D.; Rosatti, G. Dense snow avalanche modelling with Voellmy rheology: TRENT2D vs. RAMMS2D. In Proceedings of the Vistual Snow Science Workshop—VSSW 2020, Fernie, BC, Canada, 4–6 October 2020.
37. Tan, W.Y. *Shallow Water Hydrodynamics*, 1st ed.; Elsevier Science: Amsterdam, The Netherlands, 1992; ISBN 9780080870939.
38. Bladé, E.; Cea, L.; Corestein, G.; Escolano, E.; Puertas, J.; Vázquez-Cendón, E.; Dolz, J.; Coll, A. Iber: Herramienta de simulación numérica del flujo en ríos. *Rev. Int. Métodos Numér. Cálc. Diseño Ing.* **2014**, *30*, 1–10. [CrossRef]
39. Bladé, E.; Cea, L.; Corestein, G. Modelización numérica de inundaciones fluviales. *Ing. Agua* **2014**, *18*, 68.
40. Cea, L.; Bladé, E. A simple and efficient unstructured finite volume scheme for solving the shallow water equations in overland flow applications. *Water Resour. Res.* **2015**, *51*, 5464–5486. [CrossRef]
41. Cea, L.; Legout, C.; Darboux, F.; Esteves, M.; Nord, G. Experimental validation of a 2D overland flow model using high resolution water depth and velocity data. *J. Hydrol.* **2014**, *513*, 142–153. [CrossRef]

42. Uber, M.; Nord, G.; Legout, C.; Cea, L. How do modeling choices and erosion zone locations impact the representation of connectivity and the dynamics of suspended sediments in a multi-source soil erosion model? *Earth Surf. Dyn.* **2021**, *9*, 123–144. [CrossRef]
43. Aranda, J.Á.; Sánchez-Juny, M.; Sanz-Ramos, M.; Beneyto, C. Design of Drainage Downspouts Systems over a Road Embankment. *Water* **2023**, *15*, 3529. [CrossRef]
44. Aranda, J.Á.; Beneyto, C.; Sánchez-Juny, M.; Bladé, E. Efficient Design of Road Drainage Systems. *Water* **2021**, *13*, 1661. [CrossRef]
45. Cea, L.; Bermúdez, M.; Puertas, J.; Bladé, E.; Corestein, G.; Escolano, E.; Conde, A.; Bockelmann-Evans, B.; Ahmadian, R. IberWQ: New simulation tool for 2D water quality modelling in rivers and shallow estuaries. *J. Hydroinform.* **2016**, *18*, 816–830. [CrossRef]
46. Sañudo, E.; Cea, L.; Puertas, J. Modelling Pluvial Flooding in Urban Areas Coupling the Models Iber and SWMM. *Water* **2020**, *12*, 2647. [CrossRef]
47. Sanz-Ramos, M.; Martí-Cardona, B.; Bladé, E.; Seco, I.; Amengual, A.; Roux, H.; Romero, R. NRCS-CN Estimation from Onsite and Remote Sensing Data for Management of a Reservoir in the Eastern Pyrenees. *J. Hydrol. Eng.* **2020**, *25*, 05020022. [CrossRef]
48. Ruiz-Villanueva, V.; Gamberini, C.; Bladé, E.; Stoffel, M.; Bertoldi, W. Numerical Modeling of Instream Wood Transport, Deposition, and Accumulation in Braided Morphologies Under Unsteady Conditions: Sensitivity and High-Resolution Quantitative Model Validation. *Water Resour. Res.* **2020**, *56*, e2019WR026221. [CrossRef]
49. Ruiz-Villanueva, V.; Mazzorana, B.; Bladé, E.; Bürkli, L.; Iribarren-Anacona, P.; Mao, L.; Nakamura, F.; Ravazzolo, D.; Rickenmann, D.; Sanz-Ramos, M.; et al. Characterization of wood-laden flows in rivers. *Earth Surf. Process. Landf.* **2019**, *44*, 1694–1709. [CrossRef]
50. Sanz-Ramos, M.; López-Gómez, D.; Bladé, E.; Dehghan-Souraki, D. A CUDA Fortran GPU-parallelised hydrodynamic tool for high-resolution and long-term eco-hydraulic modelling. *Environ. Model. Softw.* **2023**, *161*, 105628. [CrossRef]
51. Arbat-Bofill, M.; Bladé, E.; Sánchez-Juny, M.; Niñerola, D.; Dolz, J.; Sanchez-Juny, M.; Niñerola, D.; Dolz, J.; Arbat Bofill, M.; Bladé, E.; et al. Case studies of reservoir sedimentation as a consequence of soil erosion. In *Reservoir Sedimentation*; CRC Press: Boca Raton, FL, USA, 2014; pp. 83–92. ISBN 9781138026759.
52. Bladé, E.; Sanz-Ramos, M.; Dolz, J.; Expósito-Pérez, J.; Sánchez-Juny, M. Modelling flood propagation in the service galleries of a nuclear power plant. *Nucl. Eng. Des.* **2019**, *352*, 110180. [CrossRef]
53. Roe, P. A basis for upwind differentiating of the two dimensional unsteady Euler equations. In *Numerical Methodes for Fluid Dynamics II*; Morton, B., Ed.; Oxford University Press: Oxford, UK, 1986.
54. Bingham, E.C. An investigation of the laws of plastic flow. *Bull. Bur. Stand.* **1916**, *13*, 309–353. [CrossRef]
55. Pastor, M.; Haddad, B.; Sorbino, G.; Cuomo, S.; Drempetic, V. A depth-integrated, coupled SPH model for flow-like landslides and related phenomena. *Int. J. Numer. Anal. Methods Geomech.* **2009**, *33*, 143–172. [CrossRef]
56. Chen, H.; Lee, C.F. Runout Analysis of Slurry Flows with Bingham Model. *J. Geotech. Geoenviron. Eng.* **2002**, *128*, 1032–1042. [CrossRef]
57. Naef, D.; Rickenmann, D.; Rutschmann, P.; McArdell, B.W. Comparison of flow resistance relations for debris flows using a one-dimensional finite element simulation model. *Nat. Hazards Earth Syst. Sci.* **2006**, *6*, 155–165. [CrossRef]
58. Voellmy, A. Über die Zerstörungskraft von Lawinen. *Schweizerische Bauzeitung* **1955**, *73*, 15.
59. Bryant, S.M. *Application of Tailings Flow Analyses to Field Conditions*; University of California: Berkeley, CA, USA, 1983.
60. de' Michieli Vitturi, M.; Esposti Ongaro, T.; Lari, G.; Aravena, A. IMEX_SfloW2D 1.0: A depth-averaged numerical flow model for pyroclastic avalanches. *Geosci. Model Dev.* **2019**, *12*, 581–595. [CrossRef]
61. Jeyapalan, J.K.; Duncan, J.M.; Seed, H.B. Investigation of Flow Failures of Tailings Dams. *J. Geotech. Eng.* **1983**, *109*, 172–189. [CrossRef]
62. Gens, A.; Alonso, E.E. Aznalcóllar dam failure. Part 2: Stability conditions and failure mechanism. *Géotechnique* **2006**, *56*, 185–201. [CrossRef]
63. Alonso, E.E.; Gens, A. Aznalcóllar dam failure. Part 1: Field observations and material properties. *Géotechnique* **2006**, *56*, 165–183. [CrossRef]
64. Alonso, E.E.; Gens, A. Aznalcóllar dam failure. Part 3: Dynamics of the motion. *Géotechnique* **2006**, *56*, 203–210. [CrossRef]
65. Ayala-Carcedo, J. La rotura de la balsa de residuos mineros de Aznalcóllar (España) de 1998 y el desastre ecológico consecuente del río Guadiamar: Causas, efectos y lecciones. *Bol. Geol. Min.* **2004**, *115*, 711–738.
66. Sanz-Ramos, M.; Bladé, E.; Dolz, J.; Sánchez-Juny, M. Revisiting the Hydraulics of the Aznalcóllar Mine Disaster. *Mine Water Environ.* **2022**, *41*, 335–356. [CrossRef]
67. Sanz-Ramos, M.; Bladé, E.; Dolz, J.; Sánchez-Juny, M. El desastre de Aznalcóllar: ¿lodos o aguas ácidas? *Ing. Agua* **2021**, *25*, 229–239. [CrossRef]
68. Consultec Ingenieros SL. Calibración de las tres estaciones de aforo del río Guadiamar (EA nº 56, EA nº 90 y EA nº 76) por métodos informáticos hidráulicos. *Estud. Riada Tóxica* **1999**, *35*.
69. Antón-Pacheco, C.; Arranz, J.C.; Barettino, D.; Carrero, G.; Giménez, M.; Gómez, J.A.; Gumiel, J.C.; López-Pamo, E.; Martín Rubí, J.A.; Martínez-Pledel, B.; et al. Actuaciones para el reconocimiento y retirada de los lodos depositados sobre el terreno, y su restauración edáfica y morfológica. *Bol. Geol. Min.* **2001**, *112*, 93–122.
70. CSIC Guadiamar. *Ciencia, Técnicia y Restauración—El Accidente Minero Diez Años Después*; Consejo Superior de Investigaciones Científicas (CSIC), Ministerio de Ciencia e Innovación: Madrid, Spain, 2008; ISBN 978-84-00-08644-2.

71. Turner, J.N.; Brewer, P.A.; Macklin, M.G. Fluvial-controlled metal and As mobilisation, dispersal and storage in the Río Guadiamar, SW Spain and its implications for long-term contaminant fluxes to the Doñana wetlands. *Sci. Total Environ.* **2008**, *394*, 144–161. [CrossRef] [PubMed]
72. Bladé, E.; Gómez-Valentín, M.; Sánchez-Juny, M.; Dolz, J. Preserving steady-state in one-dimensional finite-volume computations of river flow. *J. Hydraul. Eng.* **2008**, *134*, 1343–1347. [CrossRef]
73. Dimitriadis, P.; Koutsoyiannis, D.; Iliopoulou, T.; Papanicolaou, P. A Global-Scale Investigation of Stochastic Similarities in Marginal Distribution and Dependence Structure of Key Hydrological-Cycle Processes. *Hydrology* **2021**, *8*, 59. [CrossRef]
74. Ramos-Fuertes, A.; Marti-Cardona, B.; Bladé, E.; Dolz, J. Envisat/ASAR Images for the Calibration of Wind Drag Action in the Doñana Wetlands 2D Hydrodynamic Model. *Remote Sens.* **2013**, *6*, 379–406. [CrossRef]
75. Sanz-Ramos, M.; Téllez-Álvarez, J.; Bladé, E.; Gómez-Valentín, M. Simulating the hydrodynamics of sewer-grates using a 2D-hydraulic model. In Proceedings of the 5th International Conference SimHydro, Nice, France, 20–22 November 2019; p. 8.
76. Cea, L.; Puertas, J.; Vázquez-Cendón, M.-E. Depth averaged modelling of turbulent shallow water flow with wet-dry fronts. *Arch. Comput. Methods Eng.* **2007**, *14*, 303–341. [CrossRef]
77. Sanz-Ramos, M.; Olivares, G.; Bladé, E. Experimental characterization and two-dimensional hydraulic-hydrologic modelling of the infiltration process through permeable pavements. *Rev. Int. Métodos Numér. Cálc. Diseño Ing.* **2022**, *38*, 1–12. [CrossRef]
78. Msheik, K. *Non–Newtonian Fluids: Modeling and Well-Posedness*; Universite Grenoble Alpes: Saint-Martin-d'Hères, France, 2020.
79. Takahashi, T. Debris flow: Mechanics and hazard mitigation. In *Proceedings of the ROC-JAPAN Joint Seminar on Multiple Hazards Mitigation*; National Taiwan Univerisity: Taipei, Taiwan, 1985; pp. 1075–1092.
80. Laenen, A.; Hansen, R.P. *Simulation of Three Lahars in the Mount St. Helens Area, Washington, Using a One-Dimensional, Unsteady-State Streamflow Model*; U.S. Geological Survey Water-Resources Investigations Report 88-4004; U.S. Geological Survey: Portland, OR, UA, 1988.
81. Syarifuddin, M.; Oishi, S.; Hapsari, R.I.; Legono, D. Empirical model for remote monitoring of rain-triggered lahar at Mount Merapi. *J. Japan Soc. Civ. Eng. Ser. B1 (Hydraul. Eng.)* **2018**, *74*, I_1483–I_1488. [CrossRef] [PubMed]
82. Darnell, A.R.; Phillips, J.C.; Barclay, J.; Herd, R.A.; Lovett, A.A.; Cole, P.D. Developing a simplified geographical information system approach to dilute lahar modelling for rapid hazard assessment. *Bull. Volcanol.* **2013**, *75*, 713. [CrossRef]
83. Macedonio, G.; Pareschi, M.T.T. Numerical simulation of some lahars from Mount St. Helens. *J. Volcanol. Geotherm. Res.* **1992**, *54*, 65–80. [CrossRef]
84. O'Brien, J.S.; Julien, P.Y. Laboratory Analysis of Mudflow Properties. *J. Hydraul. Eng.* **1988**, *114*, 877–887. [CrossRef]
85. Bartelt, P.; Valero, C.V.; Feistl, T.; Christen, M.; Bühler, Y.; Buser, O. Modelling cohesion in snow avalanche flow. *J. Glaciol.* **2015**, *61*, 837–850. [CrossRef]
86. Kelfoun, K. Suitability of simple rheological laws for the numerical simulation of dense pyroclastic flows and long-runout volcanic avalanches. *J. Geophys. Res. Solid Earth* **2011**, *116*, 1–14. [CrossRef]
87. Frattini, P.; Crosta, G.B.; Fusi, N.; Dal Negro, P. Shallow landslides in pyroclastic soils: A distributed modelling approach for hazard assessment. *Eng. Geol.* **2004**, *73*, 277–295. [CrossRef]
88. de Bélizal, E.; Lavigne, F.; Hadmoko, D.S.; Degeai, J.-P.; Dipayana, G.A.; Mutaqin, B.W.; Marfai, M.A.; Coquet, M.; Le Mauff, B.; Robin, A.-K.; et al. Rain-triggered lahars following the 2010 eruption of Merapi volcano, Indonesia: A major risk. *J. Volcanol. Geotherm. Res.* **2013**, *261*, 330–347. [CrossRef]
89. Dumaisnil, C.; Thouret, J.; Chambon, G.; Doyle, E.E.; Cronin, S.J.; Surono. Hydraulic, physical and rheological characteristics of rain-triggered lahars at Semeru volcano, Indonesia. *Earth Surf. Process. Landf.* **2010**, *35*, 1573–1590. [CrossRef]
90. Pistolesi, M.; Cioni, R.; Rosi, M.; Cashman, K.V.; Rossotti, A.; Aguilera, E. Evidence for lahar-triggering mechanisms in complex stratigraphic sequences: The post-twelfth century eruptive activity of Cotopaxi Volcano, Ecuador. *Bull. Volcanol.* **2013**, *75*, 698. [CrossRef]
91. van Westen, C.J.; Daag, A.S. Analysing the relation between rainfall characteristics and lahar activity at Mount Pinatubo, Philippines. *Earth Surf. Process. Landf.* **2005**, *30*, 1663–1674. [CrossRef]
92. Manville, V.; Major, J.J.; Fagents, S.A. Modeling lahar behavior and hazards. In *Modeling Volcanic Processes: The Physics and Mathematics of Volcanism*; Fagents, S.A., Gregg, T.K.P., Lopes, R.M.C., Eds.; Cambridge University Press: Cambridge, UK, 2013; pp. 300–330.
93. WISE. Chronology of Major Tailings Fam Failures. WISE Uranium Project Is Part of World Information Service on Energy. 2020. Available online: https://www.wise-uranium.org/mdaf.html (accessed on 15 July 2024).
94. Penman, A.D.M.; Brook, D.; Martin, P.L.; Routh, D. *Tailings Dams. Risk of Dangerous Occurrences. Lessons Learnt from Practical Experiences*; Commission Internationale des UNEP/ICOLD: Paris, France, 2001; Volume 121, ISBN 9783540773405.
95. Capart, H.; Eldho, T.I.; Huang, S.Y.; Young, D.L.; Zech, Y. Treatment of Natural Geometry in Finite Volume River Flow Computations. *J. Hydraul. Eng.* **2003**, *129*, 385–393. [CrossRef]
96. Hou, J.; Liang, Q.; Simons, F.; Hinkelmann, R. A stable 2D unstructured shallow flow model for simulations of wetting and drying over rough terrains. *Comput. Fluids* **2013**, *82*, 132–147. [CrossRef]
97. Pitman, E.B.; Nichita, C.C.; Patra, A.; Bauer, A.; Sheridan, M.; Bursik, M. Computing granular avalanches and landslides. *Phys. Fluids* **2003**, *15*, 3638–3646. [CrossRef]
98. Mergili, M.; Fischer, J.T.; Krenn, J.; Pudasaini, S.P. r.avaflow v1, an advanced open-source computational framework for the propagation and interaction of two-phase mass flows. *Geosci. Model Dev.* **2017**, *10*, 553–569. [CrossRef]

99. Courant, R.; Friedrichs, K.; Lewy, H. On the partial difference equations of mathematical physics. *IBM J. Res. Dev.* **1967**, *11*, 215–234. [CrossRef]
100. Buttinger-Kreuzhuber, A.; Konev, A.; Horváth, Z.; Cornel, D.; Schwerdorf, I.; Blöschl, G.; Waser, J. An integrated GPU-accelerated modeling framework for high-resolution simulations of rural and urban flash floods. *Environ. Model. Softw.* **2022**, *156*, 105480. [CrossRef]
101. Carlotto, T.; Borges Chaffe, P.L.; Innocente dos Santos, C.; Lee, S. SW2D-GPU: A two-dimensional shallow water model accelerated by GPGPU. *Environ. Model. Softw.* **2021**, *145*, 105205. [CrossRef]
102. Dehghan-Souraki, D.; López-Gómez, D.; Bladé-Castellet, E.; Larese, A.; Sanz-Ramos, M. Optimizing sediment transport models by using the Monte Carlo simulation and deep neural network (DNN): A case study of the Riba-Roja reservoir. *Environ. Model. Softw.* **2024**, *175*, 105979. [CrossRef]
103. Klose, C.D. Mine Water Discharge and Flooding: A Cause of Severe Earthquakes. *Mine Water Environ.* **2007**, *26*, 172–180. [CrossRef]

water

MDPI

Article

Hydrochemical Characteristics and Water Quality Evaluation of Groundwater in the Luohe Formation of Binchang Mining Area, China

Xu Wang [1,2], Kui Sun [2,3,*], Wanchao Ma [2,3], Jie Peng [2,3], Ruiping Liu [4], Jianping Chen [2,3], Kun Zhang [2,3], Shuai Gao [2,3], Cheng Li [2,3] and Penghua Zhang [2,3]

1 Environmental Protection Management Division, CHN Energy Shendong Clal Group Co., Ltd., Yulin 719315, China; m19991565517@163.com
2 Key Laboratory of Mine Geological Hazards Mechanism and Control, MNR, Xi'an 710054, China; 19829393069@163.com (W.M.); ax15209282093@163.com (C.L.)
3 Shaanxi Institute of Geo-Environment Monitoring, Xi'an 710054, China
4 Loess and Ecological Restoration Department of Xi'an Center of China Geological Survey, Ministry of Natural Resources, Xi'an 710054, China
* Correspondence: m1037946337@163.com

Abstract: The groundwater of the Luohe Formation in Binchang mining area is the main source of water for industrial and agricultural use and for drinking water for residents in the area. In order to study the hydrochemical characteristics and water-quality status of Luohe Formation groundwater in the mining area, statistical analysis, Piper three-line diagram, ion ratio relationship, and other methods were used to study the hydrochemical characteristics and formation factors of the groundwater. The Nemerow index evaluation method and the fuzzy comprehensive evaluation method based on principal component analysis were used to evaluate the groundwater quality in the mining area. The results show that the groundwater is weakly acidic as a whole, and the content of SO_4^{2-} and Cl^- have strong variability in terms of spatial distribution. The groundwater chemical type gradually evolves from $SO_4 \bullet HCO_3 \bullet Cl$–Na, SO_4–Na and $SO_4 \bullet Cl$–Na-type water in the north of the mining area to $SO_4 \bullet HCO_3 \bullet Cl–Na \bullet Ca$, $HCO_3 \bullet SO_4–Na \bullet Mg$, and $SO_4 \bullet Cl–Na \bullet Ca \bullet Mg$-type water in the south. The formation of the hydrochemical composition of groundwater in the study area may be related to multiple factors such as cation-alternating adsorption, carbonate and sulfate dissolution, and hydraulic exchange with the groundwater of the upper Huachi Formation. Comparing the evaluation results of the Nemerow index method and the principal component analysis method, the latter's evaluation results can take into account the contribution of each indicator to the overall groundwater quality, and to a certain extent can weaken the control effect of a certain pollution indicator, exceeding the limit on the entire evaluation result. Therefore, the evaluation results based on the principal component analysis method are more credible.

Keywords: groundwater of the Luohe Formation; hydrochemical characteristics; water-quality evaluation; Nemerow index method; Binchang mining area

Citation: Wang, X.; Sun, K.; Ma, W.; Peng, J.; Liu, R.; Chen, J.; Zhang, K.; Gao, S.; Li, C.; Zhang, P. Hydrochemical Characteristics and Water Quality Evaluation of Groundwater in the Luohe Formation of Binchang Mining Area, China. *Water* **2024**, *16*, 1913. https://doi.org/10.3390/w16131913

Academic Editor: Cesar Andrade

Received: 23 May 2024
Revised: 22 June 2024
Accepted: 26 June 2024
Published: 4 July 2024

1. Introduction

Groundwater resources are the material basis for supporting social and economic development and maintaining the functional balance and diversified services of ecosystems [1]. However, recent years have seen an increase in groundwater pollution due to human activities [2]. Western China, rich in coal resources, suffers from a fragile ecological environment and scarce water resources [3,4]. The intense and concentrated exploitation of coal in recent years has damaged aquifer structures, leading to issues such as declining groundwater levels [5,6], water resource leakage [7–10], and worsening water quality [11]. This has significantly diminished the ecological functionality of coal mining areas. Under the same geological background, the concentration of the main chemical indicators of groundwater in the coal mining area is significantly higher than that in the non-mining area. Coal is rich in harmful elements such as Cl and Pb, which have a strong carcinogenic risk and threaten the drinking water safety of residents [12]. Therefore, research on the hydrochemical characteristics of groundwater in mining areas, scientific evaluation of groundwater quality, and the timely grasp of groundwater environmental conditions can effectively promote the protection and rational development of water resources.

Factor analysis and multivariate statistical analysis have been favored in the analysis of hydrochemical characteristics [13]. Principal component analysis, systematic clustering, the Piper diagram, and the Gibbs diagram are mostly used to analyze the genesis and evolution of groundwater hydrochemistry [14–16]. Spatial analysis of principal component load scores and hydrochemical types can be used to study the controlling factors of groundwater chemical formation [17].

Water-quality evaluation is primarily divided into single-factor evaluation and comprehensive evaluation methods. The single-factor evaluation method directly highlights excessive components and regions of groundwater quality [18]. Li Lijun et al. used single-factor evaluation and the superposition index method to assess groundwater pollution in Songyuan City and analyzed its impact [19]. The comprehensive evaluation method fully reflects the overall status of groundwater quality. Comprehensive evaluation methods include the water quality index method [20], fuzzy comprehensive evaluation method [21], geostatistics method [22,23], and provenance extension method [24]. The fuzzy comprehensive evaluation method is widely used. Dahiya et al. [25] used the fuzzy comprehensive evaluation model to assess water quality at 42 groundwater sampling sites in Haryana State, Southern India. The results showed that about 64% of the groundwater was of satisfactory or acceptable quality. Peng et al. [26] used the fuzzy comprehensive evaluation method to assess water quality at 34 groundwater sampling points in Zhaoyuan City, Shandong Province. The results showed that 88% of the water was polluted. However, the fuzzy comprehensive evaluation method is cumbersome. The evaluation process becomes more complex with more water sample data, and constructing the weight matrix is the most critical problem.

The analytic hierarchy process, principal component analysis [27], entropy weight method [28], rough set method [29], and factor analysis are commonly employed to identify key evaluation indicators and determine their weights. Xia et al. [30] enhanced the fuzzy comprehensive evaluation model using principal component analysis. This simplification significantly streamlined the evaluation process, yielding satisfactory results.

The Binchang mining area, situated in the hill and gully regions of the Eastern Longdong Loess Plateau, serves as the central construction zone for the Huanglong coal base. According to statistics, Binzhou City and Changwu County in the mining area have total water resources of 7687 million m^3 and 45,494 million m^3, respectively. However, in areas with severe water shortages, the per capita water resources are merely 210.00 m^3 and 249.60 m^3. The aquifer of the Lower Cretaceous Luohe Formation, rich in water, is the principal source for industrial, agricultural, and drinking water in the Binchang mining area, holding significant strategic reserve value [31]. Since the 1980s, the exploration and development of coal resources in the area have led to the opening of several modern fully mechanized mines, resulting in subsequent groundwater pollution. Based on the analysis of hydrogeological characteristics, the chemical characteristics and formation factors of groundwater in the Cretaceous Luohe Formation in the Binchang mining area of Shaanxi Province are discussed by means of statistical analysis, Piper diagram, and ion combination ratio analysis. Groundwater quality in the mining area is evaluated using the Nemerow index method and fuzzy comprehensive evaluation based on principal component analysis. The expected outcomes of this research are to offer theoretical support for the development and utilization of groundwater resources in the mining area.

2. General Situation of the Research Area

2.1. Current Situation of Coal Resources Development

The Binchang mining area, the principal site within the Huanglong coal base, serves as the main mining district in Shaanxi Province. This area is the primary energy provider for Shaanxi's Guanzhong district. Spanning 46.00 km east to west and 36.50 km north to south, the mining area covers 978.00 km^2. Geological exploration of the coalfield in the Binchang mining area started in the 1980s, with reserves reaching 7.562 billion tons by the end of 2016. Currently, the mining area houses 13 operational coal mines, comprising 12 large-scale mines with capacities of at least 1.20 Mt/a and one medium-sized mine with a capacity of 0.45–1.20 Mt/a. The mining area's total production capacity stands at 49.60 Mt/a. As shown in Table 1. The mining area's coal-bearing stratum, the Jurassic Yan'an Formation, includes eight coal seams. The entire area permits mining of the No. 4 coal seam, which varies in thickness from 0.15 to 35.04 m and averages 10.64 m.

Table 1. Capacity and status of the main coal mines in the Binchang mining area [32].

Coal Mine	Capacity (Mt/a)	Status	Coal Mine	Capacity (Mt/a)	Status
Mengcun	6.00	Production	Wenjiapo	3.20	Production
Xiaozhuang	5.00	Production	Dafosi	3.00	Production
Hujiahe	5.00	Production	Huoshizui	3.00	Production
Tingnan	5.00	Production	Xiagou	3.00	Production
Yangjiaping	5.00	Construction	Shuiliandong	1.50	Production
Gaojiabu	5.00	Production	Jiangjiahe	0.90	Production
Yadian	4.00	Construction			

2.2. Aquifer Characteristics

The Binchang mining area is situated in the southwestern margin of the Ordos Basin, within the southern hydrogeological unit of the Cretaceous groundwater basin. The area predominantly lies in the Jinghe River Basin, which is categorized as the Jinghe Hydrogeological Unit. The study area is within this unit. Groundwater development and utilization in the study area have shifted from primarily relying on Quaternary loose layer groundwater, with Cretaceous Luohe sandstone shallow groundwater as a supplement, to primarily using Cretaceous Luohe Formation sandstone groundwater, with Quaternary loose-layer groundwater as a supplement. The aquifers in the Binchang mining area, from top to bottom, include: the Quaternary Holocene alluvial–diluvial aquifer, Quaternary Pleistocene sandy loess and sandy clay aquifer, Neogene red clay impermeable layer, Cretaceous Huachi Formation sandstone aquifer, Cretaceous Luohe Formation pore-fissure aquifer group, Yijun Formation conglomerate pore-fissure aquifer group, Anding Formation impermeable layer, Zhiluo Formation fissure aquifer, Yan'an Formation fissure aquifer, Fuxian Formation mudstone impermeable layer, and Triassic sandstone aquifer (Figure 1). Among these, the Cretaceous Luohe Formation aquifer stands out for its high water abundance, excellent water quality, and significant water supply and ecological importance.

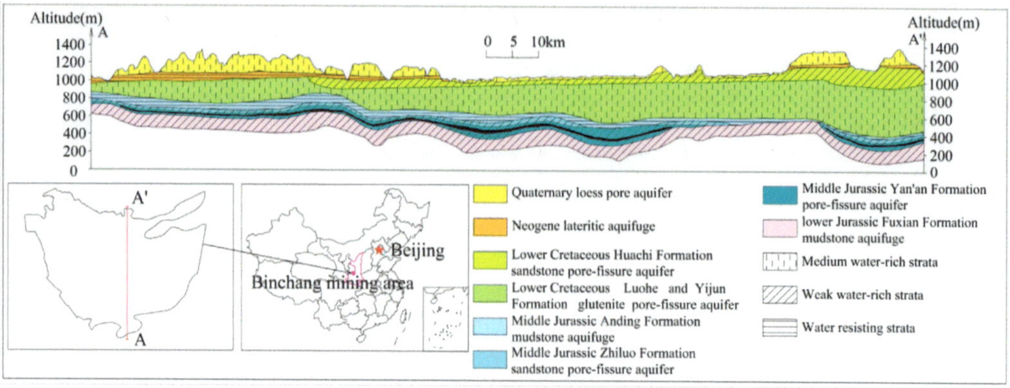

Figure 1. Hydrogeological section of the Binchang mining area [32].

2.3. Aquifer Occurrence Characteristics of Cretaceous Luohe Formation

The Luohe Formation from the Cretaceous period is mainly exposed in the East Valley of Jinghe and its tributaries. In the northern part of the minefield, this formation lies beneath the Cretaceous Huachi Formation, with thicknesses ranging from 0 to 580.84 m, and typically between 200 and 300 m. The Luohe Formation primarily consists of purple to dark purple medium- to coarse-grained sandstone, along with conglomerate, sandy conglomerate, mudstone, and sandy mudstone. The bottom interface of the Luohe Formation is a continuous plane surface, contacting the underlying Yijun Formation strata. The formation's base typically inclines northwest, forming a monocline. Due to its geological structure, the thickness of the Luohe Formation gradually decreases from northwest to southeast, disappearing at the southeastern boundary of the mining area (Figure 2).

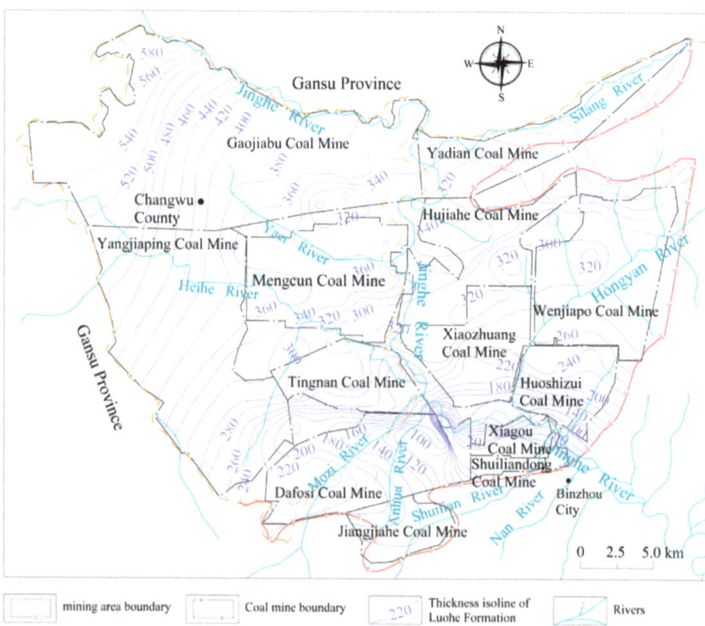

Figure 2. The contour of the Luohe Formation's thickness in the Binchang mining area [32].

Vertically, the Luohe Formation is affected by alternating sedimentation of meandering-braided-meandering river facies from the bottom to the top and shows obvious segmentation. The floodplain and channel deposits of the meandering river sedimentary section are well-developed, displaying clear positive cycles and a binary structure. The lower structure consists of edge beach deposits, characterized mainly by lateral accretion, while the upper structure is composed of floodplain deposits, primarily formed by vertical deposition. The sediment section of the middle-braided river exhibits a positive cycle progressing from bottom to top. The sediments primarily consist of gravel and sand. The sediment features relatively coarse grains, and it is more water-rich compared to the sediments of the upper and lower meandering rivers [33].

3. Sample Collection, Testing, and Analysis

3.1. Sample Collection and Testing

The groundwater of the Luohe Formation in the Cretaceous System is a vital water source for industrial, agricultural, and residential use in the Binchang mining area. To investigate water environment issues caused by coal mining in the area, water samples were collected from 10 Luohe Formation groundwater monitoring wells in the Binchang mining area. Groundwater samples were collected, stored, and sent for inspection following HJ/T164-2004 "Technical Specifications for Groundwater Environmental Monitoring". Before sampling, the pre-cleaned and sterilized 5 L high-density polyethylene bottles were rinsed 2–3 times, and samples were filtered with a low-speed vacuum pump during sampling. After collection, samples were stored in a dark, low-temperature environment and sent to the laboratory for testing within the specified time. Key detection indicators include total dissolved solids (TDS), K^+, Na^+, Ca^{2+}, Mg^{2+}, COD_{Mn}, Cl^-, SO_4^{2-}, HCO_3^-, NO_3^-, F^-, NH_3-N, and pH. pH was measured using a portable water quality analyzer, TDS by the weighing method, K^+ and Na^+ via flame atomic absorption spectrophotometry, Ca^{2+} and Mg^{2+} by ethylenediaminetetraacetic acid titration, Cl^-, SO_4^{2-}, and F^- by ion chromatography, HCO_3^- by the acid–base method, and NH_3-N by neutralization titration and sodium reagent spectrophotometry. Analysis results are presented in Table 2.

Table 2. Statistical results of groundwater hydrochemical parameters in the Binchang mining area.

Sample Date	Na^+ (mg/L)	K^+ (mg/L)	Ca^{2+} (mg/L)	Mg^{2+} (mg/L)	Cl^- (mg/L)	SO_4^{2-} (mg/L)	HCO_3^- (mg/L)	TDS (mg/L)	pH	COD_{Mn} (mg/L)	F^- (mg/L)	NH_3–N (mg/L)	NO_3^- (mg/L)
W1	284.00	3.11	42.10	21.90	128.00	476.00	230.00	1156.00	8.24	0.64	0.69	0.06	0.35
W2	306.00	2.60	20.00	69.30	131.00	288.00	232.00	960.00	8.59	0.82	0.60	0.02	0.36
W3	262.00	2.62	23.80	16.60	102.00	302.00	296.00	879.00	8.30	0.42	0.52	0.07	0.24
W4	270.00	3.02	21.40	13.30	96.70	247.00	284.00	858.00	8.25	0.45	0.76	0.09	1.59
W5	170.00	3.48	79.31	60.99	46.00	125.60	350.50	964.00	7.60	0.58	0.37	0.02	341.70
W6	177.00	2.40	52.33	9.92	93.57	160.20	330.10	680.00	7.74	1.13	0.31	0.02	15.52
W7	792.50	4.92	131.60	94.21	606.40	1467.00	189.90	3196.00	7.76	0.78	0.41	0.03	2.55
W8	50.40	1.92	63.78	23.80	8.75	48.04	382.70	399.00	7.70	0.36	0.50	0.26	6.72
W9	103.50	1.78	42.52	18.35	60.13	97.96	62.90	467.00	8.06	0.49	0.45	0.02	8.00
W10	728.00	6.90	134.10	29.75	145.90	1694.00	195.70	2843.00	7.71	0.77	1.00	0.55	2.58
Max	792.50	6.90	134.10	94.21	606.40	1694.00	382.70	3196.00	8.59	1.13	1.00	0.55	341.70
Min	50.40	1.78	20.00	9.92	8.75	48.04	62.90	399.00	7.60	0.36	0.31	0.02	0.24
Mean	314.34	3.28	61.09	35.81	141.85	490.58	255.38	1240.20	8.00	0.64	0.56	0.11	37.96
Std.	249.28	1.55	42.27	28.66	168.55	589.74	93.93	969.07	0.34	0.24	0.21	0.17	106.83
Cv	0.79	0.47	0.69	0.80	1.19	1.20	0.37	0.78	0.04	0.37	0.37	1.50	2.81

3.2. Analytical Methods

After completing the sample tests, we used SPSS19 software to calculate and correlate the results. The Piper tri-line diagram was generated using Aquachem 3.7 software, and groundwater quality in the study area was assessed using the Nemerow index and fuzzy comprehensive evaluation methods, which are based on principal component analysis.

3.2.1. Nemerow Index Evaluation Method

The Nemerow index evaluation method is as follows:

The evaluation scores of each individual index are determined according to Table 3. The classification criteria of each index are based on the water quality criteria of "Specification for regional groundwater contamination investigation and evaluation" (DZ/T0288-2015) and "Standard for groundwater quality" (GB/T14848-2017) (Table 4).

Table 3. Ratings for groundwater quality classes of a single factor.

	I	II	III	IV	V
Fi	0	1	3	6	10

Table 4. Groundwater quality classification standards.

	Na^+ (mg/L)	Ca^{2+} (mg/L)	Mg^{2+} (mg/L)	Cl^- (mg/L)	SO_4^{2-} (mg/L)	TDS (mg/L)	pH	COD_{Mn} (mg/L)	F^- (mg/L)	NH_3–N (mg/L)	NO_3^- (mg/L)
I	≤100	≤100	≤10	≤50	≤50	≤300	6.5 ≤ pH ≤ 8.5	≤1.0	≤1.0	≤0.02	≤2.0
II	≤150	≤200	≤20	≤150	≤150	≤500	6.5 ≤ pH ≤ 8.5	≤2.0	≤1.0	≤0.10	≤5.0
III	≤200	≤400	≤50	≤250	≤250	≤1000	6.5 ≤ pH ≤ 8.5	≤3.0	≤1.0	≤0.50	≤20.0
IV	≤400	≤800	≤200	≤350	≤350	≤2000	5.5 ≤ pH < 6.5; 8.5 < pH ≤ 9.0	≤10.0	≤2.0	≤1.50	≤30.0
V	>400	>800	>200	>350	>350	>2000	pH < 5.5; pH > 9.0	>10.0	>2.0	>1.50	>30.0

The comprehensive evaluation score is calculated by the following formula [34]:

$$F = \sqrt{\frac{F_{max}^2 + \overline{F}^2}{2}} \tag{1}$$

where

$$\overline{F} = \frac{1}{n}\sum F_i \tag{2}$$

F_i is the evaluation score of each individual index, dimensionless; F_{max} is the maximum value of the evaluation score Fi for each individual index, dimensionless; and n is the number of indicators [34].

The groundwater quality classification of each sample is determined according to Table 5.

Table 5. Rating scale of groundwater quality classification.

	I	II	III	IV	V
F	F < 0.80	0.80 ≤ F < 2.50	2.50 ≤ F < 4.25	4.25 ≤ F < 7.20	F ≥ 7.20

3.2.2. Fuzzy Comprehensive Evaluation Method Based on Principal Component Analysis

The fuzzy comprehensive evaluation method based on principal component analysis transforms multiple evaluation indexes into several comprehensive evaluation indexes by the principal component analysis method and then evaluates the water quality by the fuzzy comprehensive evaluation method.

Assuming that there are n samples and m indices for each sample, the original data matrix X can be expressed as an $n \times m$ order matrix:

$$X = \begin{bmatrix} x_{11} & x_{12} & \cdots & x_{1m} \\ x_{21} & x_{22} & \cdots & x_{2m} \\ x_{31} & x_{32} & \cdots & x_{3m} \\ x_{n1} & x_{n2} & \cdots & x_{nm} \end{bmatrix}$$

The correlation coefficients of each index are calculated by the following formula:

$$r_{ij} = \frac{\sum\limits_{k=1}^{n}(x_{ki} - \overline{x_i})(x_{kj} \quad \overline{x_j})}{\sqrt{\sum\limits_{k=1}^{n}(x_{ki} - \overline{x_i})^2 \sum\limits_{k=1}^{n}(x_{kj} - \overline{x_j})^2}}$$

where x_{kj} is the ion content of the j index of the k water sample, mg/L; x_i and x_j are the measured values of the ion content of the i and j index, mg/L; and $i, j = 1, 2, ..., m$.

According to the value of the correlation coefficient obtained, the correlation coefficient matrix R is obtained.

$$R = \begin{bmatrix} r_{11} & r_{12} & \cdots & r_{1m} \\ r_{21} & r_{22} & \cdots & r_{2m} \\ r_{31} & r_{32} & \cdots & r_{3m} \\ r_{n1} & r_{n2} & \cdots & r_{nm} \end{bmatrix}$$

The λ are obtained by solving the eigenvalue equation $|\lambda I - R| = 0$ and arranged in the order of $\lambda_1 > \lambda_2 > \lambda_3 > \ldots > \lambda_m > 0$. Then, the following formula is used to calculate the cumulative variance contribution rate. Principal components were extracted according to the criterion of cumulative variance contribution rate greater than 80%.

$$\alpha_i = \frac{\lambda_i}{\sum\limits_{i=1}^{m} \lambda_i}$$

where α_i is the variance contribution rate of component i, % and λ_i is the eigenvalue of component i, dimensionless.

The magnitude of principal component loads indicates their correlation with the principal component. Principal component loads are calculated using the formulas below:

$$l_{ij} = \sqrt{\lambda_i} e_{ij}$$

where e_{ij} is the eigenvector of the correlation coefficient matrix.

For each principal component, the main control factor is selected based on the load value and used as the evaluation criterion in fuzzy comprehensive evaluation.

A fuzzy relation matrix is derived from the membership degrees of evaluation factors to evaluation grades. If yip represents the membership degree of the first evaluation factor to the pth evaluation grade index, the corresponding membership function is defined as follows:

For grade I (evaluation grade p = 1), the membership function is:

$$y_{i1} = \begin{cases} 1 & x_{ki} \le c_{i1} \\ \frac{c_{i2} - x_{ki}}{c_{i2} - c_{i1}} & c_{i1} < x_{ki} < c_{i2} \\ 0 & x_{ki} \ge c_{i2} \end{cases}$$

For grade II to IV (evaluation grade p = 2, 3, and 4), the membership function is:

$$y_{ip} = \begin{cases} 1 - \frac{c_{ip} - x_{ki}}{c_{ip} - c_{ip-1}} & c_{ip-1} \le x_{ki} \le c_{ip} \\ \frac{c_{ip+1} - x_{ki}}{c_{ip+1} - c_{ip}} & c_{ip} < x_{ki} < c_{ip+1} \\ 0 & c_{ki} \ge c_{ip+1} \ or \ x_{ki} \le c_{ip-1} \end{cases}$$

For grade V (evaluation grade p = 5), the membership function is:

$$y_{i5} = \begin{cases} 0 & x_{ki} \le c_{i4} \\ 1 - \frac{c_{i5} - x_{ki}}{c_{i5} - c_{i4}} & c_{i4} < x_{ki} < c_{i5} \\ 1 & x_{ki} \ge c_{i5} \end{cases}$$

In the formulas above, x_{ki} represents the measured value of the i-th evaluation factor in k water samples (mg/L). c_{ip} is the boundary value for the i-th evaluation factor corresponding to the p-th evaluation grade (mg/L). y_{ip} denotes the degree of membership for the i-th evaluation factor relative to the p-th evaluation grade, and it is dimensionless.

$$Y = \begin{bmatrix} y_{11} & y_{12} & y_{13} & y_{14} & y_{15} \\ y_{21} & y_{22} & y_{23} & y_{24} & y_{25} \\ y_{31} & y_{32} & y_{33} & y_{34} & y_{35} \\ \cdots & \cdots & \cdots & \cdots & \cdots \\ y_{i1} & y_{i2} & y_{i3} & y_{i4} & y_{i5} \end{bmatrix}$$

The formula below determines the weight of each evaluation index in assessing groundwater quality:

$$W_{ki} = \frac{x_{ki}/s_i}{\sum\limits_{i=1}^{n} x_{ki}/s_i}$$

where s_i is the arithmetic average of the standard values of each grade of the I evaluation factor, mg/L.

The weight set of each evaluation factor of K water samples can be expressed as $W = (W_{k1}, W_{k2}, \ldots, W_{kn})$. The comprehensive evaluation model is $B = W \bullet R = (b_1, b_2, \ldots, b_m)$, where b_m is the membership degree of each water quality grade for each sample.

4. Results and Discussion

4.1. Analysis of Chemical Characteristics of Groundwater

4.1.1. Hydrochemical Characteristics of Major Ions

Table 2 shows that the groundwater pH varies from 7.60 to 8.59, with a small coefficient of variation, and the groundwater is weakly acidic as a whole. The TDS content of groundwater is 399.00–3196.00 mg/L, which is weak–medium salinity groundwater. Figure 3 shows that the TDS content of groundwater is higher in the eastern and northern mining areas but lower in the Jinghe Valley area. These relationships are mainly because of the large thickness of the loess layer and the deep burial of the aquifer in the eastern and western part of the mining area, combined with the influence of the mudstone aquifuge at the bottom of the overburden Huachi Formation, poor groundwater recharge conditions, and slow groundwater circulation, resulting in a higher TDS content of groundwater. The groundwater in the Jinghe River Basin is shallow and directly exposed on both sides of the valley. The recharge condition of atmospheric precipitation is good, the groundwater circulation is frequent, and the TDS content is high.

The variation coefficients of Cl^- and $SO_4{}^{2-}$ concentrations are both greater than 1.0. Analysis indicates that the larger variation in $SO_4{}^{2-}$ concentration is related to the distribution of Huachi Formation aquifers in the upper part of the Luohe Formation. The Huachi Formation aquifer, primarily in the northern section of the mining area, ranges from 0 to 260 m in thickness and thins out in the southern part. This aquifer contains substantial amounts of gypsum. Disturbances from coal mining weaken hydraulic exchanges between the Huachi and Luohe Formation aquifers, facilitating $SO_4{}^{2-}$ migration to the Luohe Formation. In the southern mining area, especially at sampling points W6, W7, W8, and W9, the Huachi Formation is absent, resulting in generally low $SO_4{}^{2-}$ concentrations. The high variation coefficient of Cl^- concentrations in groundwater indicates significant influence from environmental factors, such as mining activities, coal washery operations, and other anthropogenic activities.

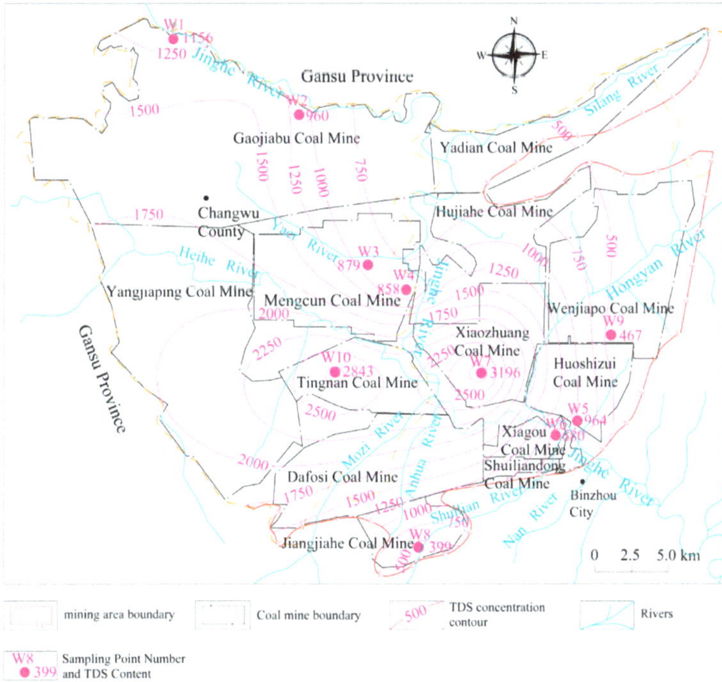

Figure 3. TDS isoline of groundwater in the Luohe Formation, Binchang mining area.

4.1.2. Chemical Types of Groundwater

Figure 4 shows that the main ions in the groundwater of the Luohe Formation are K^+ + Na^+, SO_4^{2-}, HCO_3^-, and Cl^-. In the deeper groundwater burial area of the Luohe Formation, the groundwater chemical types are mainly SO_4 • HCO_3 • Cl–Na, SO_4–Na and SO_4 • Cl–Na. In the shallower aquifer area, the groundwater chemical types are transformed into SO_4 • HCO_3 • Cl–Na • Ca, HCO_3 • SO_4–Na • Mg and SO_4 • Cl–Na • Ca • Mg. Generally, the chemical types of groundwater in the Binchang mining area are complex.

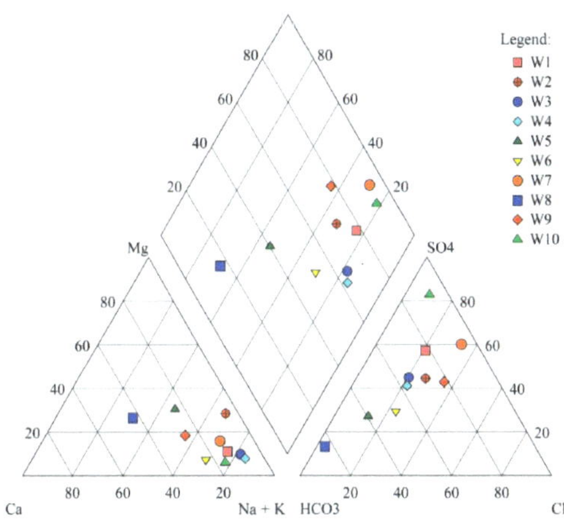

Figure 4. Piper diagram of the groundwater of the Luohe Formation in the Binchang mining area.

4.1.3. Ion Combination Ratio Analysis

The ion combination ratio serves to investigate the formation of chemical constituents in water and their ionic origins [14]. Chloride ions (Cl^-) are relatively stable in groundwater; the ratio $\rho(Na^+)/\rho(Cl^-)$ frequently helps identify the sodium (Na^+) source. If Na^+ originates from salt rock dissolution, then $\rho(Na^+)/\rho(Cl^-)$ equals 1. As depicted in Figure 5a, an increase in TDS leads to $\rho(Na^+)/\rho(Cl^-)$ significantly exceeding 1, suggesting multiple sources of Na^+ beyond rock salt dissolution. Given that Ca^{2+} and Mg^{2+} adsorb more readily to particle surfaces than Na^+, the process of cationic alternating adsorption facilitates Na^+ enrichment as TDS increases. If the primary sources of Ca^{2+} and Mg^{2+} in groundwater are carbonates and sulfates, then $\rho(Ca^{2+} + Mg^{2+})/[\rho(SO_4^{2-}) + 0.5\rho(HCO_3^-)]$ equals 1. According to Figure 5b, the value of $\rho(Ca^{2+} + Mg^{2+})/[\rho(SO_4^{2-}) + 0.5\rho(HCO_3^-)]$ is less than 1 in most samples, suggesting additional sources for Ca^{2+} and Mg^{2+}, which confirms the validity of cationic alternating adsorption. If the primary source of Ca^{2+}, Mg^{2+}, and SO_4^{2-} is sulfate dissolution, then the ratio $\rho(Ca^{2+} + Mg^{2+})/\rho(SO_4^{2-})$ equals 1. Figure 5c demonstrates that most water samples lie on both sides of the line $\rho(Ca^{2+}/Mg^{2+})/\rho(SO_4^{2-}) = 1$, suggesting that sources other than sulfate dissolution contribute to the presence of Ca^{2+}, Mg^{2+}, and SO_4^{2-} in the Luohe Formation's groundwater. When $\rho(Ca^{2+}/Mg^{2+})/\rho(SO_4^{2-}) > 1$, additional sources such as carbonate rock dissolution contribute to the levels of Ca^{2+} and Mg^{2+}. When $\rho(Ca^{2+}/Mg^{2+})/\rho(SO_4^{2-}) < 1$, other sources like hydraulic exchange with the Huachi Formation aquifer increase SO_4^{2-} levels. If carbonate dissolution primarily sources Ca^{2+}, Mg^{2+}, and HCO_3^-, then the ratio $\rho(Ca^{2+} + Mg^{2+})/0.5\rho(HCO_3^-)$ equals 1. Figure 5d reveals that most water samples have a $\rho(Ca^{2+} + Mg^{2+})/0.5\rho(HCO_3^-)$ ratio greater than 1, implying additional sources beyond carbonate dissolution for Ca^{2+}, Mg^{2+}, and HCO_3^-. The analysis of these ions indicates a complex chemical environment in the Luohe Formation's groundwater, heavily influenced by the regional geology and mining activities in the Binchang area. The chemical composition of water in the Luohe Formation is likely influenced by cationic adsorption, carbonate and sulfate dissolution, and hydraulic interactions with the Huachi Formation groundwater.

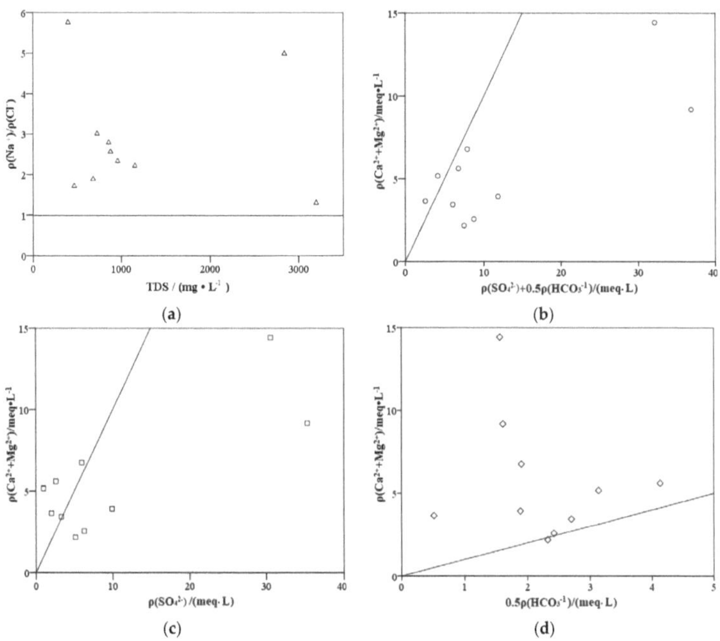

Figure 5. Correlation of major ions in the groundwater of the Luohe Formation in the Binchang mining area.

4.2. Groundwater Quality Evaluation

4.2.1. Evaluation Results of the Nemerow Index Evaluation Method

According to the different water quality standards of different indicators and the actual detection values of various indicators of water samples, the comprehensive evaluation score F of water quality of each water sample was calculated. The calculation results are shown in Table 8. Table 8 shows that there were zero water samples in Class I, three water samples in Class II, zero water samples in Class III, three water samples in Class IV, and four water samples in Class V. The groundwater quality of sampling points was mainly concentrated in Class IV and V, and the groundwater quality was generally poor.

4.2.2. Results of the Fuzzy Comprehensive Evaluation Using Principal Component Analysis

Using the introduced calculation method, we computed the variance contribution rate for each component. The calculation results are shown in Table 6. According to the cumulative contribution rate of more than 80%, three principal components were extracted, and the load matrix of principal components is shown in Table 7. Table 7 shows that TDS, Na^+, Ca^{2+}, and $SO_4{}^{2-}$ had higher loads of principal component 1; F^- had a higher load of principal component 2; and pH had a higher load of principal component 3. Therefore, the seven indicators of TDS, Na^+, Ca^{2+}, $SO_4{}^{2-}$, Cl^-, F^-, and pH were selected as the main evaluation indicators. The water quality of ten samples in the study area was assessed using fuzzy comprehensive evaluation. This yielded the membership degrees and fuzzy evaluation grades for each sample. Table 8 shows that there was one water sample in Class I, two water samples in Class II, three water samples in Class III, three water samples in Class IV, and zero water samples in Class V. The groundwater quality of sampling was mainly concentrated in Classes III and IV.

Table 6. Computation results of the variance contribution rate of each component.

Component	Characteristic Value	Contribution Rate of Variance	Accumulate (%)
1	5.200	47.273	47.273
2	2.289	20.813	68.086
3	1.713	15.576	83.662
4	0.924	8.401	92.063
5	0.600	5.452	97.515
6	0.210	1.907	99.422
7	0.043	0.391	99.813
8	0.019	0.176	99.989
9	0.001	0.011	100.000
10	4.050×10^{-17}	3.682×10^{-16}	100.000
11	-2.313×10^{-16}	-2.103×10^{-15}	100.000

Table 7. Principal component load matrix.

Index	Component		
	1	2	3
Na^+	0.964	0.080	0.203
Ca^{2+}	0.886	−0.131	−0.417
Mg^{2+}	0.564	−0.604	0.216
Cl^-	0.772	−0.369	0.425
$SO_4{}^{2-}$	0.975	0.194	0.038
TDS	0.989	0.000	0.075
pH	−0.377	0.301	0.807
COD_{Mn}	0.372	−0.291	0.185
F^-	0.344	0.857	0.016
NH_3–N	0.479	0.702	−0.474
$NO_3{}^-$	−0.118	−0.570	−0.594

Table 8. Evaluation results of groundwater quality types

Sample Date	Nemerow Index Evaluation Method		Principal Component Analysis-Based Fuzzy Comprehensive Evaluation Method.					
	F	Groundwater Quality Types	Subordination Degree					Groundwater Quality Types
W1	7.329	V	0.110	0.154	0.206	0.206	0.323	V
W2	4.609	IV	0.107	0.097	0.499	0.298	0.000	III
W3	4.452	IV	0.126	0.166	0.519	0.189	0.000	III
W4	4.398	IV	0.159	0.165	0.504	0.171	0.000	III
W5	7.266	V	0.210	0.360	0.430	0.000	0.000	III
W6	2.386	II	0.199	0.457	0.344	0.000	0.000	II
W7	7.794	V	0.030	0.028	0.038	0.000	0.905	V
W8	2.280	II	0.491	0.248	0.260	0.000	0.000	I
W9	2.236	II	0.448	0.510	0.042	0.000	0.000	II
W10	7.640	V	0.061	0.072	0.039	0.000	0.829	V

4.2.3. Comparative Analysis of Evaluation Results

The evaluation process of the Nemerow index method is relatively simple. According to the evaluation results, only three groundwater samples of potable groundwater of grade I to III are found, accounting for 30% of the total water samples. Although this method clearly indicates the over-standard status of pollution indicators, this method overemphasizes the maximum pollution factors, often leading to a higher overall pollution degree of the evaluation results because of the over-limit of one index, ignoring the overall contribution of each pollutant to groundwater pollution. Consequently, this method cannot objectively describe the continuity of environmental quality.

The fuzzy comprehensive evaluation method can show the objectively existing fuzziness and uncertainty in groundwater environmental quality, but the evaluation method is more cumbersome. By principal component analysis, the author deletes some variables that are closely related and establishes as few variables as possible that can reflect the overall water quality information, reducing the complexity of the evaluation process. Through the fuzzy comprehensive evaluation model of principal component screening, the information provided by all principal component data is fully utilized, and its contribution to the overall groundwater quality is considered. To some extent, this method can weaken the control effect of one index exceeding the limit on the whole evaluation result. According to the evaluation results, seven groundwater samples of potable groundwater samples of grade I to III are found, accounting for 70% of the water samples.

According to the location of sampling points and the analysis of evaluation results, the water samples of type V water quality are mainly distributed north of the Gaojiabao coal mine, west of the Tingnan coal mine, and south of the Xiaozhuang coal mine. The analysis indicates that all three coal mines are large-scale, each with a production capacity of 5.00 Mt/a and high mining intensity. Field investigations reveal that all three water samples originate near the mining face. Strong disturbances to the Luohe Formation's aquifer due to extensive pre-mining water exploration and drainage projects have led to sequential contamination of the aquifers. Other water samples are located in the Jinghe Wetland Reserve, Water Source Reserve, or far away from the current mining activity area. The disturbance caused by human factors such as mining activities to groundwater is relatively small, and all of them meet the potable water standards of grade III and above.

5. Conclusions

(1) The Cretaceous Luohe Formation aquifer is abundant in water with excellent quality. It is a crucial water source for industry, agriculture, and drinking water for residents in the mining area and its surroundings. Horizontally, the Luohe Formation's thickness generally decreases from northwest to southeast. Vertically, it shows distinct segmentation due to alternating sedimentation of meandering and braided river phases.

(2) Among the groundwater chemical indicators, the pH value is relatively stable and typically weakly acidic. The content levels of SO_4^{2-} and Cl^- vary significantly across different locations. The groundwater chemical type transitions from SO_4 • HCO_3 • Cl–Na, SO_4–Na, and SO_4 • Cl–Na types in the northern mining area to SO_4 • HCO_3 • Cl–Na • Ca, HCO_3 • SO_4–Na • Mg, and SO_4 • Cl–Na • Ca • Mg types in the south. The groundwater hydrochemical composition is influenced by multiple factors, including cation adsorption, carbonate and sulfate dissolution, and hydraulic exchange with the upper Huachi Formation groundwater.

(3) Significant differences exist in the groundwater quality evaluation results obtained by the Nemerow index and the fuzzy comprehensive evaluation method based on principal component analysis. The latter method provides more reasonable and reliable results. In the study area, 70% of groundwater samples meet Class III water standards or higher. Poor groundwater quality areas are mainly found to the north of Gaojiabao coal mine, the west of Tingnan coal mine, and the south of Xiaozhuang coal mine, where coal mining intensity is relatively high.

Author Contributions: Conceptualization, X.W. and R.L.; conceptualization and writing—original draft, K.S.; writing—original draft and Formal analysis, W.M.; data curation and writing—review & editing, J.P.; investigation and methodology, J.C. and K.Z.; visualization, S.G. and C.L.; software, C.L.; Investigation, P.Z. All authors have read and agreed to the published version of the manuscript.

Funding: This paper is sponsored by the Natural Science Basic Research Program of Shaanxi Province, China (2019JLZ-03). This study would not have been possible without a ministry–province cooperation-based pilot project entitled "A Technological System for Ecological Remediation Evaluation of Open-Pit Mines" initiated by the Ministry of Natural Resources in 2023.

Data Availability Statement: The data used to support the findings of this study are available from the corresponding author upon request.

Conflicts of Interest: Author Xu Wang was employed by the company CHN Energy Shendong Clal Group Co., Ltd. The remaining authors declare that the research was conducted in the absence of any commercial or financial relationships that could be construed as a potential conflict of interest.

References

1. Ouedraogo, I.; Defourny, P.; Vanclooster, M. Mapping the groundwater vulnerability for pollution at the pan African scale. *Sci. Total Environ.* **2016**, *544*, 939–953. [CrossRef] [PubMed]
2. Kazakis, N.; Voudouris, K.S. Groundwater vulnerability and pollution risk assessment of porous aquifers to nitrate: Modifying the DRASTIC method using quantitative parameters. *J. Hydrol.* **2015**, *525*, 13–25. [CrossRef]
3. Fan, L.M. On the Water–Preserved Mining. *J. China Coal Soc.* **2017**, *42*, 27–35. (In Chinese) [CrossRef]
4. Fan, L.M. Some Scientific Subjects in Water-Preserved Coal Mining. *J. China Coal Soc.* **2019**, *44*, 667–674. (In Chinese) [CrossRef]
5. Ma, X.D.; Fan, L.M.; Zhang, X.T.; Zhang, H.Q.; Zhang, Y.F.; Shen, T. Driving force analysis for water and wetlands evolution at Yushenfu mining area. *J. China Coal Soc.* **2015**, *40*, 1126–1133. (In Chinese) [CrossRef]
6. Fan, L.M.; Li, T.; Xiang, M.X.; He, W.Z.; Wu, B.Y.; Peng, J.; Li, Y.H.; Li, C.; Zheng, M.M.; Chen, J.P.; et al. Effect of Coal Mining on Springs in Yushenfu Mining Area of China. *Geofluids* **2018**, *3*, 3564360. [CrossRef]
7. Fan, L.; Ma, X. A review on investigation of water-preserved coal mining in western China. *Int. J. Coal Sci. Technol.* **2018**, *5*, 411–416. [CrossRef]
8. Sun, K.; Fan, L.M.; Xia, Y.C.; Li, C.; Chen, J.P.; Wu, B.Y.; Peng, J. Research on carrying capacity of geological environment based on the concept of coal mining under water-containing. *J. China Coal Soc.* **2019**, *44*, 830–839. (In Chinese) [CrossRef]
9. Ma, X.D.; Huang, J.T.; Li, J.X.; Ning, S.X. Groundwater level threshold under the constrain of ecology security in mining area. *J. China Coal Soc.* **2019**, *44*, 675–680. (In Chinese) [CrossRef]
10. Fan, L.; Ma, L.; Yu, Y.; Wang, S.; Xu, Y. Water-conserving mining influencing factors identification and weight determination in northwest China. *Int. J. Coal Sci. Technol.* **2019**, *6*, 95–101. [CrossRef]
11. Wu, B.Y.; Bian, H.Y.; Peng, J.; Xiang, M.X.; Li, W.L. Hydrochemical Characteristics Analysis of Groundwater in the Western Eco-Enviroment Fragility Area. *Yellow River* **2019**, *41*, 65–69. (In Chinese) [CrossRef]
12. Hussain, R.; Wei, C.; Luo, K. Hydrogeochemical characteristics, source identification and health risks of surface water and groundwater in mining and non-mining areas of Handan, China. *Environ. Earth Sci.* **2019**, *78*, 402. [CrossRef]
13. Saha, P.; Paul, B. Groundwater quality assessment in an industrial hotspot through interdisciplinary techniques. *Environ. Monit. Assess.* **2019**, *191*, 326. [CrossRef] [PubMed]

14. Cloutier, V.; Lefebvre, R.; Therrien, R.; Savard, M.M. Multivariate statistical analysis of geochemical data as indicative of the hydrogeochemical evolution of groundwater in a sedimentary rock aquifer system. *J. Hydrol.* **2008**, *353*, 294–313. [CrossRef]
15. Liu, J.; Hao, Y.; Gao, Z.; Wang, M.; Liu, M.; Wang, Z.; Wang, S. Determining the factors controlling the chemical composition of groundwater using multivariate statistics and geochemical methods in the Xiqu coal mine, North China. *Environ. Earth Sci.* **2019**, *78*, 364. [CrossRef]
16. Li, X.H.; Wang, R.; Li, J.F. Study on hydrochemical characteristics and formation mechanism of shallow groundwater in eastern Songnen Plain. *J. Groundw. Sci. Eng.* **2018**, *6*, 161–170. [CrossRef]
17. Chen, L.W.; Xu, D.Q.; Yin, X.X.; Xie, W.P.; Zeng, W. Analysis on hydrochemistry and its control factors in the concealed coal mining area in North China: A case study of dominant inrush aquifers in Suxian mining area. *J. China Coal Soc.* **2017**, *42*, 996–1004. (In Chinese) [CrossRef]
18. Huan, H.; Lian, X.Y.; Yang, Y.; Jia, Y.F.; Jiang, Y.H. Study on screening method for groundwater quality assessment based on Level Difference Method. *Res. Environ. Sci.* **2020**, *33*, 402–410. (In Chinese) [CrossRef]
19. Li, L.J.; Ma, L.; Zhang, J.; Zhao, Y.N.; Chen, C.Y.; Wang, H.Y.; Yu, Q.H.; Sun, C. Groundwater pollution assessment and pollution factor analysis in Songyuan City, Jilin Province. *Acta Geosci. Sin.* **2014**, *35*, 156–162. (In Chinese) [CrossRef]
20. Wu, C.; Wu, X.; Qian, C.; Zhu, G. Hydrogeochemistry and groundwater quality assessment of high fluoride levels in the Yanchi endorheic region, northwest China. *Appl. Geochem.* **2018**, *98*, 404–417. [CrossRef]
21. Chabukdhara, M.; Gupta, S.K.; Kotecha, Y.; Nema, A.K. Groundwater quality in Ghaziabad district, Uttar Pradesh, India: Multivariate and health risk assessment. *Chemosphere* **2017**, *179*, 167–178. [CrossRef] [PubMed]
22. Boufekane, A.; Saighi, O. Assessing groundwater quality for irrigation using geostatistical method—Case of wadi Nil Plain (North-East Algeria). *Groundw. Sustain. Dev.* **2019**, *8*, 179–186. [CrossRef]
23. Khanoranga; Khalid, S. An assessment of groundwater quality for irrigation and drinking purposes around brick kilns in three districts of Balochistan province, Pakistan, through water quality index and multivariate statistical approaches. *J. Geochem. Explor.* **2019**, *197*, 14–26. [CrossRef]
24. Wang, M.W.; Zhou, T.L.; Dong, J.Q.; Long, J.Y. A novel extension evaluation model of groundwater quality based on con-nection cloud model. *China Environ. Sci.* **2018**, *38*, 3035–3041. (In Chinese) [CrossRef]
25. Dahiya, S.; Singh, B.; Gaur, S.; Garg, V.; Kushwaha, H. Analysis of groundwater quality using fuzzy synthetic evaluation. *J. Hazard. Mater.* **2007**, *147*, 938–946. [CrossRef] [PubMed]
26. Peng, K.N.; Zhang, W.; Zhu, H.H.; Zhou, J.W.; Wan, H.J.; Zhao, J. Hydro-chemical characteristics and quality evaluation of groundwater in Zhaoyuan city. *Saf. Environ. Eng.* **2018**, *25*, 106–111, 138. (In Chinese) [CrossRef]
27. Banerji, S.; Mitra, D. Geographical information system-based groundwater quality index assessment of northern part of Kolkata, India for drinking purpose. *Geocarto Int.* **2019**, *34*, 943–958. [CrossRef]
28. Cui, Y.; Feng, P.; Jin, J.; Liu, L. Water Resources Carrying Capacity Evaluation and Diagnosis Based on Set Pair Analysis and Improved the Entropy Weight Method. *Entropy* **2018**, *20*, 359. [CrossRef] [PubMed]
29. He, H.; Tian, C.; Fang, J.; Jin, G.; An, L. Evaluation of the environmental quality associated with near-surface groundwater characteristics in coal-mining areas based on rough set and uncertainty measure theory. *Desalination Water Treat.* **2019**, *152*, 139–147. [CrossRef]
30. Xia, Y.T.; Gui, H.R.; Zhao, H.H.; Li, J.; Guan, L.S. Temporal variability of hydro-chemical characteristics and water quality assessment of collapse pond in Zhuxianzhuang coal mining area, China. *Fresenius Eviron. Bull.* **2019**, *28*, 402–409.
31. Guo, X.M.; Dong, S.N. Seepage law of bedrock aquifer and water–preserved mining technology in deep coal seam mining. *J. China Coal Soc.* **2019**, *44*, 804–811. (In Chinese) [CrossRef]
32. Sun, K.; Fan, L.; Xia, Y.; Li, C.; Chen, J.; Gao, S.; Wu, B.; Peng, J.; Ji, Y. Impact of coal mining on groundwater of Luohe Formation in Binchang mining area. *Int. J. Coal Sci. Technol.* **2020**, *8*, 88–102. [CrossRef]
33. Li, C.F.; Hu, W.Y.; Liu, Y.F. Vertical hydrogeological characteristics of Luohe aquifer and its significance of water-preserved coal mining. *J. China Coal Soc.* **2019**, *44*, 847–856. (In Chinese) [CrossRef]
34. Liu, G.; Liu, Y.H.; He, T.Y.; Zeng, Y.; Huang, Q.Q. Study on water chemistry and water quality evaluation of groundwater in a landfill. *Environ. Sci. Technol.* **2019**, *42*, 221–228. (In Chinese) [CrossRef]

water

MDPI

Article

The Hydrogeochemical Processes of Groundwater in the Bieletan Area, the Western Potash Production Region in China

Rui Duan [1,2], Liang Chang [1,2], Xiaofan Gu [1,2,*], Xiaodeng Li [1,2], Xiangzhi You [1,2], Qunhui Zhang [1,2] and Qian Wang [1,2]

1 Xi'an Center of China Geological Survey, Xi'an 710119, China; chdrui@163.com (R.D.)
2 Key Laboratory for Groundwater and Ecology in Arid and Semi-Arid Areas, China Geological Survey, Xi'an 710119, China
* Correspondence: uiaoan@126.com

Abstract: The hydrogeochemical research of groundwater in the Bieletan area, China's largest potash producing zone, used a variety of methods, including multivariate analysis, saturation index, and hydrogeochemical modeling. Water samples were collected and analyzed for physicochemical parameters, along with soluble ions from soil cores. The results showed that total dissolved solids (TDS) of groundwater exceeded 300 g/L, with the main hydrochemical characteristics being Cl-Mg type and Cl-Na type. Groundwater is recharged by lake water and canal water, with evaporation being the main factor affecting water chemistry. Hydrogeochemical modeling analyzed the processes occurring from these two different recharge sources: mineral precipitation mainly occurred with lake water recharge, while mineral dissolution mainly occurred with canal water recharge. Regarding potash dissolution, canal water and lake water recharge resulted in 8.860 mmol/L of polyhalite dissolution and 0.278 mmol/L of carnallite dissolution, respectively. This study highlights the complex hydrogeochemical processes controlling groundwater in the potash-rich Bieletan area, providing insights for water resource management and potash mining.

Keywords: hydrochemical characteristics; geochemical modeling; Qarhan Salt Lake; PHREEQC

Citation: Duan, R.; Chang, L.; Gu, X.; Li, X.; You, X.; Zhang, Q.; Wang, Q. The Hydrogeochemical Processes of Groundwater in the Bieletan Area, the Western Potash Production Region in China. *Water* **2024**, *16*, 1833. https://doi.org/10.3390/w16131833

Academic Editor: Cesar Andrade

Received: 27 May 2024
Revised: 17 June 2024
Accepted: 26 June 2024
Published: 27 June 2024

1. Introduction

Potash is considered a strategic resource because potassium is an essential nutrient for plants, as well as for animals and humans [1–3]. Potash resources are abundant worldwide, but both resources and production are primarily concentrated in a few countries such as Canada [4], Russia [5], and Belarus [6]. Due to limited land availability, China's agricultural production heavily relies on fertilizers, especially potash, to increase crop yields [7,8]. Most potash mines around the world use traditional underground mining methods to extract solid potash [9,10]. Solution mining is another method that involves extracting potash-bearing solutions from groundwater. The extracted saturated potash solution is then transferred to evaporation ponds, where the salt and potash are concentrated through evaporation. This method of potash extraction is used in countries such as Chile [11,12], the United States [13], and China [14].

Qarhan Salt Lake is the largest salt lake in China. The Bieletan area, situated in the western part of Qarhan Salt Lake, is the largest potash production region in Qarhan Salt Lake and China. The potash-bearing layers in the Bieletan area are thin and lenticular, with multiple sedimentary strata that are generally thin and discontinuous. Therefore, solution mining is used to extract potash from these layers. Understanding the hydrogeochemical processes occurring during groundwater flow is crucial for the development and utilization of potash [15]. Xiao et al. used isotopic and numerical simulation methods to study the origin of groundwater in the Qarhan Salt Lake area, finding that the contribution of modern groundwater to the salt lake area is minimal [16]. Li et al. used PHREEQC to study the factors affecting the dissolution of solid potash minerals in the Qarhan Salt Lake area [17]. However, previous studies have primarily concentrated on the groundwater origin and potash development in the study area, paying less attention to the hydrogeochemical processes of groundwater during potash extraction. Understanding these processes is crucial for comprehending the hydrogeochemical mechanisms of groundwater.

A suite of hydrogeochemical techniques, including multivariate statistics, saturation indices, and hydrogeochemical modeling, is extensively employed to investigate groundwater sources, influencing factors, and hydrogeochemical processes [18–22]. De Caritat et al. used multivariate statistical analysis and saturation indices to study the factors influencing groundwater chemistry in the arid regions of Saudi Arabia [23]. El Alfy et al. investigated the groundwater sources in the Lake Woods region of Australia using saturation indices and hydrogeochemical modeling, concluding that the weathering of potassium silicates in the main aquifer is the source of potassium enrichment [24]. Parisi et al. investigated the spatiotemporal distribution of groundwater chemistry in the arid and semi-arid regions of southern Italy using multivariate statistical analysis and hydrogeochemical methods [25]. These studies demonstrate that the combination of these methods can effectively analyze hydrogeochemical processes in groundwater.

This study combines multivariate statistics, saturation indices, and hydrogeochemical modeling methods to analyze the hydrogeochemical processes of groundwater in the Bieletan area by examining soluble salts in soil and groundwater chemical indicators. The objectives of this study are to (1) examine the hydrogeochemical characteristics of groundwater in the Bieletan area; (2) analyze the key influencing factors of groundwater; and (3) investigate the primary hydrogeochemical processes occurring during groundwater recharge.

2. Study Area

2.1. Location and Climate

The Qaidam Basin is a large, closed basin located in the northwest of China. Surrounded by mountains exceeding 4000 m in height, the basin's core plain has an average elevation of around 2800 m. This central plain is covered by dry salt flats and numerous shallow saline lakes (Figure 1). It contains roughly 27 salt lakes [26]. Eight rivers converge in the Qarhan Salt Lake area, which includes four major lakes from west to east: Senie Lake, Dabuxun Lake, South Huobuxun Lake, and North Huobuxun Lake [27]. These lakes are all terminal lakes. The Bieletan area, where Senie Lake is located, serves as the primary production area for potassium and lithium salts in the Qarhan Salt Lake area. The Bieletan area, covering approximately 950 km^2, extends between east longitudes 94°00′00″–94°55′00″ and north latitudes 36°45′00″–37°18′00″. Situated at elevations ranging from 2678 to 2681 m, it features relatively flat terrain.

The mountain ranges in the southwest of the Qaidam Basin block warm and wet air currents from the Indian Ocean, creating a climate dominated by cold air masses from Siberia, characterized as a dry continental climate. The Bieletan area has a dry environment with low relative humidity, low moisture content, strong atmospheric transparency, long daylight hours, intense solar radiation, relatively high temperatures, and a short frost-free season. The Bieletan area has an average annual temperature of 5.3 °C, with 24.1 mm of precipitation and 3500 mm of evaporation. The average relative humidity is 26%, and the average wind speed is 4.3 m/s, primarily from the northwest and southwest directions.

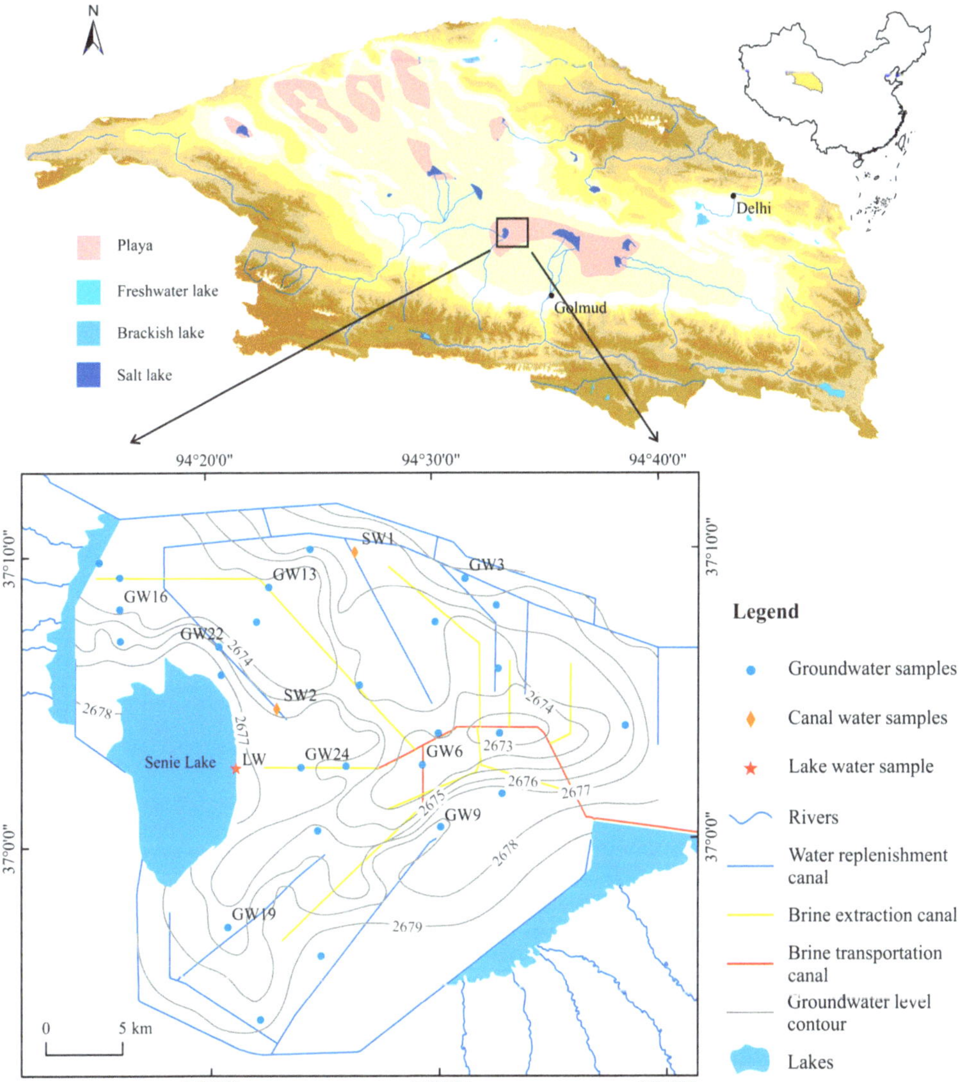

Figure 1. Location of the study area, and water sampling sites.

2.2. Geological and Hydrogeological Settings

The Qaidam Basin, situated within a complex network of compressive structures in the northeastern Qinghai-Tibet Plateau, represents the largest topographic depression on the plateau [28]. This geological environment has fostered the formation of extensive halite deposits intertwined with clastic sediment [29]. The Bieletan region, entirely covered in quaternary sediments, has sediment thicknesses of up to 2700 m. The upper portion has salt-bearing strata up to 70 m thick, mostly formed of clastic, halite, and potash layers (Figure 2). The postal layer is predominantly located in the uppermost layers. Shallow groundwater is primarily found in the upper quaternary porous sediments, which are also the focus of this study. Senie Lake and artificial water replenishment canals are the main sources of groundwater recharge. Groundwater primarily flows from the surrounding areas toward the central area, with evaporation and human extraction serving as the main discharge processes. Extracted groundwater is transported through brine transportation canals to salt farms beyond the study area. It is evident that there are two groundwater depression funnels in the research region, associated with groundwater extraction.

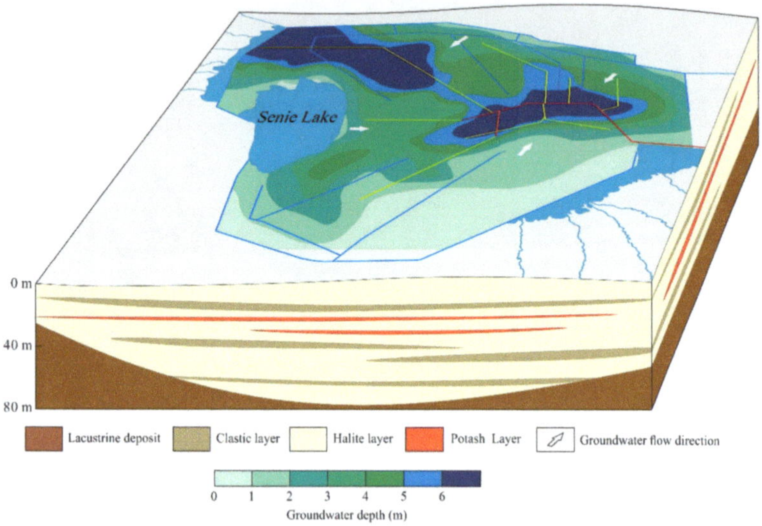

Figure 2. Geological model and groundwater depth of the study area.

3. Material and Methods

3.1. Sampling Collection

This study collected 26 groundwater samples in September 2023. Additionally, 2 replenishment canal samples and 1 Senie Lake water sample were collected to investigate their impact on groundwater recharge. The water quality data included measurements of pH, concentrations of specific cations (K^+, Na^+, Ca^{2+}, Mg^{2+}), anions (SO_4^{2-}, Cl^-, CO_3^{2-}, HCO_3^-), TDS, Li^+, and water density. In October 2023, a total of 21 soluble salt samples were collected from three boreholes (GW6, GW16, GW19). Each borehole yielded 7 soluble salt samples from depths of 0.5 m, 1 m, 2 m, 3 m, 5 m, 7 m, and 10 m below the surface, respectively. Dissolved ions in these samples were detected by analyzing the filtrate, which was filtered after the soil sample was mixed with no-CO_2 water by a ratio of 1 (soil) to 5 (water) and vibrated for 3 min. The detection of all samples included measurements of pH, concentrations of specific cations (K^+, Na^+, Ca^{2+}, Mg^{2+}), anions (SO_4^{2-}, Cl^-, CO_3^{2-}, HCO_3^-), and Li^+. All samples were analyzed by the Xi'an Center of the China Geological Survey. The concentrations of K^+, Na^+, Ca^{2+}, Mg^{2+}, and Li^+ were analyzed using an Inductively Coupled Plasma Optical Emission Spectrometer (ICP-OES) (PerkinElmer Inc., Shelton, WA, USA). SO_4^{2-} was also measured using ICP-OES. Cl^- was determined by silver nitrate titration, while CO_3^{2-} and HCO_3^- were analyzed using titration methods. TDS were measured by drying and weighing. To ensure the validity and accuracy of the data, duplicate analyses were performed for each sample, with the replicates showing an error within ±5%. The accuracy and precision of the analyses were verified by analyzing certified reference materials (CRMs) under the same conditions, thereby guaranteeing the reliability of the data. Additionally, blank samples were included in the analysis to monitor and control potential contamination during sample handling and processing. The ion charge balance error percentage (%CBE) is used to calculate the ion balance error of water samples. The ion charge balance error percentage is calculated according to Formula (1):

$$\%CEB = \frac{(|TC - TA|)}{(TC - TA)} \times 100\% \quad (1)$$

where TC and TA represent the total concentrations of cations and anions, respectively. The units for both cations and anions are milliequivalents per liter (meq/L). The ion balance error for all water samples was less than 1.69%, and the ion balance error for soluble salts was less than 9.65%. An ion balance error of less than 10% is considered to provide reliable chemical analysis results for hydrogeochemical investigations [30].

3.2. Saturation Index

The Saturation Index (SI) is crucial for evaluating the equilibrium and reactivity between minerals and groundwater [31]. The SI of minerals in water was calculated according to the following Formula (2):

$$SI = \log(\frac{IAP}{K_{sp}}) \quad (2)$$

where, IAP is the ion activity product of the dissolving mineral, and K_{sp} is the solubility product constant of the mineral. PHREEQC 3.7.3 was used to calculate the SI of possible mineral phases [32]. A positive SI indicates supersaturation (precipitation potential), a negative SI indicates undersaturation (dissolution potential), and an SI of zero indicates equilibrium.

3.3. Hydrogeochemical Modeling

Geochemical simulation is an essential tool for studying the evolution of water chemistry, and inverse modeling is one of the key modeling methods for hydrogeochemical research [33,34]. In this study, PHREEQC is employed for geochemical simulation. Determining the initial and final solutions along the groundwater flow path is a necessary step in modeling. Inverse modeling calculates the moles of minerals dissolved in or precipitated from the solution based on mass balance rules to explain the compositional differences between the initial and final endpoint solutions [35]. The mass balance of the conceptual models can be expressed as the following Formula (3):

$$\sum_{j-1}^{n} a_{ij}x_j = b_i \quad (3)$$

where a_{ij} represents the stoichiometric coefficient of element i in mineral j, x_j denotes the molar quantity of minerals or gases that have either dissolved, precipitated, or degassed, and b_i signifies the change in concentration of element i in the final water solution relative to the initial water solution.

4. Results

4.1. Characteristics of Major Indices

The statistical summary of chemical analysis results for groundwater (GW), replenishment canal water samples (SW), and lake water (LW) is presented in Table 1. The pH values of GW range from 7.76 to 8.71, with an average of 8.41. The average pH for SW is 8.85, and for LW it is 8.75. Therefore, all water bodies in the study area exhibit weak alkalinity. The average concentrations of K^+, Na^+, and Ga^{2+} in GW are 9.98 g/L, 47.95 g/L, and 0.73 g/L, respectively, whereas in SW they are 0.72 g/L, 5.62 g/L, and 0.11 g/L, respectively. For LW, the concentrations of K^+, Na^+, and Ga^{2+} are 2.52 g/L, 4.70 g/L, and 0.05 g/L, respectively. This suggests that the average concentrations of K^+, Na^+, and Ga^{2+} are higher in GW compared to SW and LW. The average concentrations of Mg^{2+} and ${CO_3}^{2-}$ in GW are 47.94 g/L and 0.08 g/L, respectively, while in SW they are 52.02 g/L and 0.37 g/L, and in LW they are 106.56 g/L and 0.78 g/L, respectively. This indicates that the average concentrations of Mg^{2+} and $CO_3{}^{2-}$ in GW are lower compared to SW and LW. The average concentrations of $SO_4{}^{2-}$, Cl^-, and total dissolved solids (TDS) in GW are 11.89 g/L, 213.24 g/L, and 332.03 g/L, respectively, while in SW they are 8.62 g/L, 152.27 g/L, and 219.73 g/L, and in LW they are 16.61 g/L, 297.79 g/L, and 428.46 g/L, respectively. The average concentrations of $SO_4{}^{2-}$, Cl^-, and TDS are higher in GW compared to SW but lower compared to LW. The coefficient of variability (CV) represents the ratio of the standard deviation (SD) to the mean, with a higher CV indicating greater variability. The CV values for the chemical components of GW range from 1% to 149%, with Ca^{2+} and $HCO_3{}^-$ having CV values greater than 100% and Cl^-, pH, TDS, and density having CV values less than 10%.

Table 1. Statistical analysis of hydrogeochemical parameters of water samples (in g/L except pH and density in g/cm^3).

Type	ID	pH	K^+	Na^+	Ca^{2+}	Mg^{2+}	$SO_4{}^{2-}$	Cl^-	TDS	$CO_3{}^{2-}$	$HCO_3{}^-$	Li^+	Density
GW	GW1	8.49	4.05	60.02	0	37.81	6.24	200.54	309.96	0.15	0.18	0.09	1.21
	GW2	8.23	32.81	66.24	1.87	28.39	3.13	214.80	347.67	0	0.32	0.11	1.23
	GW3	6.76	1.85	106.12	1.62	8.32	3.43	190.74	312.23	0	0.48	0.02	1.20
	GW4	8.56	20.97	31.23	0.44	58.08	6.74	229.06	347.22	0.09	0.70	0.17	1.24
	GW5	8.61	6.79	62.04	0.66	37.12	6.84	203.21	317.09	0.04	0.54	0.09	1.22
	GW6	8.65	5.61	56.52	0.54	39.42	8.28	202.32	313.16	0.05	0.52	0.10	1.22
	GW7	8.49	15.30	20.50	0.31	69.06	7.06	237.97	350.73	0.05	0.52	0.11	1.24
	GW8	8.57	6.70	40.53	0.34	53.01	10.65	213.02	324.77	0.04	0.77	0.09	1.23
	GW9	8.67	1.27	26.38	0.29	66.03	9.47	224.60	328.76	0.13	0.82	0.12	1.23
	GW10	8.52	15.80	19.78	0.40	68.82	5.94	238.87	350.54	0.05	0.78	0.27	1.25
	GW11	8.55	12.01	25.62	0.28	64.00	10.33	228.17	341.97	0	1.63	0.51	1.24
	GW12	8.63	8.39	42.20	0.16	51.48	19.85	206.78	329.87	0.08	0.90	0.22	1.23
	GW13	8.61	9.78	20.78	0.18	67.81	12.04	229.06	341.37	0.06	1.31	0.48	1.24
	GW14	8.30	9.47	64.11	0.17	37.09	33.19	191.05	335.32	0	0.10	0.25	1.23
	GW15	8.42	9.62	60.86	0.19	39.81	33.05	194.64	338.44	0.08	0.02	0.26	1.23
	GW16	8.44	7.78	52.24	0.59	39.50	10.48	195.09	305.91	0	0	0.21	1.21
	GW17	8.40	6.77	64.82	0.39	34.48	16.61	195.54	318.83	0.07	0	0.21	1.22
	GW18	8.47	11.48	34.67	0.37	57.01	10.77	218.86	333.31	0.19	0	0.14	1.23
	GW19	8.52	6.69	21.71	0.35	70.09	9.43	234.11	342.53	0.21	0.06	0.14	1.24
	GW20	8.61	6.62	10.82	0.13	84.04	12.27	254.74	368.80	0.32	0.09	0.18	1.26
	GW21	8.42	2.40	101.94	0.83	12.86	5.95	193.30	317.32	0.01	0	0.02	1.21
	GW22	7.85	4.55	78.08	1.31	25.79	5.04	197.78	312.65	0	0.06	0.07	1.21
	GW23	8.54	12.02	74.89	0.17	33.69	41.30	188.36	350.75	0.16	0	0.32	1.24
	GW24	8.56	11.59	40.64	0.28	51.59	14.04	209.89	328.22	0.21	0	0.18	1.23
	GW25	8.71	12.34	36.60	0.63	52.47	6.69	216.64	325.96	0.11	0.18	0.05	1.23
	GW26	8.17	14.01	27.29	5.58	58.56	0.47	232.63	339.45	0	0.52	0.31	1.23
	Min	6.76	1.27	10.82	0	8.32	0.47	188.36	305.91	0	0	0.02	1.20
	Max	8.71	32.81	106.12	5.58	84.04	41.30	254.74	368.80	0.32	1.63	0.51	1.26
	Mean	8.41	9.87	47.95	0.70	47.94	11.89	213.14	332.03	0.08	0.40	0.18	1.23
	SD	0.38	6.58	25.21	1.09	18.58	9.84	18.31	15.88	0.08	0.44	0.12	0.01
	CV(%)	5	67	53	149	39	83	9	5	97	109	68	1
SW	SW1	8.88	0.54	3.63	0.06	47.25	7.55	137.26	196.89	0.28	0.01	0.11	1.15
	SW2	8.81	0.90	7.62	0.17	56.78	9.70	167.28	242.58	0.47	0.09	0.12	1.18
	mean	8.85	0.72	5.62	0.11	52.02	8.62	152.27	219.73	0.37	0.05	0.11	1.16
LW	LW	8.75	2.52	4.70	0.05	106.5	16.61	297.79	428.46	0.78	0.11	0.22	1.31

The statistical results of core soluble salts are presented in Table 2, showing the average contents of K^+, Na^+, Ca^{2+}, and Mg^{2+} in borehole cores to be 3.31, 150.01, 5.40, and 8.10 g/kg, respectively. The average contents of $SO_4{}^{2-}$, Cl^-, $CO_3{}^{2-}$, and $HCO_3{}^-$ are 10.32, 302.70, 0.08, and 0.15 g/kg, respectively. The average content of Li^+ is 0.13 g/kg. The pH average of core leachates is 8.28, ranging from 8.03 to 8.53, indicating alkaline properties across the cores, which is one of the reasons for the alkalinity of groundwater in the study area. Na^+

and Cl^- have the largest amounts in soluble salts. K^+ and Li^+ have the largest coefficients of variation, indicating an unequal distribution throughout the strata.

Table 2. Statistical analysis of physiochemical parameters of soluble ions (in g/kg except pH).

Indices	Soluble Ions				
	Min	Max	Mean	SD	CV (%)
pH	8.03	8.53	8.28	0.16	1.93
K^+	0.36	16.62	3.31	3.72	112.19
Na^+	2.53	313.40	150.01	117.47	78.31
Ca^{2+}	0.37	11.81	5.40	3.20	59.25
Mg^{2+}	0.40	20.59	8.10	6.27	77.46
$SO_4{}^{2-}$	1.96	26.80	10.32	5.74	55.57
Cl^-	31.40	570.00	302.70	198.48	65.57
$CO_3{}^{2-}$	0.00	0.28	0.08	0.06	77.95
$HCO_3{}^-$	0.05	0.32	0.15	0.09	59.78
Li^+	0.00	0.42	0.13	0.14	110.67

4.2. Correlation Analysis

Correlation analysis is a valuable tool in groundwater chemistry studies for identifying relationships between different chemical parameters, understanding the underlying geochemical processes [36]. The findings of a Spearman correlation study performed to evaluate the possible link between various ions of the soluble salts are shown in Table 3. The analysis examined the correlation coefficients among 21 sets of 10 ions. The substantial correlation coefficients between Na^+ and Cl^-, Ca^{2+} and $SO_4{}^{2-}$, Mg^{2+} and $CO_3{}^{2-}$, Mg^{2+} and $HCO_3{}^-$, with values of 0.99, 0.78, 0.79, and 0.86, respectively, indicate widespread distribution of rock salt minerals, gypsum, and dolomite in the region. Negative correlation coefficients between K^+, Mg^{2+}, $CO_3{}^{2-}$, $HCO_3{}^-$, and Na^+, Cl^- suggest that the formation of rock salt in the area inhibits the formation of other minerals.

Table 3. Correlation matrix of the main ions in soluble salt.

	K^+	Na^+	Ca^{2+}	Mg^{2+}	Cl^-	$SO_4{}^{2-}$	$CO_3{}^{2-}$	$HCO_3{}^-$	Li^+	pH
K^+	1	−0.42	0.04	0.82 **	−0.4	0.45 *	0.4	0.62 **	0.59 **	0.12
Na^+		1	0.58 **	−0.76 **	0.99 **	0.19	−0.82 **	−0.87 **	−0.89 **	−0.71 **
Ca^{2+}			1	−0.22	0.61 **	0.78 **	−0.47 *	−0.35	−0.57 **	−0.47 *
Mg^{2+}				1	−0.73 **	0.19	0.79 **	0.86 **	0.87 **	0.57 **
Cl^-					1	0.21	−0.79 **	−0.85 **	−0.88 **	−0.71 **
$SO_4{}^{2-}$						1	−0.15	0.01	−0.11	−0.2
$CO_3{}^{2-}$							1	0.80 **	0.85 **	0.73 **
HCO^{3-}								1	0.81 **	0.63 **
Li^+									1	0.71 **
pH										1

Note: ** indicates a significant correlation at the 0.01 level; * indicates a significant correlation at the 0.05.

Table 4 analyzes the correlation coefficients of major ions in groundwater. TDS show a positive correlation with most indicators, notably significant positive correlations with K^+, Mg^{2+}, Cl^-, and Li^+. Conversely, TDS exhibit a significant negative correlation with Na^+ and Ca^{2+}. The negative correlations of Na^+ and Ca^{2+} with other indicators suggest that they are subjected to inhibition by other ions. Specifically, the correlation coefficients between Na^+ and Mg^{2+}, and between Na^+ and Cl^-, are highest at −0.98 and −0.91, respectively. This indicates that the dissolution of Na^+ is most strongly inhibited by Mg^{2+} and Cl^-. The highest correlation coefficient between Ca^{2+} and $SO_4{}^{2-}$ is −0.87, suggesting that Ca^{2+} is most strongly inhibited by $SO_4{}^{2-}$. Additionally, Mg^{2+} and Cl^- demonstrate a high degree of correlation, suggesting they share a common source, likely due to the dissolution of halite and magnesium chloride. The correlation coefficients of Mg^{2+}, $CO_3{}^-$ and $HCO_3{}^-$

with pH are 0.48, 0.53 and 0.44, respectively, indicating a positive correlation, whereas those of Na^+ and Ca^{2+} with pH are −0.49 and −0.44, indicating a negative correlation.

Table 4. Correlation matrix of GW.

	K^+	Na^+	Ca^{2+}	Mg^{2+}	Cl^-	SO_4^{2-}	CO_3^{2-}	HCO_3^-	Li^+	TDS	pH
K^+	1	−0.32	−0.07	0.28	0.38	−0.01	0.01	0.16	0.48 *	0.63 **	−0.01
Na^+		1	0.38	−0.98 **	−0.91 **	−0.06	−0.44 *	−0.47 *	−0.37	−0.55 **	−0.49 *
Ca^{2+}			1	−0.42 *	−0.12	−0.87 **	−0.52 **	−0.08	−0.55 **	−0.48 *	−0.44 *
Mg^{2+}				1	0.89 **	0.11	0.48 *	0.43 *	0.37	0.57 **	0.48 *
Cl^-					1	−0.21	0.33	0.49 **	0.19	0.59 **	0.37
SO_4^{2-}						1	0.44 *	−0.22	0.51 **	0.19	0.28
CO_3^{2-}							1	−0.13	0.18	0.35	0.53 **
HCO_3^-								1	0.13	0.21	0.44 *
Li^+									1	0.54 **	0.02
TDS										1	0.11
pH											1

Note: ** indicates a significant correlation at the 0.01 level; * indicates a significant correlation at the 0.05.

4.3. Saturation Index (SI)

The SI is crucial for studying groundwater chemistry processes, serving as a key parameter used to assess the equilibrium state between groundwater and minerals [37,38]. As illustrated in Figure 3, the distribution of saturation indices in the water samples from the survey area reveals that all minerals in the LW and SW water samples have SI values less than 0, indicating undersaturation. In most water samples from the GW body, the saturation indices of dolomite minerals are greater than 0, indicating saturation, while calcite and halite in the GW body are distributed near the SI = 0 line, indicating equilibrium. The saturation indices of the remaining minerals, polyhalite, bischofite, carnallite, and sylvite, are less than 0, indicating an undersaturated state. During the replenishment of GW by SW water, the saturation indices of most minerals increase, suggesting mineral dissolution in the aquifer. Similarly, during the replenishment of GW by LW water, the saturation indices of polyhalite, calcite, and dolomite significantly increase, indicating mineral dissolution during water transport. Conversely, the saturation indices of bischofite and carnallite minerals decrease, suggesting mineral precipitation during water transport.

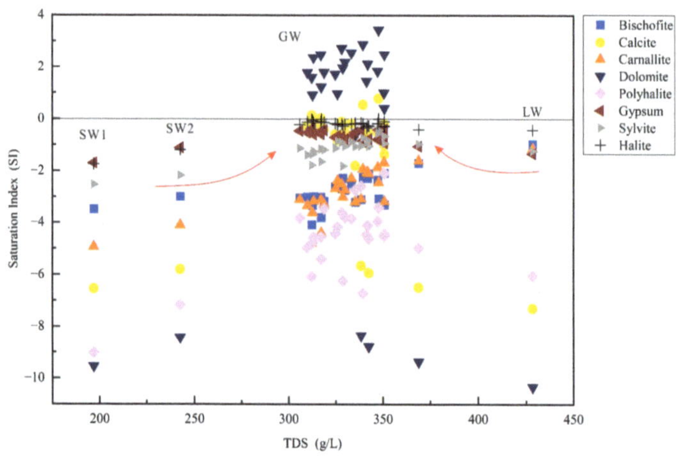

Figure 3. Relationship between Mineral Saturation Index and Total Dissolved Solids (TDS), with red arrows indicating the evolution direction of the mineral saturation index.

4.4. Hydrochemical Processes: Inverse Modeling

Reverse hydrogeochemical modeling employs mass balance models to determine the amount of mineral dissolution or precipitation occurring between two different points in groundwater flow processes [39]. In this study, PHREEQC is utilized for reverse water-rock interaction modeling, quantitatively representing water–rock interactions and validating hydrogeochemical processes occurring during groundwater recharge based on the groundwater flow field, in order to study the hydrogeochemical processes occurring along the groundwater recharge routes from lake water and canal water. The reaction simulation paths selected were LW to GW24 and SW1 to GW13, as illustrated in Figure 4. Figure 4a,b show schematic diagrams of groundwater flow patterns between points LW and GW24, and points SW1 and GW13, respectively. Figure 4c,d compare the hydrochemical characteristics between points LW and GW24, and points SW1 and GW13, displaying variations in major indices. From LW to GW24, it can be observed that the concentrations of Na^+, K^+, and Ca^{2+} increase, while the values of the other indices decrease. From SW1 to GW13, the concentration of CO_3^{2-} decreases, while the other indices increase. This indicates significant differences in the hydrochemical processes occurring along the paths from LW to GW24 and from SW1 to GW13.

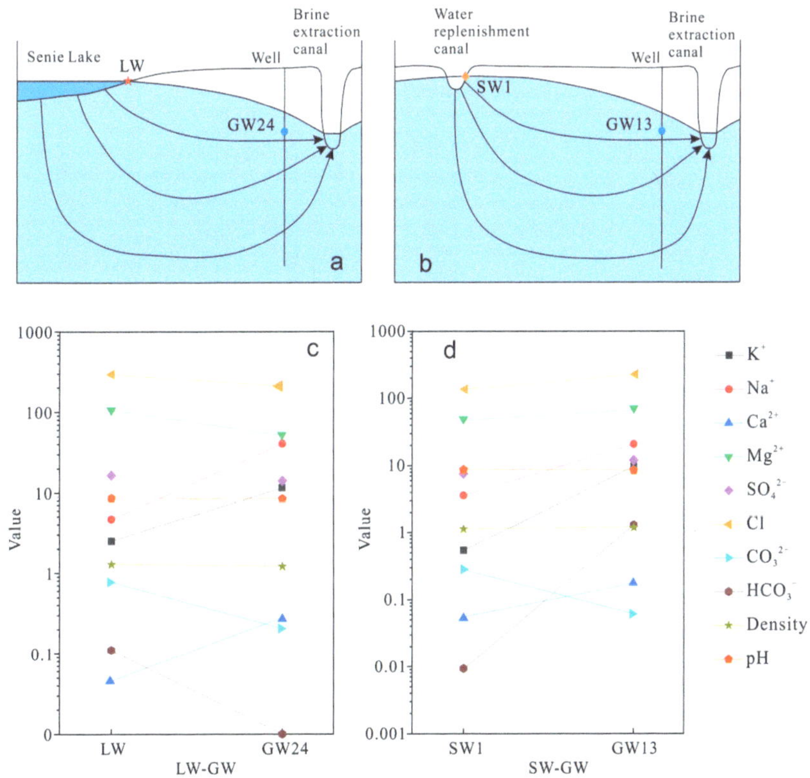

Figure 4. Schematic diagram of groundwater flow patterns between points (**a**) LW and GW24, and (**b**) SW1 and GW13, black arrows represent groundwater flowlines. Comparison of hydrochemical characteristics between points (**c**) LW and GW24, and (**d**) SW1 and GW13 (values in g/L except pH and density in g/cm^3).

The transfer of mineral components in groundwater along the flow paths is reported in Table 5. For LW to GW24, along the flow path, cation exchange leads to the entry of 18.72 mmol/L of Ca^{2+} into groundwater and 37.45 mmol/L of Na^+ out of groundwater. The primary water–rock interactions causing changes in groundwater chemistry include the precipitation of bischofite, carnallite, and gypsum, with precipitation rates of 15.980 mmol/L, 17.860 mmol/L, and 36.490 mmol/L, respectively. Additionally, polyhalite and halite dissolve, with dissolution rates of 8.860 mmol/L and 38.100 mmol/L, respectively.

Table 5. Mineral phase transfer for different pathways calculated by PHREEQC.

Phase	ρ/(mmol/L)		
	Chemical Formula	**LW-GW**	**SW-GW**
CaX_2	CaX_2	18.720	2.671
NaX	NaX	−37.450	−5.342
Bischofite	$MgCl_2 \cdot 6H_2O$	−15.980	/
Calcite	$CaCO_3$	/	−5.626
Carnallite	$KMgCl_3 \cdot 6H_2O$	−17.860	0.278
Dolomite	$CaMg(CO_3)_2$	/	2.820
Polyhalite	$K_2MgCa_2(SO_4)_4 \cdot 2H_2O$	8.860	/
Gypsum	$CaSO_4 \cdot 2H_2O$	−36.490	−0.017
Halite	NaCl	38.100	6.346

Note: positive value indicates that the mineral phase has dissolved into groundwater, whereas negative values indicate that mineral phases may precipitate out of groundwater.

For SW1 to GW13, along the flow path, cation exchange results in 2.671 mmol/L of Ca^{2+} entering groundwater and 5.342 mmol/L of Na^+ leaving groundwater. The main water–rock interactions causing changes in groundwater chemistry are the precipitation of calcite and gypsum, with precipitation rates of 5.626 mmol/L and 0.017 mmol/L, respectively. Additionally, carnallite, dolomite, and halite dissolve, with dissolution rates of 0.278 mmol/L, 2.820 mmol/L, and 6.346 mmol/L, respectively.

5. Discussion

5.1. Hydrogeochemistry

The Piper diagram is frequently utilized for analyzing the hydrochemical types of groundwater. As shown in Figure 5, the anions of GW water are primarily distributed in Zones C (Magnesium type) and D (Sodium type), while cations are mainly distributed in Zone G (Chloride type). Therefore, the hydrochemical types of GW are identified as Cl-Mg type and Cl-Na type. SW and LW exhibit anion distributions in Zone C (Magnesium type) and cation distributions in Zone G (Chloride type), indicating that the hydrochemical types of SW and LW are Cl-Mg type. All water bodies belong to the low-Ca high-Cl type, with the lowest Na^+ + K^+ content and the highest Mg^{2+} content in SW and LW waters, suggesting that during the migration process from SW and LW to GW, the concentration of Na^+ + K^+ gradually increases while the concentration of Mg^{2+} decreases. The concentration of Cl^- is highest in all water bodies, indicating that Cl^- is the dominant anion during water migration, and its variation is relatively small.

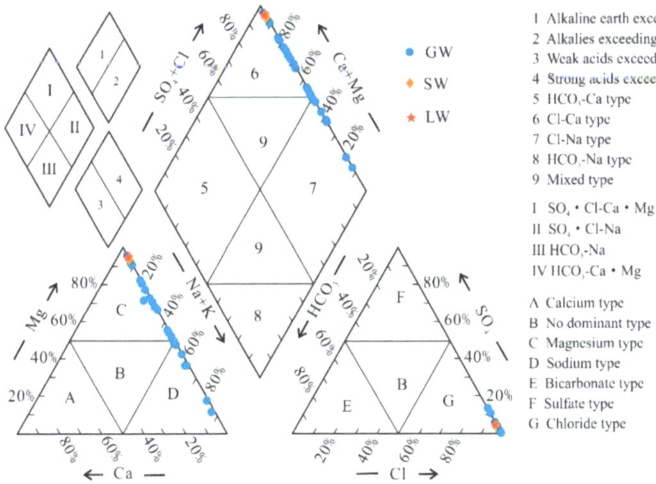

Figure 5. Piper diagram of water samples.

5.2. Water–Rock Interaction

The Gibbs diagram is a widely used tool for understanding the natural formation mechanisms of water bodies [40,41]. According to the Gibbs diagram, there are three primary mechanisms: Evaporation Dominance, Rock Dominance, and Precipitation Dominance. As shown in Figure 6, all the water sample points are clustered in the upper right corner, indicating that the formation mechanism of all the water samples is predominantly influenced by evaporation.

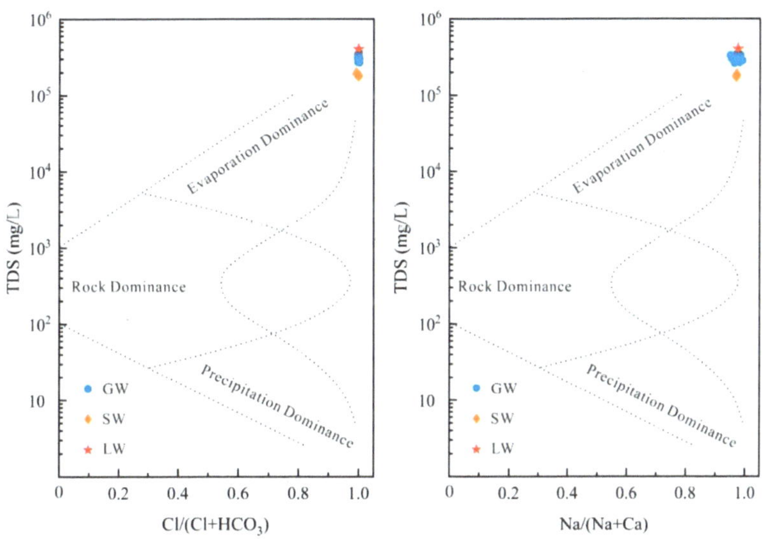

Figure 6. Gibbs diagram of water samples.

To further analyze the factors influencing the formation of water chemistry, scatter plots depicting the relationships between various components were created as shown in Figure 7. Figure 7a,b illustrate the relationships between Cl^- vs. Na^+ and Cl^- vs. Mg^{2+}, respectively. It can be observed that Na^+ is negatively correlated with Cl^-. As the Cl^- concentration increases in the GW water bodies, the Na^+ concentration decreases exponentially, with all water samples falling below the 1:1 line. This indicates that the excess Cl^- may originate from the dissolution of minerals such as carnallite and bischofite. The Na^+ concentrations in LW and SW are lower compared to those in GW, suggesting that as LW and SW recharge the groundwater, the Na^+ concentration rises, indicating the dissolution of halite. Mg^{2+} shows a positive correlation with Cl^-, with its concentration increasing as Cl^- concentration rises. This is attributed to the dissolution of carnallite and bischofite. Figure 7c shows the relationship between Ca^{2+} and $SO_4{}^{2-}$. In GW, Ca^{2+} is negatively correlated with $SO_4{}^{2-}$, with most water samples lying below the 1:1 line. This suggests that the dissolution of gypsum is not the source of Ca^{2+} and $SO_4{}^{2-}$ in the study area. Additionally, there is evidence of Ca^{2+} depletion and $SO_4{}^{2-}$ enrichment, indicating that the dissolution of polyhalite affects the sources of Ca^{2+} and $SO_4{}^{2-}$.

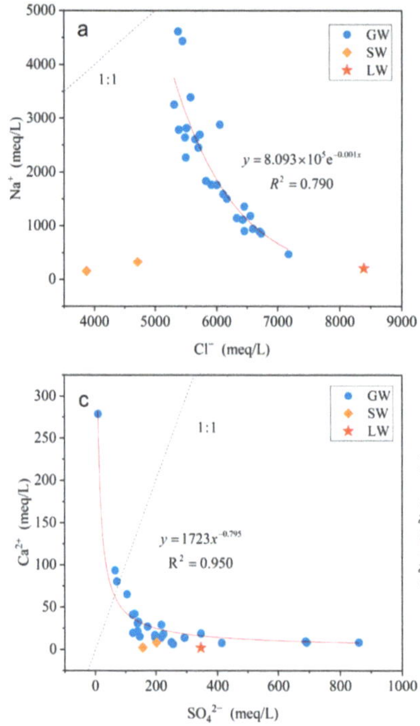

Figure 7. Scatter diagrams of (**a**) Cl^- vs. Na^+, (**b**) Cl^- vs. Mg^{2+}, (**c**) Ca^{2+} vs. $SO_4{}^{2-}$, and (**d**) (Ca^{2+} + Mg^{2+}) vs. ($HCO_3{}^-$ + $SO_4{}^{2-}$) in water samples.

Figure 7d displays the relationship between (Ca^{2+} + Mg^{2+}) and ($HCO_3{}^-$ + $SO_4{}^{2-}$). All water samples are located above the 1:1 line, indicating that the contributions of calcite, dolomite, and gypsum dissolution to the concentrations of (Ca^{2+} + Mg^{2+}) in groundwater are minimal. Instead, the dissolution of other minerals, such as polyhalite, increases the concentrations of Ca^{2+} and Mg^{2+}.

5.3. Cation Exchange Interaction

Cation exchange is another crucial mechanism that governs water chemistry [42]. Chloro-alkaline indices (CAI-I and CAI-II) serve as valuable tools for evaluating the direction of alternating cation adsorption processes. These indices are calculated as follows:

$$\mathrm{CAI-I} = \frac{\mathrm{Cl^-} - (\mathrm{Na^+} + \mathrm{K^+})}{\mathrm{Cl^-}} \tag{4}$$

$$CAI-II = \frac{Cl^- - (Na^+ + K^+)}{HCO_3^- + SO_4^{2-} + CO_3^{2-} + NO_3^-} \tag{5}$$

If both are negative values, it indicates cation exchange as shown in Equation (6) occurring in the groundwater system. Conversely, if both are positive values, it indicates reverse cation exchange as shown in Equation (7).

$$Na_2X + Ca^{2+} \rightarrow 2Na^+ + CaX_2 \tag{6}$$

$$2Na^+ + CaX_2 \rightarrow Ca^{2+} + Na_2X \tag{7}$$

As shown in Figure 8a, all sampling points in the study area lie in the first quadrant, indicating cation exchange as described in Equation (7), where adsorbed Ca^{2+} is replaced by Na^+. This process leads to an increase in Ca^{2+} concentration and a decrease in Na^+ concentration in groundwater. There is an outlier in the Figure 8a with a very high CAI-II value, indicating strong cation exchange at this point, which has led to a significant decrease in Na^+. It is hypothesized that this point may have been more strongly influenced by confined groundwater recharge. Additionally, the relationship between $[(Ca^{2+} + Mg^{2+}) - (HCO_3^- + SO_4^{2-})]$ vs. $[Na^+ + K^+ - Cl^-]$ and is commonly used to study cation exchange. $[(Ca^{2+} + Mg^{2+}) - (HCO_3^- + SO_4^{2-})]$ represents the increase or decrease in Ca^{2+} and Mg^{2+} in the water body, excluding dissolution of gypsum, calcite, and dolomite. $[Na^+ + K^+ - Cl^-]$ represents the increase or decrease in Na^+, excluding the dissolution of halite. If cation exchange is the primary process controlling groundwater mineralization, their relationship should be linear, and the slope should be close to -1. As shown in Figure 8b, there is a good linear relationship between $[(Ca^{2+} + Mg^{2+}) - (HCO_3^- + SO_4^{2-})]$ vs. $[Na^+ + K^+ - Cl^-]$ in the study area, with a linear regression slope of -1.019, which is very close to the theoretical value of -1.0. This indicates the involvement of Na^+, Ca^{2+}, and Mg^{2+} in cation exchange in groundwater.

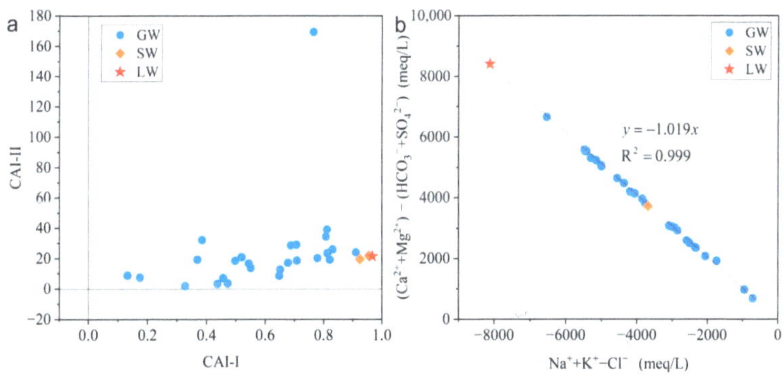

Figure 8. Plots of (**a**) CAI-II versus CAI-I and (**b**) $(Ca^{2+} + Mg^{2+}) - (HCO_3^- + SO_4^{2-})$ versus $Na^+ + K^+ - Cl^-$.

5.4. Hydrogeochemical Processes

Chloro-alkaline indices and reverse hydrogeochemical modeling show that cation exchange has a greater impact on groundwater chemistry during groundwater recharge from lake water than from canal water, with the cation exchange process being stronger in the former. Carnallite precipitates along the flow path of groundwater recharge from lake water, whereas it dissolves along the flow path of groundwater recharge from canal water. Gypsum precipitates along both lake water and canal water flow paths for groundwater recharge, but the precipitation rate is faster in lake water recharge. During the processes of groundwater recharge from both lake water and canal water, halite experiences varying degrees of dissolution, with a faster dissolution rate in lake water, possibly due to the cation exchange processes occurring along the flow paths.

6. Conclusions

This paper investigates groundwater, surface water, and soluble salts in the Bieletan region. Utilizing multivariate techniques, saturation indices, and hydrogeochemical modeling, the study identifies the hydrochemical characteristics of groundwater, the main influencing factors on groundwater, and the hydrogeochemical processes occurring during different recharge processes in the Bieletan area. The following key conclusions are summarized:

1. Groundwater (GW) cation concentrations are ranked as Na^+ > Mg^{2+} > K^+ > Ga^{2+}, while in the case of lake water (LW) and stream water (SW), cation concentrations are Mg^{2+} > Na^+ > K^+ > Ga^{2+}. The anion concentrations in GW are ranked as Cl^- > $SO_4{}^{2-}$ > $HCO_3{}^-$ > $CO_3{}^{2-}$, whereas in LW and SW, the order is Cl^- > $SO_4{}^{2-}$ > $CO_3{}^{2-}$ > $HCO_3{}^-$. The mean total dissolved solids (TDS) values for GW and SW are 332.03 g/L and 219.73 g/L, respectively, while for LW, it is 428.46 g/L. The hydrochemical types of GW are classified as Cl-Mg type and Cl-Na type, whereas LW and SW are categorized as Cl-Mg type.
2. Evaporation is the primary mechanism governing the formation of GW, LW, and SW. Cation exchange and mineral precipitation/dissolution are identified as the primary factors influencing water chemistry. Minerals in LW and SW are generally undersaturated, while in GW, apart from dolomite, calcite, and halite, which are saturated or near-saturated, other minerals remain undersaturated.
3. GW receives recharge from two distinct sources, LW and SW. These two different water bodies undergo distinct hydrogeochemical processes during groundwater recharge. The rates of cation exchange and mineral dissolution are faster in LW, where mineral precipitation predominantly occurs, whereas mineral dissolution is more prevalent in SW.

Author Contributions: Investigation, R.D., X.G. and X.Y.; writing—original draft preparation, R.D. and X.G.; writing—review and editing, L.C.; visualization, X.L., X.Y., Q.Z. and Q.W. All authors have read and agreed to the published version of the manuscript.

Funding: China Strategic Geological Survey Project: Water Resources Survey in the Salt Lake Area of the Qaidam Basin (DD20230301); Qinghai Environmental Geological Survey Bureau Science and Technology Project "Study on Water Balance and Hydrodynamic Mechanism of the Qarhan Salt Lake under Changing Environmental Conditions" (2023-ZK-01); Key Research and Development Program of Shaanxi (2022SF-327).

Data Availability Statement: Dataset available upon request from the authors.

Acknowledgments: We thank QingHai Salt Lake Industry Co., Ltd., for their assistance with the field investigation. We also extend our gratitude to Mounan Ma, Xiaoyong Wang, and Bin Luo for their hard work and support during the field investigation.

Conflicts of Interest: The authors declare no conflicts of interest.

References

1. Song, Y.; Hu, H.; Ye, C.; Zhao, Y.; Ma, L.; Zhang, J.; Yan, Q.; Lin, Y. Evaluation on liquefaction effect of potassium dissolution extraction from low-grade solid potash ore in Qarhan Salt Lake, northern of Tibetan Plateau. *Carbonates Evaporites* **2024**, *39*, 56. [CrossRef]
2. Zoroddu, M.A.; Aaseth, J.; Crisponi, G.; Medici, S.; Peana, M.; Nurchi, V.M. The essential metals for humans: A brief overview. *J. Inorg. Biochem.* **2019**, *195*, 120–129. [CrossRef] [PubMed]
3. Bhateria, R.; Jain, D. Water quality assessment of lake water: A review. *Sustain. Water Resour. Manag.* **2016**, *2*, 161–173. [CrossRef]
4. Rawashdeh, R.A.; Xavier-Oliveira, E.; Maxwell, P. The potash market and its future prospects. *Resour. Policy* **2016**, *47*, 154–163. [CrossRef]
5. Dmitrieva, D.; Ilinova, A.; Kraslawski, A. Strategic management of the potash industry in Russia. *Resour. Policy* **2017**, *52*, 81–89. [CrossRef]

6. al Rawashdeh, R.; Maxwell, P. Analysing the world potash industry. *Resour. Policy* **2014**, *41*, 143–151. [CrossRef]
7. Song, X.; Geng, Y.; Zhang, Y.; Zhang, X.; Gao, Z.; Li, M. Dynamic potassium flows analysis in China for 2010–2019. *Resour. Policy* **2022**, *78*, 102803. [CrossRef]
8. Chen, W.; Geng, Y.; Hong, J.; Yang, D.; Ma, X. Life cycle assessment of potash fertilizer production in China. *Resour. Conserv. Recycl.* **2018**, *138*, 238–245. [CrossRef]
9. Broughton, P.L. Economic geology of southern Saskatchewan potash mines. *Ore Geol. Rev.* **2019**, *113*, 103117. [CrossRef]
10. Kumar Katta, A.; Davis, M.; Kumar, A. Assessment of greenhouse gas mitigation options for the iron, gold, and potash mining sectors. *J. Clean. Prod.* **2020**, *245*, 118718. [CrossRef]
11. Cabello, J. Lithium brine production, reserves, resources and exploration in Chile: An updated review. *Ore Geol. Rev.* **2021**, *128*, 103883. [CrossRef]
12. Bustos-Gallardo, B.; Bridge, G.; Prieto, M. Harvesting Lithium: Water, brine and the industrial dynamics of production in the Salar de Atacama. *Geoforum* **2021**, *119*, 177–189. [CrossRef]
13. Wurtsbaugh, W.A.; Miller, C.; Null, S.E.; DeRose, R.J.; Wilcock, P.; Hahnenberger, M.; Howe, F.; Moore, J. Decline of the world's saline lakes. *Nat. Geosci.* **2017**, *10*, 816–821. [CrossRef]
14. Song, H.; Fan, Q.; Li, Q.; Chen, T.; Yang, H.; Han, C. Recharge processes limit the resource elements of Qarhan Salt Lake in western China and analogues in the evaporite basins. *J. Oceanol. Limnol.* **2023**, *41*, 1226–1242. [CrossRef]
15. Li, R.; Liu, C.; Jiao, P.; Hu, Y.; Liu, W.; Wang, S. Simulation study on the mining conditions of dissolution of low grade solid potash ore in Qarhan Salt Lake. *Sci. Rep.* **2021**, *11*, 10539. [CrossRef] [PubMed]
16. Xiao, Y.; Hao, Q.; Luo, Y.; Wang, S.; Dang, X.; Shao, J.; Huang, L. *Origin of Brines and Modern Water Circulation Contribution to Qarhan Salt Lake in Qaidam Basin, Tibetan Plateau*; EDP Sciences: Les Ulis, France, 2019; Volume 98, p. 12025.
17. Li, R.; Liu, C.; Jiao, P.; Liu, W.; Tang, D.; Wang, S. The effect of solvent chemistry on potassium dissolution extraction from low-grade solid potash ore in Qarhan Salt Lake, China. *Appl. Geochem.* **2020**, *115*, 104550. [CrossRef]
18. Tziritis, E.; Skordas, K.; Kelepertsis, A. The use of hydrogeochemical analyses and multivariate statistics for the characterization of groundwater resources in a complex aquifer system. A case study in Amyros River basin, Thessaly, central Greece. *Environ. Earth Sci.* **2016**, *75*, 339. [CrossRef]
19. Nematollahi, M.J.; Clark, M.J.R.; Ebrahimi, P.; Ebrahimi, M. Preliminary assessment of groundwater hydrogeochemistry within Gilan, a northern province of Iran. *Environ. Monit. Assess.* **2018**, *190*, 242. [CrossRef]
20. Chidambaram, S.; Anandhan, P.; Prasanna, M.V.; Ramanathan, A.L.; Srinivasamoorthy, K.; Senthil Kumar, G. Hydrogeochemical Modelling for Groundwater in Neyveli Aquifer, Tamil Nadu, India, Using PHREEQC: A Case Study. *Nat. Resour. Res.* **2012**, *21*, 311–324. [CrossRef]
21. Guo, W.; Li, P.; Du, Q.; Zhou, Y.; Xu, D.; Zhang, Z. Hydrogeochemical Processes Regulating the Groundwater Geochemistry and Human Health Risk of Groundwater in the Rural Areas of the Wei River Basin, China. *Expo. Health* **2024**, *16*, 291–306. [CrossRef]
22. Duan, R.; Li, P.; Wang, L.; He, X.; Zhang, L. Hydrochemical characteristics, hydrochemical processes and recharge sources of the geothermal systems in Lanzhou City, northwestern China. *Urban Clim.* **2022**, *43*, 101152. [CrossRef]
23. de Caritat, P.; Bastrakov, E.N.; Jaireth, S.; English, P.M.; Clarke, J.D.A.; Mernagh, T.P.; Wygralak, A.S.; Dulfer, H.E.; Trafford, J. Groundwater geochemistry, hydrogeology and potash mineral potential of the Lake Woods region, Northern Territory, Australia. *Aust. J. Earth Sci.* **2019**, *66*, 411–430. [CrossRef]
24. El Alfy, M.; Lashin, A.; Abdalla, F.; Al-Bassam, A. Assessing the hydrogeochemical processes affecting groundwater pollution in arid areas using an integration of geochemical equilibrium and multivariate statistical techniques. *Environ. Pollut.* **2017**, *229*, 760–770. [CrossRef] [PubMed]
25. Parisi, A.; Alfio, M.R.; Balacco, G.; Güler, C.; Fidelibus, M.D. Analyzing spatial and temporal evolution of groundwater salinization through Multivariate Statistical Analysis and Hydrogeochemical Facies Evolution-Diagram. *Sci. Total Environ.* **2023**, *862*, 160697. [CrossRef]
26. Qin, Z.; Li, Q.; Li, W.; Fan, Q.; Chen, T.; Wu, C.; Wang, J.; Shan, F. Elemental Variations and Mechanisms of Brines in the Context of Large-Scale Exploitation: A Case Study of Xitaijnar Salt Lake, Qaidam Basin. *Aquat. Geochem.* **2023**. [CrossRef]
27. Fan, Q.; Lowenstein, T.K.; Wei, H.; Yuan, Q.; Qin, Z.; Shan, F.; Ma, H. Sr isotope and major ion compositional evidence for formation of Qarhan Salt Lake, western China. *Chem. Geol.* **2018**, *497*, 128–145. [CrossRef]
28. Guo, J.; Wei, X.; Long, G.; Wang, B.; Fan, H.; Xu, S. Three-dimensional structural model of the Qaidam basin: Implications for crustal shortening and growth of the northeast Tibet. *Open Geosci.* **2017**, *9*, 174–185. [CrossRef]
29. Stober, I.; Zhong, J.; Bucher, K. From freshwater inflows to salt lakes and salt deposits in the Qaidam Basin, W China. *Swiss J. Geosci.* **2023**, *116*, 5. [CrossRef]
30. Karunanidhi, D.; Aravinthasamy, P.; Subramani, T.; Kumar, D.; Setia, R. Investigation of health risks related with multipath entry of groundwater nitrate using Sobol sensitivity indicators in an urban-industrial sector of South India. *Environ. Res.* **2021**, *200*, 111726. [CrossRef]
31. Zhu, G.; Wu, X.; Ge, J.; Liu, F.; Zhao, W.; Wu, C. Influence of mining activities on groundwater hydrochemistry and heavy metal migration using a self-organizing map (SOM). *J. Clean. Prod.* **2020**, *257*, 120664. [CrossRef]
32. Wu, J.; Li, P.; Qian, H. Hydrochemical characterization of drinking groundwater with special reference to fluoride in an arid area of China and the control of aquifer leakage on its concentrations. *Environ. Earth Sci.* **2015**, *73*, 8575–8588. [CrossRef]

33. Morán-Ramírez, J.; Ledesma-Ruiz, R.; Mahlknecht, J.; Ramos-Leal, J.A. Rock–water interactions and pollution processes in the volcanic aquifer system of Guadalajara, Mexico, using inverse geochemical modeling. *Appl. Geochem.* **2016**, *68*, 79–94. [CrossRef]
34. Pérez-Ceballos, R.; Canul-Macario, C.; Pacheco-Castro, R.; Pacheco-Ávila, J.; Euán-Ávila, J.; Merino-Ibarra, M. Regional Hydrogeochemical Evolution of Groundwater in the Ring of Cenotes, Yucatán (Mexico): An Inverse Modelling Approach. *Water* **2021**, *13*, 614. [CrossRef]
35. Guo, Y.; Lyu, Z.; Wang, G.; Ma, L.; Xu, Q.; Huang, X.; Gao, S. Hydrogeochemical simulation of groundwater in Eastern Fengfeng mining area. *Coal Geol. Explor.* **2016**, *44*, 101–105. [CrossRef]
36. Li, P.; Wu, J.; Qian, H. Assessment of groundwater quality for irrigation purposes and identification of hydrogeochemical evolution mechanisms in Pengyang County, China. *Environ. Earth Sci.* **2013**, *69*, 2211–2225. [CrossRef]
37. Adimalla, N.; Venkatayogi, S. Geochemical characterization and evaluation of groundwater suitability for domestic and agricultural utility in semi-arid region of Basara, Telangana State, South India. *Appl. Water Sci.* **2018**, *8*, 44. [CrossRef]
38. Zaidi, F.K.; Nazzal, Y.; Jafri, M.K.; Naeem, M.; Ahmed, I. Reverse ion exchange as a major process controlling the groundwater chemistry in an arid environment: A case study from northwestern Saudi Arabia. *Environ. Monit. Assess.* **2015**, *187*, 607. [CrossRef] [PubMed]
39. Xu, F.; Li, P.; Du, Q.; Yang, Y.; Yue, B. Seasonal Hydrochemical Characteristics, Geochemical Evolution, and Pollution Sources of Lake Sha in an Arid and Semiarid Region of Northwest China. *Expo. Health* **2023**, *15*, 231–244. [CrossRef]
40. Chaudhary, V.; Satheeshkumar, S. Assessment of groundwater quality for drinking and irrigation purposes in arid areas of Rajasthan, India. *Appl. Water Sci.* **2018**, *8*, 218. [CrossRef]
41. Li, P.; He, X.; Guo, W. Spatial groundwater quality and potential health risks due to nitrate ingestion through drinking water: A case study in Yan'an City on the Loess Plateau of northwest China. *Hum. Ecol. Risk Assess.* **2019**, *25*, 11–31. [CrossRef]
42. Jia, H.; Qian, H.; Zheng, L.; Feng, W.; Wang, H.; Gao, Y. Alterations to groundwater chemistry due to modern water transfer for irrigation over decades. *Sci. Total Environ.* **2020**, *717*, 137170. [CrossRef] [PubMed]

water

MDPI

Review

Recent Progress on Acid Mine Drainage Technological Trends in South Africa: Prevention, Treatment, and Resource Recovery

Jeffrey Baloyi [1,*], Nishani Ramdhani [1], Ryneth Mbhele [1] and Denga Ramutshatsha-Makhwedzha [2]

1 Smart Places Cluster, Council for Scientific and Industrial Research (CSIR), Pretoria 0001, South Africa; nramdhani@csir.co.za (N.R.); rmbele@csir.co.za (R.M.)

2 Department of Chemical, Metallurgical and Materials Engineering, Pretoria West Campus, Tshwane University of Technology, Private Bag X680, Pretoria 0183, South Africa; makhwedzhad@tut.ac.za

* Correspondence: baloyisj1@csir.co.za; Tel.: +27-82-646-3813

Abstract: South Africa is the home of major global mining operations, and the acid mine drainage (AMD) contribution has been attributed to abandoned mine sites and huge pyrite-bearing tailings from coal and gold mines. Determining the true economic impact and environmental liability of AMD remains difficult. Researchers have been looking into several treatment technologies over the years as a way to reduce its possible environmental impact. Different methods for active and passive remediation have been developed to treat AMD. The aim of this review was to describe the AMD-impacted environments and critically discuss the properties of AMD and current prediction and preventative methods and technologies available to treat AMD. Furthermore, this study critically analysed case studies in South Africa, gaps in AMD research, and the limitations and prospects offered by AMD. The study outlined future technological interventions aimed at a pattern shift in decreasing sludge volumes and operational costs while effectively improving the treatment of AMD. The various treatment technologies have beneficial results, but they also have related technical problems. To reduce the formation of AMD, it is recommended that more preventive methods be investigated. Moreover, there is a current need for integrated AMD treatment technologies that result in a well-rounded overall approach towards sustainability in AMD treatment. As a result, a sustainable AMD treatment strategy has been made possible due to water reuse and recovery valuable resources such sulphuric acid, rare earth elements, and metals. The cost of AMD treatment can be decreased with the use of recovered water and resources, which is essential for developing a sustainable AMD treatment process. More study is required in the future to improve the effectiveness of the various strategies used, with a focus on reducing the formation of secondary pollutants and recovery of valuable resources.

Keywords: acid mine drainage; sustainable remediation technologies; prevention and prediction; resource recovery

Citation: Baloyi, J.; Ramdhani, N.; Mbhele, R.; Ramutshatsha-Makhwedzha, D. Recent Progress on Acid Mine Drainage Technological Trends in South Africa: Prevention, Treatment, and Resource Recovery. *Water* **2023**, *15*, 3453. https://doi.org/10.3390/w15193453

Academic Editor: Liliana Lefticariu

Received: 4 August 2023
Revised: 27 September 2023
Accepted: 28 September 2023
Published: 30 September 2023

1. Introduction

Acid mine drainage (AMD) is a persistent pollutant as a result of current and past mining activities, which is currently one of the critical environmental challenges in South Africa and globally. AMD presents a challenge for operational and deserted mines, in shafts below ground, open holes, waste rock mounds, and powder tailings [1–4]. AMD is more serious in deserted and inactive mines, where there is no pumping occurring and the water table recoils, in contrast to active mines where the water table levels are kept to a minimum through the use of pumps [5–7]. Besides the environmental impacts, AMD also impacts sustainability, which includes environmental, community, and financial concerns. AMD has an effect on the removal of resources, which has an effect on economic activity, infrastructure, and people's livelihoods. This has a substantial effect on the guidelines of developing countries for climate change and their efforts to switch to becoming 'green' economies. New sustainable technologies, efficient management plans, and AMD treatment procedures are currently required. In South Africa, AMD, mainly from gold mine tailings dams/slime dams, is one of the most serious environmental and socio-economic challenges [8,9]. Acidity is formed when pyrite, in the gold-bearing ores, is oxidised and encounters atmospheric oxygen. This oxidation goes through several steps, as follows:

$$4FeS_2 + 15O_2 + 14H_2O \rightarrow 4Fe(OH)_3 + 8H_2SO_4 \quad (1)$$

On the left side of the above equation, pyrite is ferrous sulphide and the right has ferric oxide and sulphate. Thus, both the iron and sulphur components of pyrite have been oxidised. Acid mine drainage develops when water permeates through the zone of oxidised pyrites forming sulphuric acid, which drains out into the surrounding environment [8,10].

AMD causes significant environmental problems that are both locally and globally intractable in the near future. It takes development of effective, innovative, and affordable approaches for addressing and overcoming these issues. Mining for minerals, such as gold, copper, and nickel, has been linked to AMD issues, which could have long-term consequences for streams and biodiversity. Some metal mining effluents contain high levels of poisonous cyanides and heavy metals, which have major human health and environmental consequences [3,11,12]. To remediate AMD, many wastewater treatment technologies have been used, including neutralization [13], selective precipitation [14], use of membranes [15], exchange of ions [16], and the removal of sulphate biologically [17]. The challenge, however, is that the constituents of AMD, while hazardous, may be collected and turned into valuable materials that can be commercialized. Sulphuric acid, for example, has a major demand in the chemical and metallurgical industries [18,19]. While the need for critical minerals and erratic metals in the ground is probably set to increase going forward, there is a desire to create innovative results that blend ground earth metal reclamation with AMD remediation [20,21]. The financial advantages could then be utilized to offset the entire expense of AMD treatment.

To that purpose, the most recent research on AMD development, prevention, and treatment is summarized and critically reviewed. The central focus of this article is to evaluate studies on the prevention, prediction, impact of AMD on the water quality, management, and prospective reclamation of beneficial by-products from mine drainage. The specific objectives are to (1) briefly present the past impacts and possible future effects of AMD on the water condition in South Africa, (2) discuss the developments on the prediction and prevention of AMD, and (3) review of the main large-scale AMD treatment processes applied locally for AMD elimination and simultaneous recovery of the valuable by-products.

2. Past Impact of AMD on the Condition of Water in South Africa

Mining has been the backbone of the South African economy for many years. Coal and gold mining are the major sources of AMD as sulphide-bearing materials are concentrated in geological environments containing coal and gold ore deposits [22]. Gold mining pollution is associated mainly with the release of harmful elements from tailings and other types of mining waste [23]. The infiltration of water through sulphide-containing tailings piles and ponds, surface and underground workings, waste, and development rocks leads to the leaching of large volumes of metals like zinc, nickel, lead, copper, and sulphate ions into streams and river ecosystems [24]. This results in acid mine drainage (AMD) with severe detrimental effects on the receiving water bodies. Heavy metal pollution and acid mine drainage is a very important environmental concern where waste materials containing metal-rich sulphides from mining activity have been stored or abandoned. Tailings and rock dumps are associated with surface impacts which greatly affect surface and ground water quality. The underground impacts are caused by the influx of water into the underground workings and the subsequent dewatering of the aquifer [25].

To maintain safe mining conditions, over 120 mines would have to pump out the water that had been collected in them. However, as the mines became derelict, drainage of the mine voids became less frequent, and the voids started filling up with water which resulted in AMD [26]. AMD can contaminate shallow aquifers, and if the underground mine water reaches the near surface, it starts to decant and flows down to wetlands, streams, and rivers. Since these rivers are used as a water source for agriculture, recreation, and drinking purposes, AMD potentially affects the quality of this water. The two main types of mining occurring in South Africa contribute significantly to the generation of AMD, as listed below in Table 1. The resultant water bodies affected by this AMD pollution are also shown in Table 1 together with the affected areas.

The Western, Central, and Eastern Basins have been highlighted since they have been affected the most by AMD. The mining shafts started decanting in 2002, polluting the Tweelopiespruit that drains into the Krugersdorp Game Reserve [27]. There are abandoned mines which are not operational which have become sources of AMD and water ingress into the shafts. AMD decants through three old mine shafts [28]. The volume being decanted was noted at an average of 20 ML/d, and up to 60 ML/d during the wet season. Approximately 12 ML/d is treated partially, and the remainder flows into the Tweelopiespruit. About 27 ML/d of AMD needs to be treated in this basin to maintain the water below the environmental critical level (ECL). The long-term effects of decanting into this stream result in the pollution of the Hartbeespoort Dam, the Crocodile River, and the transboundary Limpopo River [29].

Table 1. Types of mining in South Africa that predominantly contribute to AMD pollution and their areas of impact.

	Provinces	Key Areas	Water Resources Impacted	References
Gold	Northwest, Gauteng, Mpumalanga, Limpopo	Within Gauteng: Witwatersrand gold spans the Central and Eastern Basins Within the Witwatersrand Eastern Basin: Brakpan, Springs, Nigel Klerksdorp Kloof, Driefontein, Western Deep Levels	Tweelopiespruit, Hartbeespoort Dam, Crocodile River, Limpopo River, Vaal River, Klip River, Blesbokspruit, Barrage, Vaal Dam	[30–32]
Coal		Witbank, Delmas, Secunda	Boesmanspruit, Blesbokspruit, Vaal River	[33–35]

Since the quarrying of mines began, dumps containing the wasteful end-products from gold mining have become a common sight around mining towns and have been releasing contaminated water for decades. Tailings dumps flourish in the upper catchments of springs at the Blesbokspruit and Klip Rivers, where this pollution is predominant [36]. The Witwatersrand gold mines have shut down over several years, and water began to fill in the voids as the pumping of the mines had stopped. This accumulated water travelled into neighbouring mines since all the mines are connected. This process forced these neighbouring mines to take on the responsibility of pumping. A subsidy was initiated by the government to help the mines cover the costs involved in pumping the additional high volumes of water which filled up and requires treatment to an acceptable quality. This treatment involves the use of lime to increase the pH levels and the pumping of oxygen into the water to trigger the iron to oxidize and precipitate along with several other heavy metals. The precipitated iron will settle out, separate, and be discarded in dumps of tailings and the remainder of the water is diverted into local rivers. However, these discharges increased the number of pollutants previously transported via the rivers through the mining towns.

The salt levels present in the Vaal River indicate the effects of long-winded and point source pollutants stemming from the gold mines in the Central and Western Basins, which has increased to greater than twice the amount between the Barrage and Vaal Dams. This is due to the incoming water from the Klip and the Blesbokspruit Rivers (via the Sukerbos River) [37]. The water composition is not good at the Barrage which makes it necessary to periodically let go off water from the Vaal Dam to decrease the salt content for the users downriver of the Vaal. During the rainy season this is not a problem, but during a drought, this could pose a challenge, when the water upriver of the Vaal system, which is mainly for Gauteng, must be discharged for the aim of dilution [38]. The void began to fill once the last Goldfield mine shut down and stopped pumping water. Decantation of the Western Basin began in 2002. When draining of the Central Basin ceased in 2008, the water levels continued to rise at a rate of 12 m per month. Pumping at the Eastern Basin started to slow down in late 2010, and eventually stopped at the beginning of 2011. The decanted water emanating from the void is of very bad quality, as is noted from the water draining from the Western basin. The level of sulphate is typically approximately 3500 mg/L, with a pH range of 2–3. There are also elevated iron and other heavy metal concentrations present in the water. Oxidation occurs when the iron is exposed to air which leaves a bright orange stream of precipitate on the banks and beds of rivers. It is expected that the basins in Boksburg (Central) and Nigel (Eastern), where the bottommost shafts are located, will decant in approximately three years if there are no interventions put into place [39].

These points of decantation are created on the assumption that the water is spontaneously flowing via the cavities and that mine quarries are the only gaps to these voids. However, this may not be true. For example, in the Western Basin, it was found that water was draining from a borehole on a farm, and thereafter from a longstanding mine quarry which may have or may not have been linked up to the central void. Many decant points can arise if the flow rate through the void is not enough to allow for the inflow.

3. Possible Future Impacts of AMD on the State of Water in South Africa

The Olifants River catchment is in a state of deterioration [40]. There was an idea to connect a plant to treat the high number of pollutants in Brugspruit which is close to Witbank, but the idea had limited efficacy [40]. The main purpose was to address the pH challenge and it did not impact the salt concentrations in the water. A plant to treat water was commissioned in the area (eMalahleni Water Reclamation Plant) which operates by reverse osmosis [41]. This plant showed the potential to treat highly polluted water to an acceptable standard for drinking purposes. The setback is that this water costs more than (approximately triple) the water that is distributed to this area from the Vaal River by the local water agency [42]. This plant, although beneficial in producing drinking water for the public, has a limitation of not being beneficial for the complete improvement of rivers that are polluted in the area [28]. The state of the water of the Olifants River will remain in a state of deterioration in the future [43]. This is due to the massive quantities of coal that are found in the Olifants Catchment, which are not mined [44], thereby leading to a rise in pollutants in the future.

Coal mining was occurring for a vast number of years in the upper catchment of the Vaal River [45]. These mines are mostly deep and are still actively operated and overseen. Nevertheless, it was found that there is a high inflow of applications for permits for new mines in that catchment. Funds were set aside by the government in the past to tackle the issue of decanting from the gold mines in Witwatersrand, which will involve draining as well as the straightforward operational treatment being reinitiated (lime being added and iron being removed) in the goldfields which are presently affected [46]. This will aid in stopping the unrestrained draining of the basins (Western, Central, and Eastern) [47]. However, as much as this intervention will vastly help to improve the Western Basin, it will not affect the state of the water of the Vaal River. Instead, it will take the system back to its original state at a time when the mines were still being pumped and treated and water was drained from the mine cavities.

Several different technologies have been developed for desalination of contaminated water from local mines. Only one of these has been commercialised and is being implemented, reverse osmosis treatment technology (used at Witbank), which has shown that although this type of reverse osmosis treatment may help overcome the challenges, it is not feasible [42]. It is possible that although most of the suggested treatment methods are appropriate for remedying water with contaminants at the point source (e.g., pumped from old mines), it is unlikely that polluted water from diffuse sources such as waste dumps, could be treated as well [48]. For gold mines, the water accumulated in the cavity can most often be accessed and treated as a point source. Coal mining is more composite however, and it may not be possible to avoid uninhibited draining of AMD from restored opencut mines [49]. Thus, the state of water in these regions should be anticipated to decline.

Since the treatment techniques used for prevention do not need to be continuous (for maintenance as well), they are more viable than conventional treatment techniques. However, most preventative treatments, such passivation and microencapsulation, are still under experimentation and focus on pure pyrite systems [50] The results were encouraging for microencapsulation techniques; however, this is restricted to batch, single-metal systems. Therefore, it is unclear if they may be used for waste rocks and mine tailings that contain a variety of minerals such silicates and aluminosilicates in complex environmental conditions. This suggests that continuous tests should be conducted to determine the efficacy of various microencapsulation techniques using real or synthetic tailings, including waste rocks high in pyrite [51]. It is important to thoroughly examine the long-term stability of treated waste rocks and tailings under environmental conditions, such as drying–wetting cycles. Utilizing or recycling mine waste for use in building and geopolymer materials is another choice to consider while attempting to limit the generation of AMD. Finding value in and managing mine waste has become incredibly important [52].

4. AMD Prevention

The complexity of the treatment system that is required to guarantee that effluent standards will be satisfied depends on a number of variables [53]. These include the chemical properties of the AMD, the volume of water that needs to be treated, the local climate, the topography, the properties of the sludge, and the anticipated lifespan of the plant [53]. Various treatment techniques have been developed and can be categorised as either 'abiotic' or 'biotic', the former of which does not rely on biological activities while the latter does [5].

Passive treatments involve the passage of mine water through a controlled environment, rather than a receiving water body, where naturally occurring geochemical and biological reactions take place and improve the mine water quality [10,54,55]. Examples of passive abiotic treatment include anoxic limestone drains, open limestone channels, limestone leach beds, slag leach beds, diversion wells, limestone sand, and oxidation channels [12]. These materials all generate alkalinity which help to neutralise the AMD and raise the pH, while at the same time oxidising and precipitating out metals. Passive biotic treatments, on the other hand, include wetlands and bioreactors where natural biological processes work, either in aerobic or anaerobic conditions, to neutralise the AMD and precipitate the hazardous concentrations of contaminants (e.g., metals) over time [12,56]. In line with this, Ramla and Sheridan [57] proved the efficacy of utilizing indigenous South African grass as a suitable organic substrate for sulphate-reducing bacteria to reduce sulphate to sulphides during the passive biotic treatment of AMD. In this experiment, Hyparrhenia hirta grass supplemented with soil containing microbes produced the best outcomes.

Both passive biotic and abiotic AMD treatments require relatively little resource input, tend to be more useful for AMD flows of less than 2 to 5 ML/d with low acidity, e.g., <800 mg/L as $CaCO_3$, and which require little metal and sulphate removal [54,58,59]. However, in comparison to active treatments, passive treatments need larger areas of land and additional time to neutralise AMD and precipitate the contaminants. Thus, passive options are more applicable when AMD treatment needs to be accomplished at closed mine sites with low AMD flows as they are a potentially lower-cost, longer-term sustainable option. An example of one that has been studied in some detail is the passive system set up to treat AMD mine seepage from a long-abandoned mine near the town of Red Oak in eastern Oklahoma [56]. The benefits of passive systems are their self-sufficiency, infrequent maintenance requirements, and extremely low operating and capital expenses. However, the quality of the resultant effluent is poorer than that produced by active treatment systems [54].

Unlike passive treatments which depend mostly on naturally occurring reactions, active AMD treatment is performed in a constructed plant where processes are controlled and sustained via the continuous input of resources [60]. It involves the utilisation of alkaline substances to increase the pH of the drainage and precipitate heavy and toxic metals from the AMD [55]. In line with this, the operating and capital costs of sustaining effective and efficient functioning of the plant can be high as it requires a continuous supply of chemicals, electrical and mechanical power sources, and the employment of operations and maintenance staff [60]. Therefore, active AMD treatment is more suited for application at operational mine sites where the necessary resources are more readily available [61]. Additionally, the kind of neutralizing agent utilized affects the effectiveness, cost, and potential environmental effects of using an active treatment system [62]. The selection of the neutralizing agent is based on the chemical composition of the AMD, site-specific conditions, and expected outcomes with the understanding that some level of cost and benefit trade-off will be required. For instance, sodium hydroxide is more effective in AMD treatment than lime but is approximately 1.5 times more expensive and must be handled in line with specific health and safety requirements due to its hazardous nature [62]. Likewise, anhydrous ammonia requires safe handling and if excessively used can spur nitrification or denitrification in receiving water bodies [62]. In some cases, the split treatment of AMD may yield the most desirable results, e.g., using lime and limestone [63].

The benefits of active AMD treatment, however, can be considered as great advantages over passive treatment techniques [64]. These include that active treatment can be applied to all AMD flow rates, it is fast and effective, it produces good quality effluent with a potential for cost recovery via the sale of the resulting water, metals, and by-products and involves a lower cost in the handling and disposal of generated sludge [10,60,62,65]

Traditional abiotic active treatment of AMD is characterised by the use of alkaline chemicals to neutralise acids, deactivate metals, and precipitate salts [66,67]. Calcium-, sodium-, ammonium-, or magnesium-based chemicals that have been, and are, used to neutralise AMD include calcium carbonate (limestone, $CaCO_3$), calcium hydroxide (slaked lime, $Ca(OH)_2$), calcium oxide (lime or quick/burnt lime, CaO), sodium hydroxide (caustic soda, NaOH), sodium carbonate (soda ash, Na_2CO_3), ammonium hydroxide (NH_4OH), and magnesium hydroxide ($Mg(OH)_2$) [68]. Using calcium hydroxide, calcium oxide, or calcium carbonate can result in large amounts of sludge that retain water as calcium will bond with sulphates and then precipitate out of solution, together with the metals (usually as hydroxides), at higher pH levels [69]. Recycling this sludge is difficult, leaving disposal to landfill or sludge dams as the main option for handling this waste [69]. Furthermore, some metal hydroxides are amphoteric, which presents the probability for dissolution of potentially harmful chemicals from the sludge both during and after disposal [70]. Thus, the sludge disposed at landfill sites or sludge dams would need to be properly managed and regularly monitored to ensure that no long-term negative environmental impacts occur [69]. In addition, landfills and sludge dams can occupy large areas of land, especially when considerable amounts of sludge are produced and disposed of [69]. Neutralising chemicals such as magnesium hydroxide, ammonium hydroxide, and sodium hydroxide have proven to be comparatively more useful as they tend to precipitate metals (e.g., as hydroxides) while leaving the sulphate in solution. This sulphate can subsequently be treated to produce gypsum, which may be valuable in other economic and industrial sectors.

An area of rising importance in the active abiotic treatment of AMD is the use of waste by-products to treat other wastes. An example of this is the use of calcium-containing waste such as dust from cement and lime kilns to neutralise AMD and precipitate metals. Another example is the use of coal combustion by-products to partially treat AMD. Coal-based by-products generally are very good adsorbents because they have a high surface area, microporous structure, and high surface reactivity [71]. However, these by-products seem to be best at removing trace concentrations of more toxic metals such as radioactive thorium, uranium, radium, and lead. Kaur et al. [67] also describes the use of an alkaline waste material from the alumina refining industry as a possible alternative neutralising material. The costs to obtain and use one waste to treat another is often much lower, making AMD neutralisation and other treatment processes (e.g., metal extraction) potentially much cheaper. However, the feasibility of these types of options will vary depending on the type, availability, and location of the various wastes as well as the properties of the AMD. In addition, it must be noted that calcium-based neutralisation chemicals tend to produce considerable quantities of waste sludge which would still need to be dealt with if cement waste or lime kiln dust is used. Other examples of good absorbents that have shown potential to treat AMD include bauxite and naturally occurring bentonite clay. Bentonite (primarily aluminium phyllosilicate) has been used to neutralise AMD and remove metals. However, the suitability of these type of adsorbents needs to be investigated on a large, long-term scale to prove that they can work as well as the current technologies whilst also being more sustainable and cost-effective.

Active biotic AMD treatment involves the use of off-line sulfidogenic bioreactors where the hydrogen sulphide produced by sulphate-reducing bacteria (SRB) is used both to add alkalinity to neutralise the acidic waste streams and to precipitate metals as insoluble sulphide precipitates, which may then be recovered and reprocessed [5,72]. Table 2 below summarizes the various technologies used in the treatment of AMD.

Table 2. List of the AMD treatment methods [12,66].

Active/Passive	Biotic/Abiotic	Treatment Methods	Advantages
Passive	Abiotic	• Anoxic limestone drains • Open limestone channels • Limestone leach beds • Slag leach beds • Diversion wells • Limestone sand • Oxidation channels	• Self-sustaining. • Needs sporadic maintenance. • Very low operating and capital costs.
	Biotic	• Wetlands • Bioreactors	• Applicable at closed mine sites with low AMD flows with potentially lower costs and longer-term sustainability.
Active	Abiotic	• Use of alkaline chemicals such as calcium, sodium, calcium hydroxide (slaked lime), calcium carbonate (limestone), calcium oxide (lime or quick/burnt lime), sodium carbonate (soda ash), sodium hydroxide (caustic soda), magnesium hydroxide, and ammonium hydroxide	• Waste by-products can be used to treat other wastes. • Can be applied to all AMD flow rates. • Fast and effective. • Produces good quality effluent. • Cost recovery via the sale of resulting water, metals, and by-products. • Has a lower cost in the handling and disposal of generated sludge. • Suitable for operational mine sites where the necessary resources are readily available.
	Biotic	• Off-line sulfidogenic bioreactors	• Can be applied to all AMD flow rates. • Fast and effective. • Produces good quality effluent. • Potential for cost recovery via the sale of resulting water, metals, and by-products. • Lower cost in handling and disposal of generated sludge.

5. AMD Impact

High quantities of dissolved metals and acid constitute AMD, which is extremely harmful to groundwater, streams, and rivers. AMD also damages ecosystems, corrodes infrastructure, and poses a number of environmental challenges for aquatic life. This often results in contaminated water supplies to areas where freshwater is not easily accessible [73].

For AMD pollutants to affect humans, they need to be exposed and several AMD pollutants are dangerous to humans [74]. According to Orlović-Leko et al. [75], heavy metals have an adverse effect on both people and the environment, and they can linger for a very long time in natural ecosystems where they build up at higher and higher levels of the food chain. This will lead to acute and chronic diseases where metabolic functioning is disrupted by accumulation of heavy metals in vital organs and glands [65].

The water from AMD inflicts terrible damage since it starts out clear and quickly turns brilliant orange when iron oxides and hydroxides precipitate due to the high acidity levels. By becoming embedded on the river, stream, or ocean bed, this fine precipitate, known as ochre, cements substrates that serve as a food supply for benthic creatures, which eventually go extinct [74], affecting the higher levels of the food chain. Because of these indirect effects, AMD still has an effect on people and wildlife further downstream even if the acidity and heavy metals are reduced.

Heavy metals also contaminate soil which poses serious environmental issues where plant growth is affected by oxidative stress [8]. According to Li et al. [76], this causes cellular damage and disturbs homeostasis, which affects the physiology and morphology of plants. Calcium and magnesium are unavailable to plants as well as nitrogen, phosphorus, and potassium when the pH of the soil is low. At a low pH, soil particles also release aluminium, iron, and manganese, enhancing their toxicity. Furthermore, low soil pH affects how well plants use nutrients, establish roots, and tolerate drought by reducing the activity of soil organisms that break down organic materials. According to Jiao et al. [9], heavy metals are accumulated by aquatic creatures like fish both directly from tainted water and indirectly through the food chain. Since they are highly persistent and poisonous at trace levels, cadmium, copper, lead, and zinc have the potential to cause severe oxidative stress in aquatic organisms [77]. While chronic exposure can cause mortality or stunted growth, limited reproduction, malformations, or lesions, acute exposure to them can directly kill organisms. As a result, aquatic organisms' typical physiological processes—including ion exchange with the water and respiration—are influenced by the pH of the water.

6. Current Treatment Technologies and Resource Recovery

Many studies and investigations have been conducted by academics to treat AMD. The following treatment techniques are frequently applied: neutralization [78], precipitation [79], and sedimentation [80]; nevertheless, additional techniques such anaerobic bioreactors [81], sorption [82], coagulation [83], flocculation [84], and crystallization [85] may also be employed. Although these effluents are typically treated, these techniques may not be sufficient to treat the effluent characteristics to fulfil standards for discharge and/or reuse; high levels of chemical product consumption can produce significant amounts of sludge polluted with metal [9]. The membrane separation process has emerged in the treatment of AMD with an ability to have salts and metals retained from aqueous media using membrane separation methods, particularly reverse osmosis (RO), membrane distillation (MD), forward osmosis (FO), and nanofiltration (NF). The NF process is a third option between RO and UF that can retain multivalent ions and dissolved compounds with molar masses between 200 and 1000 g/mol [86]. Numerous studies have demonstrated the effectiveness of NF as a secondary or tertiary treatment system [87,88]. This is due to its low consumption of power, high efficiency, and ease of operation. Meanwhile, RO technology has been reported as a promising AMD option for producing high-quality water while minimizing the discharge.

For example, Andalaf et al. [89] developed an AMD treatment process to treat and predict the behaviour of AMD; an NF pilot-scale system with two different membranes (NF270 and NF90, France) was used. All ions were rejected with a high rejection rate (100%); however, fouling was found at a water recovery rate of 75%. Wadekar et al. [90] compared ceramic and polymeric nanofiltration membranes in the treatment of abandoned coal mine drainage. The AMD was sampled from a site in Pennsylvania and was treated with NF (ceramic and polymeric NF270 membranes, pressure: 35 bar) which was pretreated with aeration and microfiltration. Over 96% of multivalent ions were rejected by NF270. Approximately 55 to 67% of ceramic membrane rejections resulted from its use. Membrane fouling occurred with a water recovery rate of 75%. Masindi et al. [91] looked into recovering drinking water from acid mine drainage from a South African coal mine, and they found that a reverse osmosis device can successfully prepare the water. The drinking water produced by this procedure had a pH of about 6.5 and a metal removal rate of about 100%, which met the SANS 241 criteria for drinking water quality. With the use of forward osmosis, acid mine drainage can be concentrated in a way that promotes the growth of enrichment sludges and the subsequent selective metal precipitation. León-Venegas et al. [92] studies the potential for water and metal recovery from acid mine drainage from the Iberian Pyrite Belt, Southwestern Spain, by combining hybrid membrane processes with selective metal precipitation. Forward osmosis, reverse osmosis, and osmotically assisted reverse osmosis mixed with selective metal precipitation was used to treat AMD, obtaining high water recovery and an enriched metal sludge. They reported that two steps of FO using draw solutions based on sodium chloride could recover about 80% of the water from AMD. Moreover, selective metal precipitation can be used to produce sludges rich in Fe, Al, Cu, Zn, and Mn from AMD. Asif et al. [93] showed the efficacy of a direct contact (DC)-MD system for the treatment of AMD. They found that the DCMD achieved 100% removal of AMD and produced high-quality effluent. However, the permeate flux was reduced by 76% due to membrane fouling induced by membrane scaling, and this flux reduction was based on the metal content as well as the presence of bulk organics in the feed water.

This section reviews some examples of South African AMD treatment projects that are either commercially developed and in operation, in the pilot stage, or under evaluation. The alkali–barium–calcium (ABC) method, developed by the Council for Scientific and Industrial Research (CSIR) in South Africa, consists of three phases. The first stage is the addition of lime and calcium sulphide to remove metals and acids. The second stage involves treating most of the remaining water with barium carbonate to remove the remaining sulphate as barium sulphate. The barium sulphate and some sludge wastes are reduced in a coal-fired kiln to recover some of the alkaline compounds used for neutralisation as well as barium and calcium, some of which can be recycled back into the treatment process [94–96]. This method is a potentially cost-effective treatment for AMD because of the potential reuse opportunities from recycling [97]. However, it still produces a significant amount of waste sludge that must be disposed of. In addition, this process also involves high capital and operating costs, especially with running a coal-fired kiln.

A possible modification/improvement to the above is the Tshwane University of Technology's magnesium–barium–alkali (MBA) treatment process [98]. This process uses barium hydroxide for two purposes, that is, to precipitate and remove sulphate as barium sulphate and to precipitate and remove magnesium as magnesium hydroxide.

The CSIR recently developed and patented a sustainable AMD treatment technology called Magnesite–Softeners–Reverse-Osmosis–Eutectic (MASRO) Freeze Crystallisation that uses magnesite slurry to neutralise the AMD and precipitate metal hydroxides [99]. The advantage of using magnesite is that most of the gypsum, which usually forms a large portion of AMD sludge if it is first treated with a calcium compound like lime as a neutralising agent, does not precipitate with the metal hydroxide sludge. This allows for the possibility of easily concentrating and treating the metal hydroxide sludge to remove more of the valuable metals. Following this step, a lime slurry is added to the AMD, and this precipitates a gypsum sludge (70% gypsum) containing brucite (i.e., $Mg(OH)_2$). Next, the AMD is treated with a soda ash slurry to recover any residual calcium (65% as calcium carbonate) and magnesium and finally, the remaining water is treated using RO to improve its quality so it can be fit for human consumption. Masindi et al. [100] published results of a pilot plant of the MASROE process which was designed and built to treat 20 kL of AMD per day. They also calculated that the direct field costs to treat the AMD by this process was ZAR 65.60/kL.

However, if the by-products such as metal hydroxides could be treated to produce iron pigment, gypsum, and lime which could be purified to be sellable, then Masindi et al. [100] calculated that the sale of these products could perhaps yield an overall saving of approximately ZAR 9.00/kL (it is also noted that the possible sale of treated water did not appear to be accounted for and this may further increase these potential savings). Thus, presuming the above, the direct field costs could possibly be reduced to approximately ZAR 56.60/kL (i.e., a ±14% cost saving). However, it must be noted that AMD transportation costs were deemed negligible as it was assumed that the plant would be close to the AMD source. The treatment costs were also calculated presuming that the plant would operate for most of the year (i.e., 95% of the time, 24 h/day) and that electrical power and cleaning water were the only required utilities. The treatment costs also excluded any operational labour. In addition, the study did not account for any costs of processing the sludges to prepare iron pigments, gypsum, or calcium carbonate of marketable quality. Therefore, it is possible that the costs not accounted for could lead to less of a cost saving than predicted. Until further research is conducted, these additional costs might potentially increase the overall treatment costs to greater than ZAR 65.60/kL. However, one of the key findings from this research was the fact that setting up a process to effectively claim by-products for potential resale is possible if the process is planned and correctly constructed. And, the removal and resale of valuable components of the sludge such as metals, that would otherwise be potentially toxic if left in the sludge, leads to better environmental protection and potentially has less requirements for newly mined resources.

The SAVMIN process, developed by Mintek, involves five stages of AMD treatment. Firstly, lime is added to precipitate metals [101,102]. Secondly, using gypsum seed, all the remaining gypsum is removed. Thirdly, aluminium hydroxide is added to the remaining AMD water, and this produces ettringite (a calcium–aluminium sulphate mineral) which removes any remaining dissolved calcium and sulphate. Fourthly, the ettringite is removed and remixed with sulphuric acid which causes decomposition into aluminium hydroxide (which is recycled back into the process) and gypsum (some of which is recycled back into the process). Finally, in the fifth stage, the remaining water from stage four is treated with carbon dioxide to lower the pH and remove calcite by precipitation. The advantages of the SAVMIN process include high-quality by-products such as metal hydroxides, gypsum, and calcite which can potentially be resold to enhance the economic feasibility of this treatment [17,101]. However, again, it too produces significant amounts of waste sludge that would need to be disposed of.

A process developed in the United States of America is the slurry precipitation and recycling reverse osmosis (SPARRO) process. SPARRO uses membrane desalination to treat AMD and produce water at variable recoveries depending on, for instance, its chemical properties. Membrane fouling is and remains a challenge with this process and thus developing membranes to improve their performance may increase the economic feasibility of SPARRO to treat AMD [72].

Another process developed in South Africa is gypsum–continuous ion exchange (GYP-CIX). It is a continuous fluidised bed ion exchange process designed to remove calcium and sulphate from gypsum-saturated waters such as AMD [103]. During the first stage of the GYP-CIX process, cations can be removed from the AMD by cation exchange resins. After cation removal, anions are then removed by anion exchange resins. When required, the anion exchange resin is regenerated by lime while the cation exchange resin is regenerated by sulphuric acid. The advantages of this process include calcium and sulphate precipitates of relatively high quality that have the potential to be reused [104]. The use of inexpensive chemicals and efficient water recoveries are further benefits. However, as would seem to be customary, a significant amount of sludge is generated during the renewal of the ion exchange resins, and this typically necessitates an expensive disposal method.

THIOPAQ is a biotechnological approach to AMD treatment [105,106]. It involves two stages, the first being the addition of hydrogen gas to the AMD to produce sulphide from sulphate which precipitates out metal sulphides. Any excess hydrogen sulphide produced is oxidised to elemental sulphur in the second stage using sulphide-oxidising bacteria. The hydrogen gas used in the first stage of the THIOPAQ process is generated using ethanol and butanol. Recently, however, these chemicals have become quite expensive which has reduced the attractiveness of this approach.

The Rhodes BioSURE process, developed at Rhodes University in South Africa, is a biological treatment used to remove acid from AMD using waste such as sewage sludge or organic wastes [107–109]. While using these wastes will make treatment cheaper, they are also a potentially limiting reagent if a sufficient stock of them is not available. However, the advantages are re-use of sewage sludge or organic waste which would result in lower landfill loads and costs. Interestingly, this process was used by the East Rand Water Care Company in Grootvlei [72].

High-pressure reverse osmosis (HiPRO) was developed by Nafasi Water, then known as Aveng Water, in South Africa and applied at the eMalahleni treatment plant. The recovery of water was generally very good while brine and solid waste were also produced [110]. By-products from the solid waste that could potentially be sold were various purities of calcium sulphate and metal sulphates. However, the main challenge was the treatment/disposal of the remaining waste sludge and brine.

Luo et al. [111] provide details of a treatment that they successfully used to recover metals and produce hydrogen gas using microbial electrolysis cells to treat AMD. This electrolysis technology is one of the more promising methods to be developed. Through microbiologically assisted electrolysis, these researchers were able to remove copper, nickel, and iron from simulated AMD solutions while concurrently producing hydrogen to potentially offset some of the energy inputs during treatment.

Nleya et al. [64] published their research about the possible production of sulphuric acid from AMD. They summarised that, although not currently economically profitable, methods such as freeze crystallisation and acid retardation may potentially be the most promising technologies for acid recovery. These authors also indicated that the investigation and possible use of lower cost energy sources would assist to make these alternative treatments of AMD more economically viable.

The research details of staged electrochemical treatment in a laboratory to neutralise the pH and remove metals from AMD have recently been published by Brewster et al. [66]. Briefly, an electrochemical system was set up and a current applied between a cathode and anode. This caused the pH of the cathodic solution to increase while the anolyte solution pH decreased. The anions, including sulphate, were drawn across a membrane from the cathodic solution into the anolyte solution because of the decrease in pH. The increasing pH in the cathodic solution caused the dissolved metals to concurrently precipitate out at specific pH endpoints. The results indicate that metals like aluminium, iron, manganese, zinc, nickel, lead, and others were successfully removed. The advantages of this approach include the use of virtually no chemicals and the production of lower sludge volumes. Other advantages, as indicated by these authors, are the co-precipitation of most AMD metals in a controllable manner which will assist with possible recovery and recycling and the possible recovery and sale of sulphuric acid. The significant disadvantages include high initial capital costs and membrane fouling. Figure 1 summarises the current technologies used for AMD treatment.

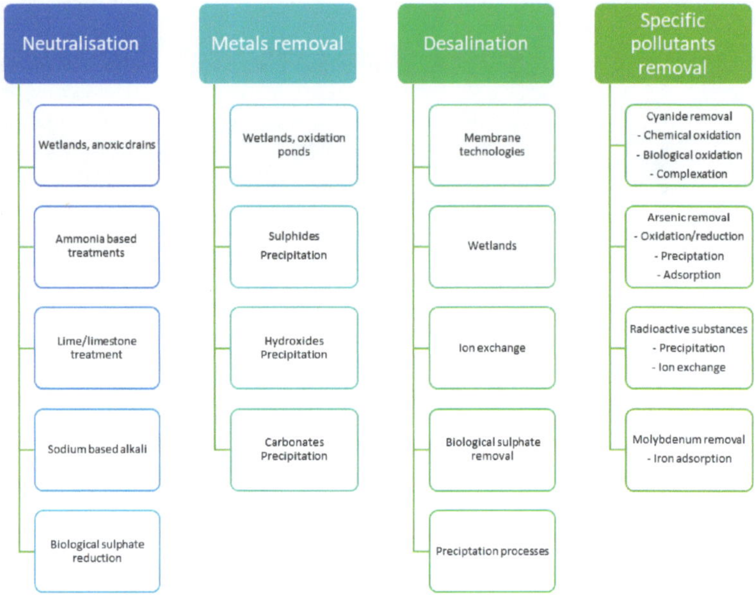

Figure 1. Current technologies used for AMD treatment.

7. Conclusions and Future Prospects

AMD, in South Africa and the rest of the world, is the cause of serious environmental and social concern and requires urgent attention. There are several active, passive, abiotic, and biotic treatments that have been investigated and, in some instances, implemented. The most widely used AMD treatment throughout the world is active chemical neutralisation which has been modified recently to be more efficient by continuously recycling most of the sludge, using fewer neutralisation chemicals, and producing less waste sludge. However, chemical neutralisation remains very expensive and produces significant amounts of potentially toxic waste. Various active and passive treatments have been investigated and generally look promising at the lab bench scale and a few at pilot plant scale. However, none of the recent techniques to treat large volumes of AMD have been implemented and proven to work at full scale for an extended period. Thus, the next step remains full-scale implementation and successful, long-term operation. Often, however, constraints like costly initial capital expenses are a factor that needs to be overcome. Besides these challenges, there is the real potential to produce by-products that could be sold to offset some of the initial capital costs and the on-going operational treatment costs. Some new technologies have successfully reclaimed metals, sulphur-based products such as gypsum and sulphuric acid, and other alkali chemicals including calcium carbonate. These alternative treatments offer several advantages including the requirement for less new materials because of recycling and a reduction in the amount and treatment and disposal costs of potentially toxic sludge waste. However, the emphasis must now be on full-scale implementation of the most promising techniques that recover and sell viable, marketable by-products to significantly offset the cost of traditional AMD treatment technologies and promote the circular economy. Future research is needed to increase the efficiency of all the methods used, with a focus on reducing the generation of secondary pollutants and recovering valuable resources.

Author Contributions: Conceptualization, J.B. and N.R.; formal analysis, J.B. and N.R.; investigation, J.B., D.R.-M., and N.R.; resources, R.M.; writing—original draft preparation, J.B. and N.R.; writing—review and editing, J.B., R.M., D.R.-M., and N.R.; visualization, J.B. and N.R.; supervision,

R.M. and J.B.; project administration, R.M. and J.B. All authors have read and agreed to the published version of the manuscript.

Funding: This research received no external funding.

Data Availability Statement: No new data were created or analysed in this study. Data sharing is not applicable to this article.

Acknowledgments: The authors acknowledge the support received for this study from Council for Scientific and Industrial Research (CSIR).

Conflicts of Interest: The authors declare no conflict of interest.

References

1. Baloyi, J.; Seadira, T.; Raphulu, M.; Ochieng, A. Preparation, Characterization and Growth Mechanism of Dandelion-like TiO2 Nanostructures and Their Application in Photocatalysis towards Reduction of Cr(VI). *Mater. Today Proc.* **2015**, *2*, 3973–3987. [CrossRef]
2. Seadira, T.; Baloyi, J.; Raphulu, M.; Moutloali, R.; Ochieng, A. Acid Mine Drainage Treatment Using Constructed Wetland. In Proceedings of the International Conference on Chemical, Integrated Waste Management and Environmental Engineering, Johannesburg, South Africa, 15–16 April 2014.
3. Ighalo, J.O.; Kurniawan, S.B.; Iwuozor, K.O.; Aniagor, C.O.; Ajala, O.J.; Oba, S.N.; Iwuchukwu, F.U.; Ahmadi, S.; Igwegbe, C.A. A Review of Treatment Technologies for the Mitigation of the Toxic Environmental Effects of Acid Mine Drainage (AMD). *Process Saf. Environ. Prot.* **2022**, *157*, 37–58. [CrossRef]
4. Nepfumbada, C.; Tavengwa, N.T.; Masindi, V.; Foteinis, S.; Chatzisymeon, E. Recovery of Phosphate from Municipal Wastewater as Calcium Phosphate and Its Subsequent Application for the Treatment of Acid Mine Drainage. *Resour. Conserv. Recycl.* **2023**, *190*, 106779. [CrossRef]
5. Johnson, D.B.; Hallberg, K.B. Acid Mine Drainage Remediation Options: A Review. *Sci. Total Environ.* **2005**, *338*, 3–14. [CrossRef] [PubMed]
6. Daraz, U.; Li, Y.; Ahmad, I.; Iqbal, R.; Ditta, A. Remediation Technologies for Acid Mine Drainage: Recent Trends and Future Perspectives. *Chemosphere* **2023**, *311*, 137089. [CrossRef]
7. Larochelle, T.; Noble, A.; Ziemkiewicz, P.; Hoffman, D.; Constant, J. A Fundamental Economic Assessment of Recovering Rare Earth Elements and Critical Minerals from Acid Mine Drainage Using a Network Sourcing Strategy. *Minerals* **2021**, *11*, 1298. [CrossRef]
8. Laker, M.C. Environmental Impacts of Gold Mining—With Special Reference to South Africa. *Mining* **2023**, *3*, 205–220. [CrossRef]
9. Jiao, Y.; Zhang, C.; Su, P.; Tang, Y.; Huang, Z.; Ma, T. A Review of Acid Mine Drainage: Formation Mechanism, Treatment Technology, Typical Engineering Cases and Resource Utilization. *Process Saf. Environ. Prot.* **2023**, *170*, 1240–1260. [CrossRef]
10. Bai, S.J.; Li, J.; Yuan, J.Q.; Bi, Y.X.; Ding, Z.; Dai, H.X.; Wen, S.M. An Innovative Option for the Activation of Chalcopyrite Flotation Depressed in a High Alkali Solution with the Addition of Acid Mine Drainage. *J. Cent. South Univ.* **2023**, *30*, 811–822. [CrossRef]
11. Azapagic, A. Developing a Framework for Sustainable Development Indicators for the Mining and Minerals Industry. *J. Clean. Prod.* **2004**, *12*, 639–662. [CrossRef]
12. Rezaie, B.; Anderson, A. Sustainable Resolutions for Environmental Threat of the Acid Mine Drainage. *Sci. Total Environ.* **2020**, *717*, 137211. [CrossRef]
13. Iakovleva, E.; Mäkilä, E.; Salonen, J.; Sitarz, M.; Wang, S.; Sillanpää, M. Acid Mine Drainage (AMD) Treatment. Neutralization and Toxic Elements Removal with Unmodified and Modified Limestone. *Ecol. Eng.* **2015**, *81*, 30–40. [CrossRef]
14. Vaziri Hassas, B.; Shekarian, Y.; Rezaee, M. Selective Precipitation of Rare Earth and Critical Elements from Acid Mine Drainage—Part I: Kinetics and Thermodynamics of Staged Precipitation Process. *Resour. Conserv. Recycl.* **2023**, *188*, 106654. [CrossRef]
15. Al-Zoubi, H.; Rieger, A.; Steinberger, P.; Pelz, W.; Haseneder, R.; Härtel, G. Optimization Study for Treatment of Acid Mine Drainage Using Membrane Technology. *Sep. Sci. Technol.* **2010**, *45*, 2004–2016. [CrossRef]
16. Felipe, E.C.B.; Batista, K.A.; Ladeira, A.C.Q. Recovery of Rare Earth Elements from Acid Mine Drainage by Ion Exchange. *Environ. Technol.* **2021**, *42*, 2721–2732. [CrossRef]
17. van Rooyen, M.; van Staden, P.J.; du Preez, K.A. Sulphate Removal Technologies for the Treatment of Mine-Impacted Water. *J. S. Afr. Inst. Min. Metall.* **2021**, *121*, 523–530. [CrossRef]
18. Mondaca, S.L.; Leiva, C.A.; Acuña, C.A.; Serey, E.A. Flow Enhancement of Mineral Pastes to Increase Water Recovery in Tailings: A Matlab-Based Imaging Processing Tool. *Sci. Program.* **2020**, *2020*, 5607242. [CrossRef]
19. Chen, G.; Ye, Y.; Yao, N.; Hu, N.; Zhang, J.; Huang, Y. A Critical Review of Prevention, Treatment, Reuse, and Resource Recovery from Acid Mine Drainage. *J. Clean. Prod.* **2021**, *329*, 129666. [CrossRef]
20. Hassas, B.V.; Rezaee, M.; Pisupati, S.V. Precipitation of Rare Earth Elements from Acid Mine Drainage by CO2 Mineralization Process. *Chem. Eng. J.* **2020**, *399*, 125716. [CrossRef]

21. Cicek, Z. Selective Recovery of Rare Earth Elements from Acid Mine Selective Recovery of Rare Earth Elements from Acid Mine Drainage Treatment Byproduct Drainage Treatment Byproduct Recommended Citation Recommended Citation. Master's Thesis, Statler College of Engineering and Mineral Resources, Morgantown, WV, USA, 2023.
22. Elghali, A.; Benzaazoua, M.; Taha, Y.; Amar, H.; Ait-khouia, Y.; Bouzahzah, H.; Hakkou, R. Prediction of Acid Mine Drainage: Where We Are. *Earth Sci. Rev.* **2023**, *241*, 104421. [CrossRef]
23. Chen, Y.; Liu, G.; Zhou, C.; Zhou, H.; Wei, Y.; Liu, Y. The Influence of Gold Mining Wastes on the Migration-Transformation Behavior and Health Risks of Arsenic in the Surrounding Soil of Mined-Area. *Front. Earth Sci.* **2023**, *10*, 1068763. [CrossRef]
24. Mukolu, N. The effect of waste management of oil drilling and gold mining extraction in the yakutia arctic (Russian). *Am. J. Humanit. Soc. Sci. Res. (AJHSSR)* **2023**, *7*, 59–72.
25. Cacciuttolo, C.; Marinovic, A. Experiences of Underground Mine Backfilling Using Mine Tailings Developed in the Andean Region of Peru: A Green Mining Solution to Reduce Socio-Environmental Impacts. *Sustainability* **2023**, *15*, 12912. [CrossRef]
26. Yuan, S.; Sui, W.; Han, G.; Duan, W. An Optimized Combination of Mine Water Control, Treatment, Utilization, and Reinjection for Environmentally Sustainable Mining: A Case Study. *Mine Water Environ.* **2022**, *41*, 828–839. [CrossRef]
27. Gonah, T. Impact of Acid Mine Drainage on Water Resources in South Africa. In *Management and Mitigation of Acid Mine Drainage in South Africa: Input for Mineral Beneficiation in Africa*; Africa Institute of South Africa: Pretoria, South Africa, 2016; pp. 41–65. [CrossRef]
28. Abiye, T.A.; Ali, K.A. Potential Role of Acid Mine Drainage Management towards Achieving Sustainable Development in the Johannesburg Region, South Africa. *Groundw. Sustain. Dev.* **2022**, *19*, 100839. [CrossRef]
29. Windisch, J.; Gradwohl, A.; Gilbert, B.M.; Dos Santos, Q.M.; Wallner, G.; Avenant-Oldewage, A.; Jirsa, F. Toxic Elements in Sediment and Water of the Crocodile River (West) System, South Africa, Following Acid Mine Drainage. *Appl. Sci.* **2022**, *12*, 10531. [CrossRef]
30. Minnaar, A. Water Pollution and Contamination from Gold Mines: Acid Mine Drainage in Gauteng Province, South Africa. In *Water, Governance, and Crime Issues*; Springer: Cham, Switzerland, 2020; pp. 193–219. [CrossRef]
31. Nofal, A.P.; Dos Santos, Q.M.; Jirsa, F.; Avenant-Oldewage, A. Camallanid Nematodes from Clarias Gariepinus (Burchell, 1822) in the Crocodile River, Gauteng, South Africa: Exploring Diversity and Divergence in an Acid-Mine Drainage Impacted Environment. *Int. J. Parasitol. Parasites Wildl.* **2022**, *19*, 196–210. [CrossRef]
32. Ouma, K.O.; Shane, A.; Syampungani, S. Aquatic Ecological Risk of Heavy-Metal Pollution Associated with Degraded Mining Landscapes of the Southern Africa River Basins: A Review. *Minerals* **2022**, *12*, 225. [CrossRef]
33. Atangana, E. Evaluation of the Impact of Coal Mining on Surface Water in the Boesmanspruit, Mpumalanga, South Africa. 2023. Available online: https://doi.org/10.21203/RS.3.RS-3184680/V1 (accessed on 3 August 2023).
34. Simpson, G.B.; Badenhorst, J.; Jewitt, G.P.W.; Berchner, M.; Davies, E. Competition for Land: The Water-Energy-Food Nexus and Coal Mining in Mpumalanga Province, South Africa. *Front. Environ. Sci.* **2019**, *7*, 422006. [CrossRef]
35. Sakala, E.; Novhe, O.; Kumar Vadapalli, V.R. Application of Artificial Intelligence (AI) to Predict Mine Water Quality, a Case Study in South Africa. In Proceedings of the Mine Water Association Conference: Technological and Ecological Challenges, International Mine Water Association Annual Conference, Perm, Russia, 15–19 July 2019.
36. McCarthy, T.S. The Impact of Acid Mine Drainage in South Africa. *S. Afr. J. Sci.* **2011**, *107*, 1–7. [CrossRef]
37. Naidoo, S. *Social Constructions of Water Quality in South Africa: A Case Study of the Blesbokspruit River in the Context of Acid Mine Drainage Treatment*; Springer Nature: Berlin/Heidelberg, Germany, 2022; pp. 1–219. [CrossRef]
38. Lourenco, M.; Curtis, C. The Influence of a High-Density Sludge Acid Mine Drainage (AMD) Chemical Treatment Plant on Water Quality along the Blesbokspruit Wetland, South Africa. *Water SA* **2021**, *47*, 35–44. [CrossRef]
39. Scott, R. *Flooding of the Central and East Rand Gold Mines*; WRC Report 486/1/95; Water Research Commission: Pretoria, South Africa, 1995.
40. Addo-Bediako, A. Comparative Spatial Assessment of Trace Metal(Loid) Pollution in the Sediments of the Lower Olifants River Basin in South Africa. *Front. Environ. Sci.* **2022**, *10*, 882393. [CrossRef]
41. Masindi, V.; Foteinis, S.; Renforth, P.; Ndiritu, J.; Maree, J.P.; Tekere, M.; Chatzisymeon, E. Challenges and Avenues for Acid Mine Drainage Treatment, Beneficiation, and Valorisation in Circular Economy: A Review. *Ecol. Eng.* **2022**, *183*, 106740. [CrossRef]
42. Wolkersdorfer, C. *Mine Water Treatment-Active and Passive Methods*; Springer: Berlin/Heidelberg, Germany, 2022; ISBN 3662657694.
43. Marr, S.M.; Swemmer, A.M. Hydrological Characteristics of Extreme Floods in the Klaserie River, a Headwater Stream in Southern Africa. *J. Limnol.* **2023**, 82. [CrossRef]
44. Netshitungulwana, K.R.T.; Gauert, C.; Vermeulen, D.; Yibas, B.; Shai, M.; Lusunzi, R. Geochemical Characterisation of the Witbank Coalfield Geological Strata and Assessment of Potential Metal Impact on the Receiving Environment. In Proceedings of the International Mine Water Association 2022 Conference Reconnect, Christchurch, New Zealand, 6–10 November 2022.
45. Obaid, A.; Adam, E.; Ali, K.A. Land Use and Land Cover Change in the Vaal Dam Catchment, South Africa: A Study Based on Remote Sensing and Time Series Analysis. *Geomatics* **2023**, *3*, 205–220. [CrossRef]
46. Alexander, A.C.; Ndambuki, J.M. Impact of Mine Closure on Groundwater Resource: Experience from Westrand Basin-South Africa. *Phys. Chem. Earth Parts A/B/C* **2023**, *131*, 103432. [CrossRef]
47. du Plessis, A. Progressive Deterioration of Water Quality Within South Africa. In *South Africa's Water Predicament: Freshwater's Unceasing Decline*; Springer: Berlin/Heidelberg, Germany, 2023; pp. 109–141.

48. Wood, D.L.; Cole, K.A.; Herndon, E.M.; Singer, D.M. Lime Slurry Treatment of Soils Developing on Abandoned Coal Mine Spoil: Linking Contaminant Transport from the Micrometer to Pedon-Scale. *Appl. Geochem.* **2023**, *151*, 105617. [CrossRef]
49. Bondarenko, V.I.; Kovalevska, I.A.; Podkopaiev, S.V.; Sheka, I.V.; Tsivka, Y.S. Substantiating Arched Support Made of Composite Materials (Carbon Fiber-Reinforced Plastic) for Mine Workings in Coal Mines. In *Proceedings of the IOP Conference Series: Earth and Environmental Science*; IOP Publishing: Bristol, UK, 2022; Volume 1049, p. 012026.
50. Camenzuli, D. Development of Orthophosphate and Silica Treatments for the Management of Environmental Contaminants at Wilkes Landfill, East Antarctica. Ph.D. Thesis, Macquarie University, Sydney, Australia, 2015.
51. Mulopo, J. Active Physical Remediation of Acid Mine Drainage: Technologies Review and Perspectives. *J. Ecol. Eng.* **2022**, *23*, 148–163. [CrossRef]
52. Alekseyev, V.A. Reasons for the Formation of Acidic Drainage Water in Dumps of Sulfide-Containing Rocks. *Geochem. Int.* **2022**, *60*, 78–91. [CrossRef]
53. Markovic, R.; Bessho, M.; Masuda, N.; Stevanovic, Z.; Bozic, D.; Trujic, T.A.; Gardic, V. New Approach of Metals Removal from Acid Mine Drainage. *Appl. Sci.* **2020**, *10*, 5925. [CrossRef]
54. Humphries, M.S.; McCarthy, T.S.; Pillay, L. Attenuation of Pollution Arising from Acid Mine Drainage by a Natural Wetland on the Witwatersrand. *S. Afr. J. Sci.* **2017**, *113*, 9. [CrossRef]
55. Seervi, V.; Yadav, H.L.; Srivastav, S.K.; Jamal, A. Overview of Active and Passive Systems for Treating Acid Mine Drainage. *IARJSET* **2017**, *4*, 131–137. [CrossRef]
56. Porter, C.M.; Nairn, R.W. Ecosystem Functions within a Mine Drainage Passive Treatment System. *Ecol. Eng.* **2008**, *32*, 337–346. [CrossRef]
57. Ramla, B.; Sheridan, C. The Potential Utilisation of Indigenous South African Grasses for Acid Mine Drainage Remediation. *Water SA* **2015**, *41*, 247. [CrossRef]
58. Qian, G.; Li, Y. Acid and Metalliferous Drainage–A Global Environmental Issue. *J. Min. Mech. Eng.* **2019**, *1*, 1–4. [CrossRef]
59. Dama-Fakir, P.; Sithole, Z.; van Niekerk, A.M.; Dateling, J.; Maree, J.P.; Rukuni, T.; Mthombeni, T.; Ruto, S.; Zikalala, N.; Hughes, C.; et al. *Mine Water Treatment Technology Selection Tool: Users' Guide (TT 711/17)*; Water Research Commission: Pretoria, South Africa, 2017.
60. Bwapwa, J.K. A Review of Acid Mine Drainage in a Water-Scarce Country: Case of South Africa. *Environ. Manag. Sustain. Dev.* **2017**, *7*, 1. [CrossRef]
61. Trumm, D. Selection of Active and Passive Treatment Systems for AMDflow Charts for New Zealand Conditions. *N. Z. J. Geol. Geophys.* **2010**, *53*, 195–210. [CrossRef]
62. RoyChowdhury, A.; Sarkar, D.; Datta, R. Remediation of Acid Mine Drainage-Impacted Water. *Curr. Pollut. Rep.* **2015**, *1*, 131–141. [CrossRef]
63. Akcil, A.; Koldas, S. Acid Mine Drainage (AMD): Causes, Treatment and Case Studies. *J. Clean. Prod.* **2006**, *14*, 1139–1145. [CrossRef]
64. Nleya, Y.; Simate, G.S.; Ndlovu, S. Sustainability Assessment of the Recovery and Utilisation of Acid from Acid Mine Drainage. *J. Clean. Prod.* **2016**, *113*, 17–27. [CrossRef]
65. Yuan, J.; Ding, Z.; Bi, Y.; Li, J.; Wen, S.; Bai, S. Resource Utilization of Acid Mine Drainage (AMD): A Review. *Water* **2022**, *14*, 2385. [CrossRef]
66. Brewster, E.T.; Freguia, S.; Edraki, M.; Berry, L.; Ledezma, P. Staged Electrochemical Treatment Guided by Modelling Allows for Targeted Recovery of Metals and Rare Earth Elements from Acid Mine Drainage. *J. Environ. Manag.* **2020**, *275*, 111266. [CrossRef]
67. Kaur, G.; Couperthwaite, S.J.; Hatton-Jones, B.W.; Millar, G.J. Alternative Neutralisation Materials for Acid Mine Drainage Treatment. *J. Water Process Eng.* **2018**, *22*, 46–58. [CrossRef]
68. Acharya, B.S.; Kharel, G. Acid Mine Drainage from Coal Mining in the United States—An Overview. *J. Hydrol.* **2020**, *588*, 125061. [CrossRef]
69. Kefeni, K.K.; Msagati, T.A.M.; Mamba, B.B. Acid Mine Drainage: Prevention, Treatment Options, and Resource Recovery: A Review. *J. Clean. Prod.* **2017**, *151*, 475–493. [CrossRef]
70. Pohl, A. Removal of Heavy Metal Ions from Water and Wastewaters by Sulfur-Containing Precipitation Agents. *Water Air Soil. Pollut.* **2020**, *231*, 1–17. [CrossRef]
71. Saleem, J.; Bin Shahid, U.; Hijab, M.; Mackey, H.; McKay, G. Production and Applications of Activated Carbons as Adsorbents from Olive Stones. *Biomass Convers. Biorefinery* **2019**, *9*, 775–802. [CrossRef]
72. Simate, G.S.; Ndlovu, S. Acid Mine Drainage: Challenges and Opportunities. *J. Environ. Chem. Eng.* **2014**, *2*, 1785–1803. [CrossRef]
73. Ruihua, L.; Lin, Z.; Tao, T.; Bo, L. Phosphorus Removal Performance of Acid Mine Drainage from Wastewater. *J. Hazard. Mater.* **2011**, *190*, 669–676. [CrossRef]
74. Kumari, M.; Bhattacharya, T. A Review on Bioaccessibility and the Associated Health Risks Due to Heavy Metal Pollution in Coal Mines: Content and Trend Analysis. *Environ. Dev.* **2023**, *46*, 100859. [CrossRef]
75. Orlović-Leko, P.; Farkaš, B.; Galić, I. A Short Review of Environmental and Health Impacts of Gold Mining. *Reliab. Theory Appl.* **2022**, *4*, 242–248.
76. Li, S.; Yu, L.; Jiang, W.; Yu, H.; Wang, X. The Recent Progress China Has Made in Green Mine Construction, Part I: Mining Groundwater Pollution and Sustainable Mining. *Int. J. Environ. Res. Public. Health* **2022**, *19*, 5673. [CrossRef] [PubMed]

77. Zhu, M.; Li, B.; Liu, G. Groundwater Risk Assessment of Abandoned Mines Based on Pressure-State-Response—The Example of an Abandoned Mine in Southwest China. *Energy Rep.* **2022**, *8*, 10728–10740. [CrossRef]
78. Weinberg, R.; Coyte, R.; Wang, Z.; Das, D.; Vengosh, A. Water Quality Implications of the Neutralization of Acid Mine Drainage with Coal Fly Ash from India and the United States. *Fuel* **2022**, *330*, 125675. [CrossRef]
79. Li, Q.; Ji, B.; Honaker, R.; Noble, A.; Zhang, W. Partitioning Behavior and Mechanisms of Rare Earth Elements during Precipitation in Acid Mine Drainage. *Colloids Surf. A Physicochem. Eng. Asp.* **2022**, *641*, 128563. [CrossRef]
80. Marove, C.A.; Sotozono, R.; Tangviroon, P.; Tabelin, C.B.; Igarashi, T. Assessment of Soil, Sediment and Water Contaminations around Open-Pit Coal Mines in Moatize, Tete Province, Mozambique. *Environ. Adv.* **2022**, *8*, 100215. [CrossRef]
81. Thisani, S.K.; Von Kallon, D.V.; Byrne, P. A Fixed Bed Pervious Concrete Anaerobic Bioreactor for Biological Sulphate Remediation of Acid Mine Drainage Using Simple Organic Matter. *Sustainability* **2021**, *13*, 6529. [CrossRef]
82. Lozano, A.; Ayora, C.; Fernández-Martínez, A. Sorption of Rare Earth Elements on Schwertmannite and Their Mobility in Acid Mine Drainage Treatments. *Appl. Geochem.* **2020**, *113*, 104499. [CrossRef]
83. Song, G.; Wang, X.; Romero, C.; Chen, H.; Yao, Z.; Kaziunas, A.; Schlake, R.; Anand, M.; Lowe, T.; Driscoll, G. Extraction of Selected Rare Earth Elements from Anthracite Acid Mine Drainage Using Supercritical CO2 via Coagulation and Complexation. *J. Rare Earths* **2021**, *39*, 83–89. [CrossRef]
84. Hu, X.; Yang, H.; Fang, X.; Shi, T.; Tan, K. Recovery of Bio-sulfur and Metal Resources from Mine Wastewater by Sulfide Biological Oxidation-Alkali Flocculation: A Pilot-Scale Study. *Sci. Total Environ.* **2023**, *876*, 162546. [CrossRef]
85. Vo, T.D.H.; Nguyen, B.S.; Vu, C.T.; Shih, Y.J.; Huang, Y.H. Recovery of Iron (II) and Aluminum (III) from Acid Mine Drainage by Sequential Selective Precipitation and Fluidized Bed Homogeneous Crystallization (FBHC). *J. Taiwan Inst. Chem. Eng.* **2020**, *115*, 135–143.
86. Maroufi, N.; Hajilary, N. Nanofiltration Membranes Types and Application in Water Treatment: A Review. *Sustain. Water Resour. Manag.* **2023**, *9*, 142. [CrossRef]
87. Zhao, S.; Chen, Y.; Wu, G.; Li, J.; Ren, Y.; Duan, X. Investigation on Nanofiltration Membrane Fouling Behaviour of Cation-Induced Apam in Strontium-Bearing Mine Water. *J. Environ. Chem. Eng.* **2023**, *11*, 110940. [CrossRef]
88. Ang, W.L.; Mohammad, A.W.; Ahmad, N.N.R.; Teow, Y.H. Role of Nanofiltration Process for Sustainability in Industries: Reuse, Recycle, and Resource Recovery. In *Nanofiltration for Sustainability*; CRC Press: Boca Raton, FL, USA, 2023; pp. 1–13.
89. Andalaft, J.; Schwarz, A.; Pino, L.; Fuentes, P.; Bórquez, R.; Aybar, M. Assessment and Modeling of Nanofiltration of Acid Mine Drainage. *Ind. Eng. Chem. Res.* **2018**, *57*, 14727–14739. [CrossRef]
90. Wadekar, S.S.; Vidic, R.D. Comparison of Ceramic and Polymeric Nanofiltration Membranes for Treatment of Abandoned Coal Mine Drainage. *Desalination* **2018**, *440*, 135–145. [CrossRef]
91. Masindi, V. Recovery of Drinking Water and Valuable Minerals from Acid Mine Drainage Using an Integration of Magnesite, Lime, Soda Ash, CO2 and Reverse Osmosis Treatment Processes. *J. Environ. Chem. Eng.* **2017**, *5*, 3136–3142. [CrossRef]
92. León-Venegas, E.; Vilches-Arenas, L.F.; Fernández-Baco, C.; Arroyo-Torralvo, F. Potential for Water and Metal Recovery from Acid Mine Drainage by Combining Hybrid Membrane Processes with Selective Metal Precipitation. *Resour. Conserv. Recycl.* **2023**, *188*, 106629. [CrossRef]
93. Asif, M.B.; Price, W.E.; Fida, Z.; Tufail, A.; Ren, T.; Hai, F.I. Acid Mine Drainage and Sewage Impacted Groundwater Treatment by Membrane Distillation: Organic Micropollutant and Metal Removal and Membrane Fouling. *J. Environ. Manag.* **2021**, *291*, 112708. [CrossRef]
94. Mulopo, J.; Zvimba, J.N.; Swanepoel, H.; Bologo, L.T.; Maree, J. Regeneration of Barium Carbonate from Barium Sulphide in a Pilot-Scale Bubbling Column Reactor and Utilization for Acid Mine Drainage. *Water Sci. Technol.* **2012**, *65*, 324–331. [CrossRef]
95. Motaung, S.; Maree, J.; De Beer, M.; Bologo, L.; Theron, D.; Baloyi, J. Recovery of Drinking Water and By-Products from Gold Mine Effluents. *Int. J. Water Resour. Dev.* **2008**, *24*, 433–450. [CrossRef]
96. Swanepoel, H.; de Beer, M.; Liebenberg, L. Complete Sulphate Removal from Neutralised Acidic Mine Drainage with Barium Carbonate. *Water Pract. Technol.* **2012**, *7*, wpt2012003. [CrossRef]
97. De Beer, M.; Maree, J.P.; Wilsenach, J.; Motaung, S.; Bologo, L.; Radebe, V. Acid Mine Water Reclamation Using the ABC Process. In Proceedings of the International Mine Water Association Symposium, Sydney, NS, Canada, 4–9 September 2010.
98. Bologo, V.; Maree, J.P.; Carlsson, F. Application of Magnesium Hydroxide and Barium Hydroxide for the Removal of Metals and Sulphate from Mine Water. *Water SA* **2012**, *38*, 23–28. [CrossRef]
99. Masindi, V.; Chatzisymeon, E.; Kortidis, I.; Foteinis, S. Assessing the Sustainability of Acid Mine Drainage (AMD) Treatment in South Africa. *Sci. Total Environ.* **2018**, *635*, 793–802. [CrossRef] [PubMed]
100. Masindi, V.; Osman, M.S.; Shingwenyana, R. Valorization of Acid Mine Drainage (AMD): A Simplified Approach to Reclaim Drinking Water and Synthesize Valuable Minerals—Pilot Study. *J. Environ. Chem. Eng.* **2019**, *7*, 103082. [CrossRef]
101. Petterson, D. Addressing Legacy Challenges. *Inside Min.* **2018**, *11*, 22–23.
102. van Rooyen, M.; van Staden, P.J. Deriving Value from Acid Mine Drainage. In *Recovery of Byproducts from Acid Mine Drainage Treatment*; Scrivener Publishing: Beverly, MA, USA, 2020; pp. 235–261. [CrossRef]
103. Robertson, A.M.; Everett, D.J.; Du Plessis, N.J. Sulfates Removal by the GYP-CIX Process Following Lime Treatment. In Proceedings of the Superfund XIV Conference and Exhibition, Washington, DC, USA, 30 November–2 December 1993.
104. Fernando, W.A.M.; Ilankoon, I.M.S.K.; Syed, T.H.; Yellishetty, M. Challenges and Opportunities in the Removal of Sulphate Ions in Contaminated Mine Water: A Review. *Miner. Eng.* **2018**, *117*, 74–90. [CrossRef]

105. Sullivan, D.; Arena, B.; de Vegt, A.; Buisman, C.; Jannsen, A. Converting Sulfide Biologically. In Proceedings of the PETSOC Annual Technical Meeting, Calgary, AB, Canada, 8–11 June 1997.
106. Dhir, B. Biotechnological Tools for Remediation of Acid Mine Drainage (Removal of Metals from Wastewater and Leachate). In *Bio-Geotechnologies for Mine Site Rehabilitation*; Elsevier: London, UK, 2018; pp. 67–82. [CrossRef]
107. Rose, P. Review: Long-Term Sustainability in the Management of Acid Mine Drainage Wastewaters Development of the Rhodes BioSURE Process. *Water SA* **2013**, *39*, 582. [CrossRef]
108. Rose, P.; Corbett, C.; Neba, A. Sewage Sludge as an Electron Donor in Biological Mine Wastewater Treatment: Development of the Rhodes BioSURE Process®. In Proceedings of the Mine Water 2004–Proceedings International Mine Water Association Symposium, Newcastle upon Tyne, UK, 20–25 September 2004; pp. 111–118.
109. Corbett, C.J. The Rhodes BioSURE Process in the Treatment of Acid Mine Drainage Wastewaters. Ph.D. Thesis, Rhodes University, Makhanda, South Africa, 2001.
110. Hutton, B.; Kahan, I.; Naidu, T.; Gunther, P. Operating and Maintenance Experience at the Emalahleni Water Reclamation Plant. In Proceedings of the International Mine Water Conference, Pretoria, South Africa, 19–23 October 2009.
111. Luo, H.; Liu, G.; Zhang, R.; Bai, Y.; Fu, S.; Hou, Y. Heavy Metal Recovery Combined with H2 Production from Artificial Acid Mine Drainage Using the Microbial Electrolysis Cell. *J. Hazard. Mater.* **2014**, *270*, 153–159. [CrossRef]

water

MDPI

Article

Risk Assessment and Control of Geological Hazards in Towns of Complex Mountainous Areas Based on Remote Sensing and Geological Survey

Weicui Ding [1,†], Gaofeng Wang [2,*,†], Qiang Yang [3,4,*,†], Youning Xu [1,5], Youlong Gao [2], Xuanhua Chen [1], Shenglin Xu [1], Lele Han [1] and Xinru Yang [1]

1 SinoProbe Center, Chinese Academy of Geological Sciences and China Geological Survey, Beijing 100084, China; dingweicuigis@163.com (W.D.); ksdzhj@sohu.com (Y.X.); xhchen@cags.ac.cn (X.C.); xushenglindizhi@163.com (S.X.); hanllgis@163.com (L.H.); yangxr422@163.com (X.Y.)
2 Center for Hydrogeology and Environmental Geology Survey, CGS, Baoding 071051, China; gemcgao@163.com
3 Institute of Geotechnical Engineering, Xi'an University of Technology, Xi'an 404100, China
4 China Institute of Geo-Environment Monitoring, Beijing 100081, China
5 Xi'an Geological Survey Center of China Geological Survey, Xi'an 710061, China
* Correspondence: wgf_cgcs303@163.com (G.W.); yang5359535@126.com (Q.Y.); Tel.: +86-150-9742-5164 (G.W.); +86-177-1002-8388 (Q.Y.)
† These authors contributed equally to this work.

Abstract: Mountainous areas have become among the most developed areas of geological hazards due to special geological environmental conditions and intensive human engineering activities. Geological hazards are a main threat to urbanization, rural revitalization, and new rural construction in complex mountainous areas. It is of great strategic significance to conduct large-scale geological hazard investigation and risk assessment in urban areas, control the risk of geological hazards at the source and propose risk control measures. In this paper, we established the technical methods of geologic hazard risk assessment and control in complex mountain towns by taking Longlin Town in the mountainous region of Gansu Longnan, China as the study area, with the Quanjia bay debris flows and Panping Village landslides as the typical pilot investigation and assessment. The methods consist of six stages—risk identification, hazard disaster model investigation, risk analysis, vulnerability assessment, risk evaluation and risk management and control measures and proposals. On this basis, the results of geological hazards with different precipitation frequencies (5%, 2%, 1%) are presented. The results show that 75.23% of the regions remained at low risk levels; 24.38% of the regions increased a risk level with decreasing precipitation frequency, and 0.39% of the regions remained at extremely high risk levels under different precipitation frequency conditions. For the Quanjia bay debris flows and Panping Village landslides case, we discussed the geological hazards risk source control contents, management and control technologies, engineering and non-engineering measures of disaster prevention and control for urban disasters and specific disaster areas. This research can provide technical support and reference for disaster prevention and mitigation, and territorial spatial planning.

Keywords: the Longnan Mountain area; geological hazard; risk assessment; risk management and control; remote sensing

check for updates

Citation: Ding, W.; Wang, G.; Yang, Q.; Xu, Y.; Gao, Y.; Chen, X.; Xu, S.; Han, L.; Yang, X. Risk Assessment and Control of Geological Hazards in Towns of Complex Mountainous Areas Based on Remote Sensing and Geological Survey. *Water* **2023**, *15*, 3170. https://doi.org/10.3390/w15183170

Academic Editors: Roger Falconer, Paolo Fabbri and Andrzej Witkowski

Received: 30 April 2023
Revised: 17 August 2023
Accepted: 29 August 2023
Published: 5 September 2023

1. Introduction

Many cities have been built on the region of canyon terraces, debris flow accumulation fans and slopes or landslides with fragile ecological environments due to the special natural geography and geological environmental conditions in the mountainous areas of western China. The occurrence and development of geological hazards in the mountains area were affected by human engineering activities in the process of city construction and development, resulting in frequent geological hazards and huge casualties and economic losses. In addition, the threat and danger of geological hazards are further increased by the increase and intensification of extreme weather, earthquakes, and large-scale engineering and economic activities. In the process of urban planning and development in mountainous areas, practical and reliable geological hazard risk zoning and control measures are an urgent need to reduce the casualties and economic losses caused by geological hazards. In particular, a scientific large-scale geological hazard survey and quantitative multi-hazard risk assessment is urgently needed to provide a technical basis for land use planning, disaster prevention and mitigation, and implementation of prevention and control projects.

The evaluation and management control of geological hazard risk is an effective way to prevent and reduce disasters, which mainly includes five major steps—risk identification, risk analysis and evaluation, risk countermeasure decision, implementation decision and risk supervision [1]. Geological hazard risk management control has become an important part of the disaster prevention and reduction strategy system in the world. At present, the technical methods and theoretical systems of geological hazard risk control have been formed basically according to the actual situation of the country or region, and geological hazard risk management has gradually changed from a semi-quantitative to a quantitative direction [2–7]. In recent years, many researchers have focused on the geological hazard risk assessment method system and the risk management system. In the mountainous areas of western China, the assessment of urban geological hazard risk, the formation of a series of ideas and technical methods for geological hazard investigation and risk assessment, and the proposition of a series of risk reduction programs have been undertaken [8–14]. Considering the possible damage of monolithic geologic hazards to population gathering areas and cities, many scholars have carried out the risk assessment of monolithic geologic hazards based on dynamic processes. For example, Xiao Lili et al. analyze the motion accumulation process of Sunjia landslide used by the numerical simulation and profiled the landslide risk level under extreme scenarios quantitatively [15]; Du Juan et al. established a computational model based on the finite volume method by considering the erosion effect and frictional resistance of the lower surface of the landslide during the motion of the landslide debris flow, and predicted the hazard and risk of the El Picacho landslide disaster in El Salvador [16]; Cui Peng et al. took the debris flow in Qingping town as an example, explained the formation mechanism of flash flood debris flow and the methods and contents of risk analysis and management, and proposed a series of risk assessment and risk management theory and method system based on the dynamic process of flash flood debris flow [17]; Shu Heping constructed the area, thickness and morphological characteristics of debris flow accumulation in Sanyanyu through physical simulation tests, and classified the hazard degree [18].

However, most of the above urban geological hazard risk assessments are based on factor analysis of statistical methods; there are many questions in the sufficient reflection of the detailed information of specific hazard sites, the application of assessment results in the risk control of each specific hazard site, and meeting the new requirements of taking the new road of urbanization and comprehensively improving the quality of urbanization [19–21]. In addition, there has been a lot of research about static risk assessment focused on the regions or single units, which are mainly applied to risk management planning, while there is a lack of research on the dynamic risk assessment of geological hazards at the town scale [22,23]. For the risk control of urban agglomerations and specific disaster sites, it is necessary to further carry out geological hazard risk source identification, hazard analysis, and vulnerability analysis based on the physical and mechanical characteristics of specific hazards and the dynamic process of hazards. Additionally, on this basis, calculate the risk value and classification of each geological hazard and hidden spot.

In this paper, we build a procedure and method for evaluating the risk of geological hazards in complex mountainous towns with reference to the technical methods and theory of geological hazard evaluation at home and abroad in Longlin Town, which is a typical mountainous town in the Lixian County, Gansu Province. Then, we analyze the potential hazards of typical geological hazards around the Longlin Town area, study the vulnerability of the town area and the surrounding disaster-bearing bodies, and divide the risk areas of geological hazards under different rainfall frequencies. Finally, the proposals for urban geological hazard risk control measures were discussed and used to provide technical support for disaster prevention and mitigation and territorial spatial planning of complex mountainous towns.

2. Materials and Methods

2.1. Overview of the Geological Environment of the Study Area

Longlin Town is located in the south of Lixian County in the West Hanshui basin of Longnan City, Gansu, and with a serious geological hazard. Its river system range includes the first-level watershed that is located on both sides of the West Hanshui River valley and takes mudstone ditch watershed as the boundary. The geographical range of the town area is between 105°02′29.5″ E~105°08′47.4″ E, 33°53′29.7″ N~33°57′50.1″ N and with the area of 38.4 km^2. Approximately 11,550 people and 9 administrative villages were in the town area. Because the climate of the town is temperate continental monsoon with a mild and humid climate, the average annual rainfall in the area is 499.4 mm, and the rainfall is concentrated from June to September, often in the form of heavy and continuous rain.

The Longlin Town area is characterized by the complex landscape of medium to high mountains and valleys that is situated in the western part of the West Lishan interrupted basin of the West Qinling Mountains in the Longnan Mountains of China. The slopes of high and steep creating favorable topographic conditions for the occurrence of geological hazards. The study area is located at the eastern Tibetan plateau active block, between the north margin of the West Qinling left-slip fault and the East Kunlun left-slip fault. The main faults in the study area are active fault zone with the NW and NE trending thrust and strike-slip characterized by complex structural styles and intense activity. Longlin Township was shaped by a wedge confined by the Lintan–Tangchang fault, the Lixian–Luojiabao fault, the Feng–Tai fault and the Liangdang–Jiangluo fault [24–26]. The stratigraphy in the region is mainly the Middle Devonian West Hanshui Group fifth and sixth lithologic section ($D_2^2xh_5$ and $D_2^2xh_6$) and includes light gray shale and slate with a small amount of chert and siltstone, etc. The rock mass in the study area is broken, weathered highly and weaker competency that is affected by the surrounding active fault. The features of rock mass are collapse and landslide, which includes the black carbonaceous shale and schist fragments with significant rheological properties.

Further, the study area experienced historically strong earthquakes frequently with a VIII degree of regional seismic intensity due to its location in the north-central part of the north–south seismic zone of China. There are as many as 15 earthquakes of Ms 7.0 magnitude or higher recorded in history, among which the 8-magnitude earthquake in Lixian County, Gansu Province, China on 21 July 1654, caused the most intense landslide. Historical earthquakes lead to the formation of geological hazards because of their age in the geotechnical structures and other structures in the area and reduce mechanical strength [26,27].

In conclusion, the study area is a site in the mid-alpine canyon area on the eastern margin of the Tibetan Plateau formed by strong erosion and cutting. The area suffered frequent geological disasters due to active neotectonics movements, complex and fragmented rock structures, and the development of weak rock layers (Figure 1). Further, in recent years, the increasing intensity of human engineering activities also aggravated the occurrence of geological hazards in the area, mainly including land use, road construction, and urbanization.

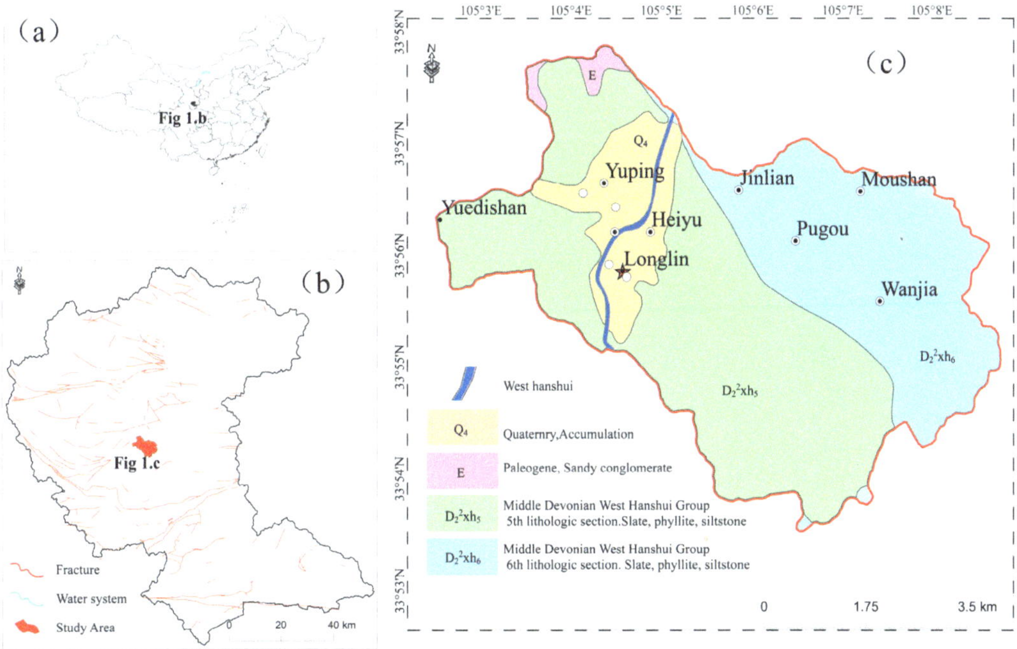

Figure 1. Geological background map of study area. (**a**). The location of the West Hanshui Basin in China. (**b**). The structural outline map of the West Hanshui Basin. (**c**). Development of weak rock layers in Longlin Town.

2.2. *Study Data Sources*

The main data (Table 1) in this study were obtained from the information listed in Table 1, mainly including ① basic feature information obtained from geological hazard site survey; ② 1:10,000 topographic map data, DEM and 1:200,000 geological map; ③ 1:10,000 land-use type data; ④ 1:10,000 accuracy of physical source feature data; ⑤ Pléiades satellite remote sensing data with 0.5 m accuracy that obtained on 3 May 2016, and UAV mapping data with 0.1 m accuracy obtained on 15 October 2020; ⑥ historical geological hazard rainfall data and 406 meteorological observation data points in the Longnan Mountains in the study area; ⑦ major geological hazard body survey and geotechnical body experimental test data.

Table 1. Data type and source.

Basic Data	Data Source and Production	Data Format
Geohazard data	From the Longnan West Hanshui Basin Disaster Geological Survey (2019–2021) project database	1:10,000 precision vector data
DEM	Geospatial data to extract slope, gully density, debris flow gully bed ratio drop, etc.	5 m × 5 m raster data
DOM/DLG	Land use type data	5 m × 5 m raster/vector data
Remote Sensing Data	Interpretation for risk source identification, carrier types, etc.	P-star and UAV data, raster data
Rainfall information	Lanzhou Central Weather Station, Longnan town geohazard Professional Monitoring Network	Vector data
Geological data	Lithological zoning, fracture structure	1:200,000 regional geological map, vector data
Survey and test data	Physical and mechanical indicators such as geotechnical density/capacity, water content/permeability coefficient, and angle of internal friction, cohesion, etc., for model calculation and analysis	Text Data Format

3. Methodology

3.1. Town Risk Assessment Process

Geohazard risk assessment is a research work that is aimed at a certain area or a typical geohazard body. The urban geohazard risk assessment investigates the threat of geohazard potential and its cascading hazards based on the geohazard data of the urban study area. In this paper, the following steps and methods are used to realize the study of geohazard risk assessment and control for towns in typical middle and high mountain valley areas (Figure 2).

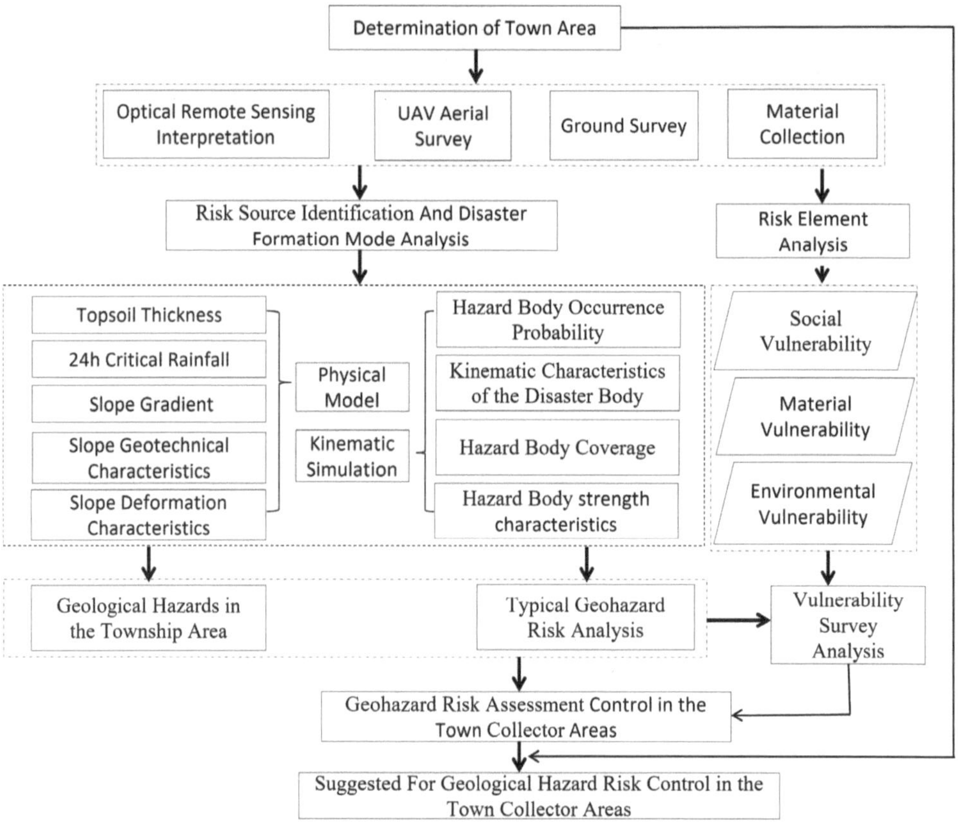

Figure 2. Flowchart of geohazard risk assessment and control for cities and towns in complex mountainous areas.

- Geological hazard risk identification.

Hazard source identification focuses on the r of the parcels within the township that may cause geological hazards, the major geological hazards and the degree of population concentration, the general impact area is within 1~3 km upstream and downstream along the river valley. The content of geohazard risk identification is analyzing the lots that may be destabilized to produce collapse and landslide mainly in the first-level slope zone on both sides of the river valley and the area below the circulation area of each branch gully, and analyzing whether the type of debris flow gully is slope or flash flood type for the mountainous gully. Additionally, we consider the extent of secondary effects after the occurrence of chain geologic hazards under extreme conditions [28].

- Research on the formation mode of geological hazards.

It is including the analysis of the conditions of geohazard potential disasters and the process of disaster formation, the study of the basic formation conditions, development characteristics, development process and characteristics of geohazards and early detection signs, and establishing geological detection signs and indicators of geohazards of different types and development stages [29–32]. The study region is a fault-controlled hillslope area, and the large landslides caused by the soft and hard laminated rock groups composed of shallow phyllite rocks such as micro phyllite rocks and schists, debris flow have multiple, multi-level block activities characteristics, and their disaster modes are featured by the chain such as landslide-debris flow or barrier lake. The large debris flow has obvious branch gully grouping to block and collapse step by step, and the flow increases significantly, which is easy to block the river valley to form a chain disaster.

- Geohazard Risk Analysis.

The overlaid analysis was carried out with the critical rainfall distribution map in the study that was obtained used by the slope hydrology model and the infinite slope stability model and the existing disaster-inducing rainfall data to obtain the spatial distribution of slope stability under different rainfall conditions in 24 h [33,34]. Then, the future development trend of the slope is qualitatively discriminated according to the surface deformation law, which is represented by the slope development rate. Finally, the two are combined for comprehensive analysis to calculate the damage probability of the slope [35–39]. In the first-class slope zone around the township area, the potential hazard degree under different rainfall conditions is predicted and analyzed one by one for landslide/debris flow that may cause disasters. Based on the geological hazard generation model, a reasonable mathematical model of geotechnical movement and fluid-solid coupling movement is used to analyze the geological hazard movement and accumulation characteristics under each condition and predict its hazard range. According to the three-level superposition of slope destabilization, landslide damage and debris flow occurrence, the hazard blocks or strip slope units composed of raster cells in the study area are reasonably delineated and form the geological hazard zoning map of the collector area.

- Vulnerability assessment of potential disaster-bearing bodies.

The vulnerability analysis of geological hazards is a comprehensive analysis way to the resilience of disaster-bearing bodies, and its degree depends on the sensitivity of disaster-bearing bodies to the effects of geological hazards, which is usually expressed by the value (or number) of disaster-bearing bodies and their vulnerability index [40]. Based on the geological hazard zoning map, it is mainly focused on the potential disaster-bearing bodies exposed to medium or higher level geological hazards, which includes permanent and temporary buildings exposed to the threat of geological hazards, linear projects such as roads/highways/pipelines, population distribution and age structure, etc., and ecological environment conditions. The comprehensive value of disaster-bearing bodies is obtained by calculating the average unit value of the above-mentioned disaster-bearing bodies and the actual number of disaster-bearing bodies, and then carrying out qualitative or quantitative vulnerability index analysis according to the form of damage by spatial movement of geological hazards and the structural strength of the disaster-bearing bodies themselves, to obtain the vulnerability assessment zoning map of potential disaster-bearing bodies of geological hazards in the study area.

- Geological hazard risk assessment in the township area.

The probability of occurrence of geological hazards in different risk areas was obtained by using historical geohazard cases and corresponding trigger rainfall record data in the study area. The quantitative and qualitative methods were used to evaluate the risk of geological hazards in urban towns by combined with the degree of risk of geological hazards and the vulnerability of hazard-affected bodies and formed the risk zoning map of geological hazards in the study area.

- Recommend countermeasures for risk control.

Proposing specific measures such as risk source elimination, risk area reduction, and integrated risk control by a comprehensive consideration of the risk distribution of disaster-bearing bodies in the study area.

3.2. Geohazard Risk Identification and the Disaster Generation Model

Geological disaster risk identification focuses on the identification of hazard sources such as landslides, landslides and debris flow in the areas that may produce geological disasters in the planning area of the township. For example, we analyze the lots that may be destabilized to produce collapse and landslide in the first-class slope zone on both sides of the river valley and the area below the circulation area of each branch ditch and analyze whether the debris flow valley is a slope type or flash flood valley in the mountainous valley. First, effectively identify the surface deformation according to the spectral and texture change characteristics of multi-period optical remote sensing images, to circle the major hidden hazard and potential geological hazard hidden danger lots combined with the topographic features and census the old landslides that had occurred and the areas with obvious signs of deformation [41–45]. Secondly, forming three-dimensional images in the exposed areas of bedrock such as weathered phyllite rock and shale are most prone to geological hazards such as collapse and landslide using an aerial survey of UAV, so that various features and signs of slope deformation can be visually analyzed, and deformation sections or blocks can be circled in detail [46]. Finally, the identification results obtained by the first two means are supplemented and verified, with emphasis on selecting potential geological hazard sites or typical geological hazard sites with obvious deformation characteristics that pose a threat to people's lives and property safety and are visually blinded by remote sensing interpretation [47]. After a general, detailed survey and verification, there are 71 geological hazards in the study area, including 4 landslides, 53 rockfalls, 7 gully-type debris flows and 7 slope debris flows (Figure 3). The total area of landslide geological hazards is approximately 3.23 km^2, accounting for 8.4% of the total area of the study area, and the density of hazard development is 1.85 places/km^2. The total area of provenance area of landslide, landslide and debris flow development is 1.67 km^2, and the area of their corresponding provenance areas are 7.2%, 57.5% and 35.3%, respectively.

Landslide is one of the most widely distributed types of geological hazards in the study area, the amount is accounting for 74.6% of the total number of hazard sites in the study area. It is densely distributed on both sides of West Hanshui and Hanjia Rivers and on the slopes of both sides of each branch ditch, mostly developed in the loose accumulation of the Quaternary sediments, Devonian carbonaceous shale, micaceous rocks and tuffs with soft and hard intervals, and easily sliding engineering rock groups. It is concluded that the large-scale landslides in the study area are ancient landslides formed under the movement of earthquake or tectonic activities with a total of 13 landslides developed by combined with the UAV images, field survey and literature. Some other small and medium-sized landslides are developed on older landslide accumulations and are characterized by the multi-period sliding that is mostly under the influence of rainfall, human engineering activities, or river erosion.

The disaster mode is follows 3 types: Type I, the old landslide back wall unloading effect produces small avalanches and slides under the collapse, the slide body forward movement, the formation of rush cover damage. Type II, the landslide in the front edge of the excavation and erosion, the leading edge of the front body forward movement, the formation of pushover damage. Type III, the side edge of the landslide under the erosion of the cutting gully, the side edge of the local sliding to form the secondary landslides, the formation of further sliding damage.

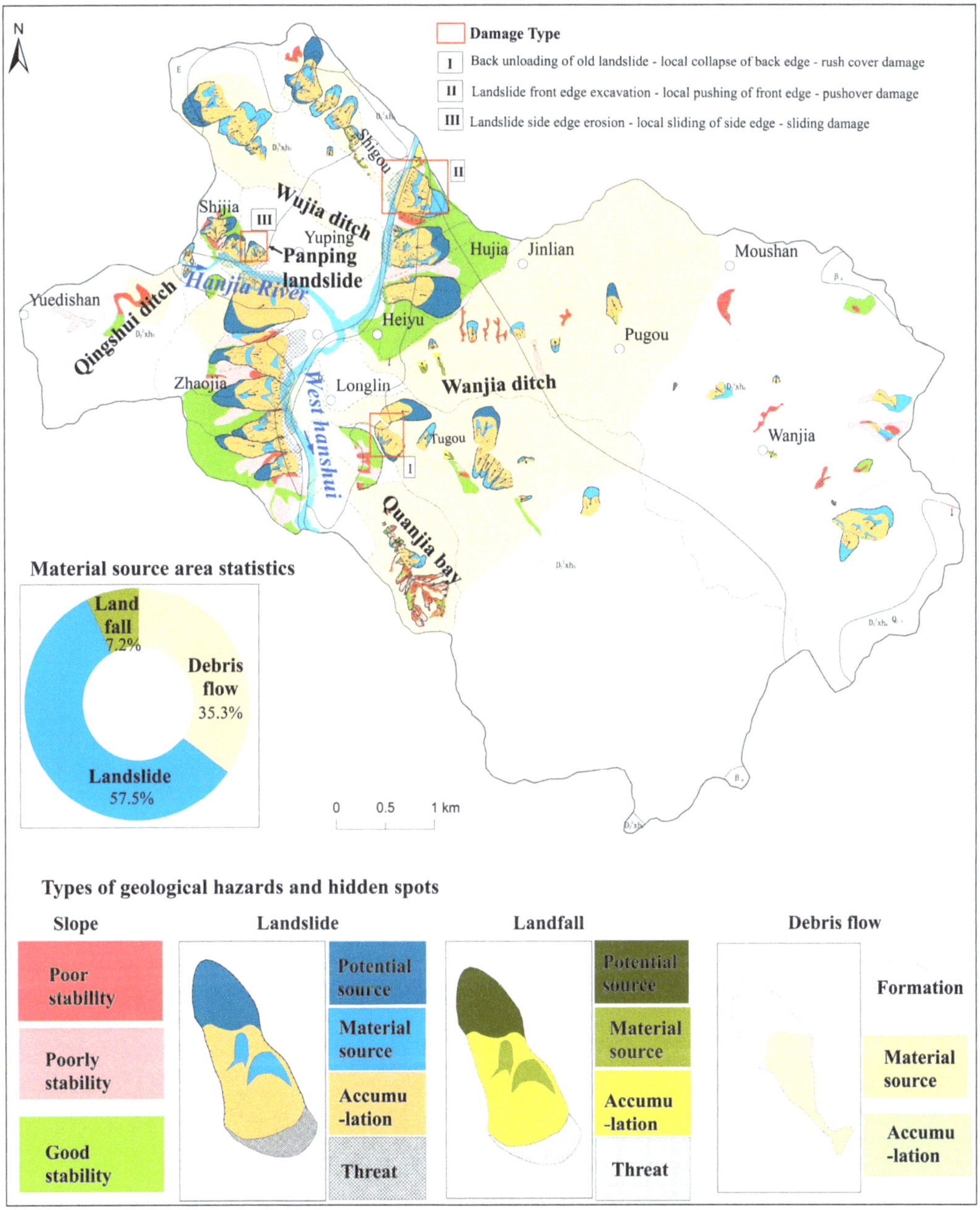

Figure 3. Distribution map of geological hazards and hidden dangers in the study area.

The basic formation conditions, development characteristics, development and evolution processes and characteristics, and early identification marks of the three types of landslides were further analyzed and studied. Finally, an intuitive map of the geological identification marks of landslides with different disaster formation modes was constructed, and geological identification marks and indicators of landslides corresponding to different types and development stages were established (Table 2).

Table 2. Landslide and hidden danger point disaster pattern identification mark statistics.

Type	Initial State	Ageing Deformation Stage	Progressive Deformation Damage Stage	Post-Damage State	Description of Model Elements
Type I	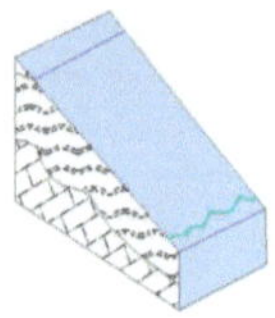	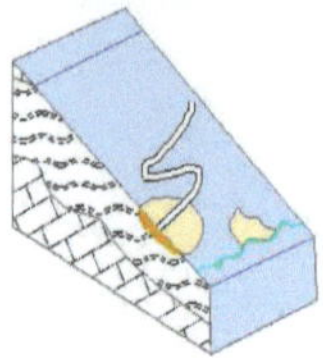	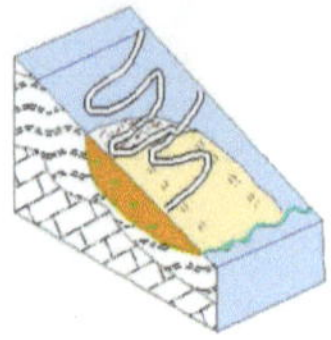	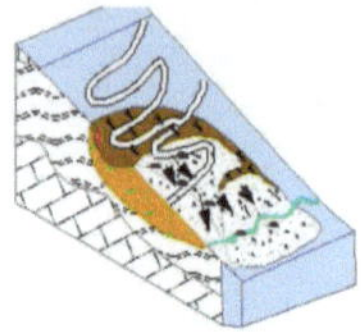	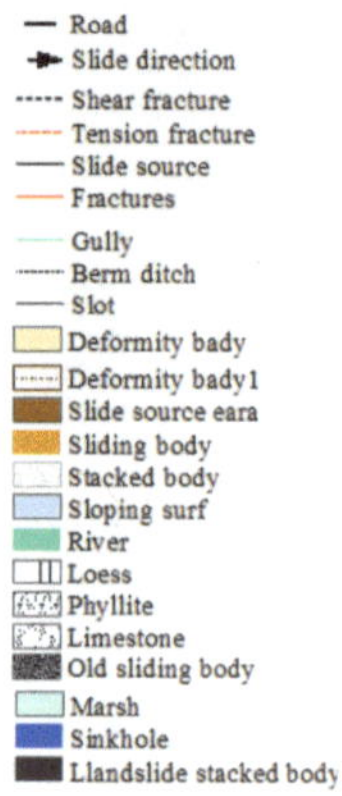
Type II	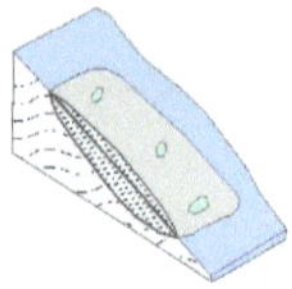	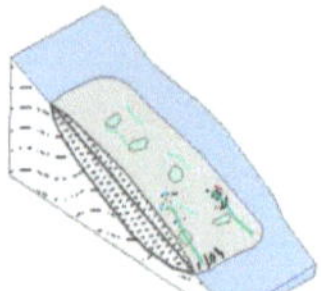	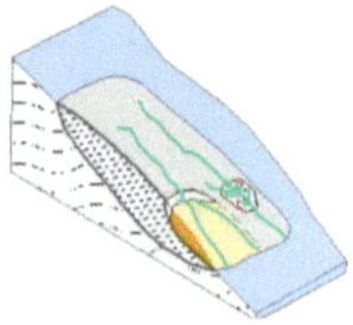	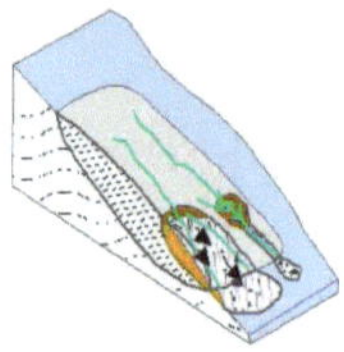	
Type III	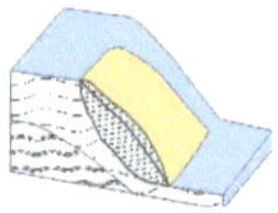	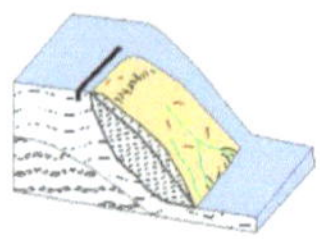	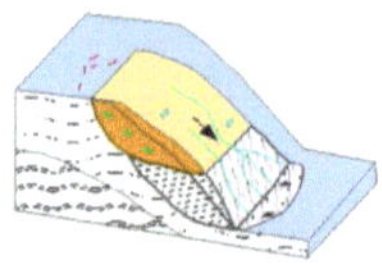	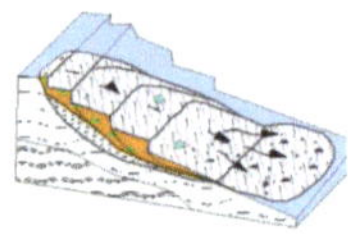	

The amount of debris flow is account for 19.7% of the total number of hazards in the study area. The overall gully debris flow is shaped narrow, and the source, circulation and accumulation area are relatively complete. The source area is mostly developed with small and medium-sized avalanche slides and other sources, with obvious signs of local blockage of the gully; the circulation area has clear traces of erosion and scouring by flowing water; the accumulation area has obvious fan-shaped land. Slope debris flow is mostly developed in the slope triangle area with the deep channel. The sources are mostly developed in the middle and lower reaches of the watershed and are small avalanche slides, the slope gradient of the gully bed is large, and the fan shape of the accumulation area is not obvious. The main disaster mode includes 2 types, the gully uncovered bottom-siltation type damage and slope runoff erosion slip type—diffuse flow type damage.

The number of collapses is relatively small, accounting for 5.7% of the total number of disasters. They mostly occur in the steeper slope areas on both sides of river gullies and highways, and the slope structure is mainly bedrock and loess-soft rock slopes with good prominence, developed structural surface, and unclear texture.

3.3. Town Risk Assessment Methods and Models

3.3.1. The Slope Stability Evaluation Model

The infinite slope model proposed by Skempton and Delory [48] is to consider the response of rainfall erosion slope surface, infiltration slope body and geotechnical properties to the slope under different rainfall frequencies, which can achieve quantitative stability evaluation of the slope. Montgomery et al. [49] combined it with the slope hydrology model to obtain a critical rainfall calculation formula for shallow landslide initiation. The slope instability in the study area is mainly influenced by rainfall, and the depth of the landslide is much smaller than the width and length of the slope. Therefore, according to the limiting equilibrium theory, the slope stability coefficient within each raster cell is F_s:

$$F_s = \frac{c' + [(\rho_s g D - \rho_w g h)cos\theta]tan\varphi'}{\rho_s g D sin\theta} \quad (1)$$

where c' is the effective cohesion of the slope (kPa). φ' is the effective angle of internal friction of the slope (°). ρ_s is the natural weight of the geotechnical body (kg/m^3). t is the potential slip thickness (m). θ is the slope inclination (°). h is the slope groundwater level (m). ρ_w is the weight of water (kg/m^3).

Under certain rainfall intensity conditions, the water table height in the slope. h is:

$$h = \frac{IA}{Tbsin\theta} \tag{2}$$

where I is the equivalent rainfall intensity (m/d). A is the watershed area (m^2). T is the hydraulic conductivity of the saturated soil (m^2/d). b is the width of the considered water flow cross-section (grid accuracy) (m).

Combining Equation (1) with Equation (2), such that F_s = 1, the critical rainfall for rainfall-induced slope initiation can be obtained as

$$I_c = T\left(\frac{b}{A}\right)\sin\theta\left(\frac{\rho_s}{\rho_w}\right)\cdot\left[\left(1-\frac{tan\theta}{tan\varphi'}\right)+\frac{c'}{\rho_w g D cos\theta tan\varphi'}\right] \tag{3}$$

3.3.2. The FLO-2D Fluid Model

Most of the debris flow hazards in the study area are rainfall controlled in nature, and few of the debris flow gullies have been subjected to engineering control measures. Therefore, this paper simulates the future debris flow hazard in the study area based on the FLO-2D model [50,51]. The FLO-2D model was proposed by O'Brien [52] in the early 1990s based on a non-Newtonian fluid model and a finite difference method to solve the motion control procedure, which can be used for two-dimensional flood hazard management and debris flow motion. In the FLO-2D model, the debris flow control equation is

$$\frac{\partial h}{\partial t} + \frac{\partial (uh)}{\partial x} + \frac{\partial (vh)}{\partial y} = I \tag{4}$$

$$(S_{ox} - S_{fx})g = \frac{\partial h}{\partial x}g + u\frac{\partial (uh)}{\partial x} + v\frac{\partial (uh)}{\partial y} + \frac{\partial u}{\partial t} \tag{5}$$

$$(S_{oy} - S_{fy})g = \frac{\partial h}{\partial y}g + u\frac{\partial (vh)}{\partial x} + v\frac{\partial (vh)}{\partial y} + \frac{\partial v}{\partial t} \tag{6}$$

where t is the evolution time (s), h is the depth (m), I is the rainfall intensity (mm/h), u is the velocity in the x-direction (m/s), and v is the velocity in the y-direction (m/s), S_{ox} and S_{oy} are the streambed slope drops in the x-direction and y-direction (%), S_{fx} and S_{fy} are the frictional slope drops in the x-direction and y-direction (%). FLO-2D provides dynamic wave mode and diffusion seeding mode to simulate the process of movement and accumulation. Equation (2) is the continuity equation, which is the volume mass conservation equation. Equations (5) and (6) are the equations of motion of the force balance. In this model, the expression of the shear stress gradient of the fluid:

$$S_f = S_y + S_v + S_{td} = \frac{\tau_y}{\gamma_m h} + \frac{K\eta u}{8\gamma_m h^2} + \frac{n^2u^2}{h^{4/3}} \tag{7}$$

where S_f is frictional decline (%), S_y is yield decline (%), S_v is viscous decline (%), S_{td} is turbulent-dispersion decline (%), τ_y is yield stress (MPa), γ_m is specific gravity of fluid (t/m^3), K is laminar drag coefficient, η is fluid viscosity coefficient, n is Manning coefficient, and v is flow velocity (m/s). The parameters τ_y and η are calculated from the equation $\eta = \alpha_1 e^{\beta_1 \cdot C_v}$ and $\tau_y = \alpha_2 e^{\beta_2 \cdot C_v}$, α_1, α_2, β_1 and β_2 are set by rheological tests or table setting.

3.3.3. The River-Flow 2D Rheological Model

The landslides in the study area mostly developed on the slopes of loose accumulations that were composed of weak and shallow metamorphic phyllite and slate zone. The damage process is usually discontinuous, which is characterized by the high concentration of non-Newtonian fluid. The River-Flow2D [15] numerical model is a multidimensional simulation software that adopts a finite volume method to integrate hydrodynamic and hydrologic elastic mesh [53,54]. In the process of landslide simulation, it mainly considers the change in frictional resistance on the bottom surface of the debris body. It can realize the two-dimensional or three-dimensional simulation and display the accumulation characteristics of landslide-clastic flow movement (slip velocity, slip distance and accumulation body thickness).

Different sliding friction calculation models are given in the River-Flow 2D numerical model calculation process for different properties of the sliding material, which can simulate the landslide motion process reasonably and determine the stress boundary conditions accurately. It mainly includes the Bingham model [55,56], the Voellmy model [57,58], and the friction flow model [59]. Among them, Bingham is a friction resistance model considering the plasticity and viscosity of the slide, when the starts to flow after reaching the critical value of shear stress, and its bottom friction value is calculated as Equation (8). Additionally, the formula of the Vowelly flow model mainly includes the turbulence term and frictional resistance term, and the formula is Equation (9). The frictional flow is calculated as Equation (10).

$$v = \frac{h}{6\eta}(2\tau - 3\tau' + \frac{\tau'^3}{\tau^2}) \tag{8}$$

$$\tau = \gamma \mathrm{h}\left(\cos\alpha + \frac{a_c}{g}\right)\tan\phi + \gamma\frac{v_i^2}{\xi} \tag{9}$$

$$\tau = \gamma \mathrm{h}\left(\cos\alpha + \frac{a_c}{g}\right)(1 - r_u)\tan\phi \tag{10}$$

where η is the Bingham viscosity coefficient. v is the slip velocity (m/s); according to the Cullen viscosity theory, τ' can be expressed by the positive slip surface stress. γ is the slip body capacity (kg/m^3). a is the slip surface inclination angle (°). ϕ is the internal friction angle (°). $a_c = {v_i}^2/R$ is the centrifugal acceleration of the curved slip surface. r_u is the cavity pressure coefficient, the ratio of the cavity pressure to the normal stress at the bottom of the calculation unit. ξ is the turbulence coefficient (m^2/s).

During the River-Flow 2D simulation, when the slide stress is gradually dispersed near the slide bed and there is a turbulence effect, the frictional resistance characteristics mainly depend on the shear stress change in the slide, and viscous stress, yield stress, dispersion stress, and inelastic collision of solid particles in the debris-fluid soil and rock mixture, etc. The standard Bingham frictional resistance model is

$$f_1(\tau_0, \tau_1) = 2\tau_b^3 - 3(\tau_y + 2\tau_\mu)\tau_b^2 + \tau_y^3 = 0 \tag{11}$$

where τ_b is the slip stress (MPa), $\tau_b = g\rho h cos\theta tan\theta_b$, τ_y is the yield stress (MPa), $\tau_y = 0.181 \cdot \exp(25.7C_V)/10$ is the yield stress (MPa), τ_u is the viscous stress (MPa), $\tau_u = 0.036 \cdot \exp(22.1C_V)/10$. $\rho = \rho_w\ (1 + 1.65C_V)$ and θ is the slope of the landslide (°). θ_b is the internal friction angle of the landslide (°), and ρ is the fluid density of the landslide debris (kg/m^3). ρ_w is the water weight (kg/m^3). C_V is the volume concentration.

According to the above evaluation model and method, this paper further elaborates on the technical framework and related technical means for geological hazard risk evaluation and control in complex mountainous towns in the typical relocation and resettlement area of the Longnan mountainous area, Longlin Town collector town. The results show that there are many potential major geological hazards in the first-class slope zone on both sides of the river valley in the planning area of the townships. We select a typical landslide and debris flow for demonstration by the FLO-2D model and River-Flow 2D because the multiple major geological hazards and the calculation and analysis process is similar to the single body. In the other areas of the township area, the 5 m × 5 m resolution raster cells are used as the basic evaluation units of geological hazards for geological hazard evaluation and risk analysis by selecting the slope stability evaluation model.

4. Model Validation and Results

4.1. Model Validation

Geological hazard risk assessment is based on the analysis of geological conditions and triggering factors of geological hazards, and analyzes the time probability and movement accumulation characteristics of geological hazards occurrence, and its core content is to determine the probability of destabilization and coverage of geological hazard bodies under different working conditions [60–63]. The types of geological hazards in the study area are mainly landslides and hazards, and geological hazards have rainfall-controlled characteristics, so a comprehensive evaluation of the hazard of different rainfall conditions is mainly conducted for slopes and major geological hazard bodies in the catchment area. Finally, the hazard levels are divided into four levels: very high, high, middle, and low.

4.1.1. Town Slope Hazard Analysis

The slope structure in the study area is mostly a laminar slope structure characterized by loose accumulation layer slope and overlying loess or residual slope accumulation rubble layer with underlying highly weathered soft bedrock, which can meet the assumptions of the infinite slope model, and the catchment area is generally small. The required physical and mechanical indices of the geotechnical body are easy to obtain and can meet the requirements of the model calculation. Firstly, the survey and investigation data of the slopes in the study area show that most of the landslide damage modes are multi-phase shallow sliding, while the thickness of the overlying gravel soil layer of the slope (such as poor stability and good stability in the slope unit) is investigated and counted. The total number of slope thickness points in this investigation is 463, the maximum overburden thickness was 26.1 m, and the minimum overburden thickness is 0.1 m. Based on these data, ArcGIS spatial analysis is used to interpolate the slope thickness and obtain the distribution of potential slope slip thickness (Figure 4). Secondly, according to the series of spatial layers such as stratigraphic structure type, material composition, and fragmentation degree from the slope refinement survey, the study area distinguishes six types of slope structure types including loose accumulation layer, loess-soft rock, loess, soft rock, soft rock-hard rock and hard rock. Based on the field survey and investigation data, the geotechnical samples of various slopes were analyzed to determine the basic physical and mechanical parameters such as effective cohesion, effective internal friction angle, natural weight of the geotechnical body, and hydraulic conductivity of saturated soil body. The slopes in the study area were rasterized into a 5 m × 5 m grid, and then the critical rainfall for each grid in the study area was obtained according to Equation (3). According to the field survey and collected historical data, the geological hazard outbreak in Longnan Mountains usually reaches 20~50 mm in 24 h rainfall, and when the 24 h continuous rainfall is more than 100 mm, it often induces a cluster geological hazard of uneven scale. The maximum 24 h rainfall ever occurred in the study area is 116.3 mm, combined with the rainfall intensity classification standard promulgated by the National Meteorological Bureau, the three 24 h rainfall amounts set under different rainfall conditions, 20-year, 50-year, and 100-year scenarios, are 25, 50 and 100 mm, respectively. With the critical rainfall distribution map, the overlay analysis is carried out under the ArcGIS platform, which would obtain the spatial distribution maps of slope stability under different rainfall conditions.

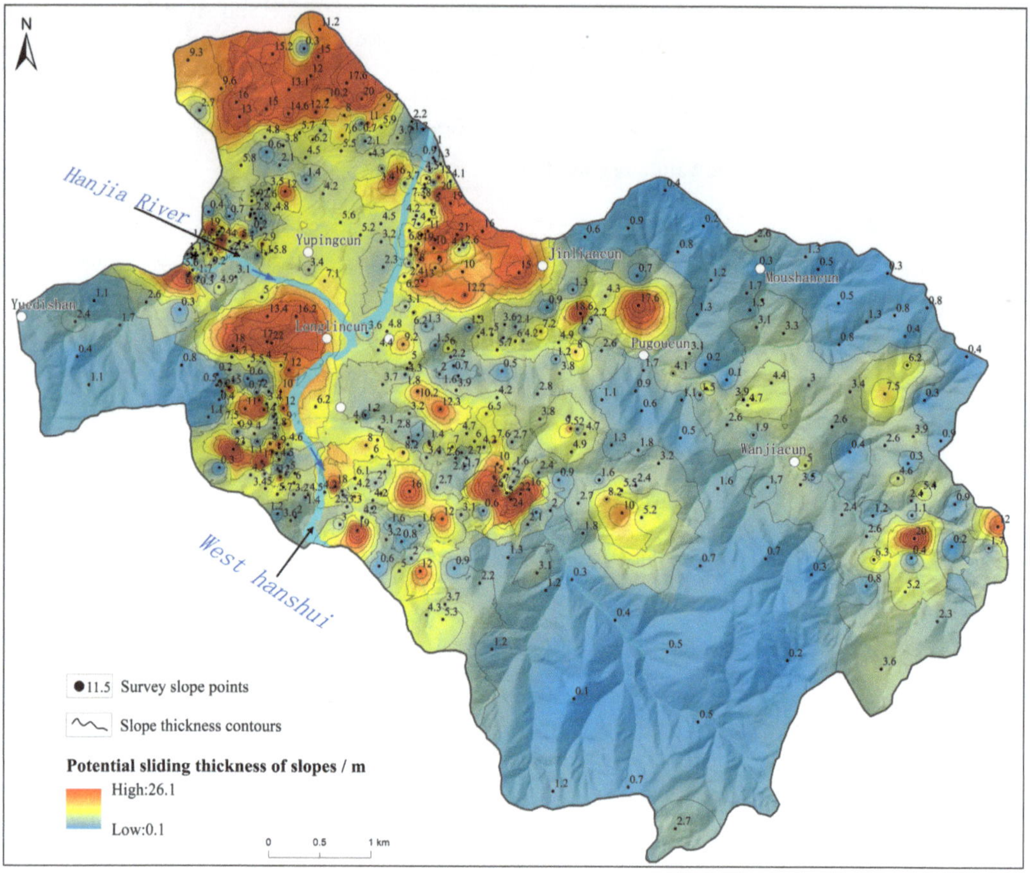

Figure 4. Distribution of potential slope damage thickness in the study area.

However, the slope stability calculation only represents the existence of a possible internal state and is only a quantitative evaluation of the stability state at different stages, but cannot represent the future development trend of the slope. The slope deformation and damage characteristics, i.e., slope development rate, is an assessment of the future slope development trend based on the surface deformation law [64]. Therefore, the individual slope damage analysis should combine the stability calculation results with the development rate judgment, and comprehensively judge the possibility of future development trend of the slope, i.e., the slope damage probability. The slope damage probability calculation method not only includes the slope stability calculation results but also considers the macroscopic deformation development state of the slope surface, combining microscopic analysis and macroscopic judgment, which can reflect the landslide damage situation more accurately. Its characteristic is to determine the development trend of slope damage in a targeted way based on the realism of slope surface deformation. Thus, it avoids focusing only on the artificial random adjustment of physical and mechanical parameters of slopes and ignores the damage probability calculation results of slope surface deformation.

Finally, according to the slope damage probability evaluation reference Table 3, the slope stability under different rainfall conditions is categorized into four levels, which are superimposed with the resulting map of landslide development rate grading to obtain the hazard zoning map of slopes in the catchment area under different rainfall conditions.

Table 3. Reference for slope failure probability evaluation.

Grading	Landslide Stability Calculation Concerning Surface Macro Deformation			Landslide Development Rate Evaluation Reference		
	Surface Macro Deformation Characteristics	**Stable State**	**Reference Value of Stability Coefficient**	**Developmental Status**	**Landslide Development Characteristics**	**Fertility Reference Values**
Extremely high	Signs of overall landslide sliding can be clearly observed on the surface, and the slide body can be separated from the slide bed	Landslide initiation	<0.9	Full developmental maturity	Landslide has been initiated and overall sliding is highly probable	0.9~1
High	Landslides can be initiated when there is localized damage to the ground surface, and overall sliding precursors appear	Unstable	0.9~1.00	Developmental maturity	Slippery slope can be started, the overall sliding possibility is high	0.7~0.9
Medium	Signs of surface deformation begin to intensify and the landslide progresses rapidly toward the initiation phase; or the surface shows significant local deformation, but the rate of deformation is slow	Critical state or less stable	1.00~1.10	Developmental immaturity or onset of development	Accelerated deformation of the landslide, with the possibility of overall sliding; or local deformation of the slope, with the possibility of forming a landslide	0.3~0.7
Low	There are only local signs of minor deformation on the surface, and there is no development trend for the time being, or no signs of deformation are observed on the surface for the time being	Basically, stable or stable	1.10~1.20	Not yet developed or not developed	The slope deformation range is very small and the possibility of landslide formation is minimal; or no landslide	<0.3

4.1.2. Typical Evaluation Demonstration

Quanjia Bay debris flow is located in the Longlin Town Quandu village group, West Hanshui left bank, its watershed area is 1.26 km^2, the relative height difference in the area is 693 m, the main channel length is 2.53 km, the average longitudinal ratio drop of the ditch bed is 273.9‰, and the total amount of loose solids source in the watershed reaches 242.38×10^4 m^3. The field investigation data show that the debris flow dynamic process is: soft and hard lithology combination of medium and shallow landslide start $\rightarrow$ blockage body instantaneous collapse flow amplification $\rightarrow$ along the course of channel erosion $\rightarrow$ more intense bend wash silt $\rightarrow$ stop silt accumulation or blockage of the river, is a typical collapse—channel erosion mixed debris flow. In the process of movement, the flow is immediately amplified 3~10 fold after experiencing the blockage and collapse of loose accumulation source or coarse and large particle source, forming a super large-scale mudflow.

The debris flow in Quanjia Bay is a rainfall type debris flow gully, and without engineering control, measures are taken. Therefore, this paper mainly deals with the simulation of the debris flow movement characteristics of the debris flow in Quanjia Bay under 100-year (1% frequency), 50-year (2% frequency) and 20-year (5% frequency) rainfall conditions based on the FLO-2D model under rainfall conditions. The characteristic values of its movement were quantitatively obtained, and the spatial distribution of the intensity values of the mud level depth and flow velocity of the debris flow in each raster cell is used as an expression of the debris flow hazard.

FLO-2D model parameters are obtained from 0.5 m DEM, four sets of particle gradation at different cross sections, and other watershed characterization parameters. Combined with the debris flow dynamic characteristics and the Gansu Province small watershed storm flood model, the debris flow flows were calculated under different capacities of 1.77 t/m^3, 1.89 t/m^3 and 1.97 t/m^3. In the actual investigation, the upstream channel in Quanjia Bay is severely blocked, and the formation after the collapse of the middle and shallow landslide weir will produce a certain amplification effect, so the input of the FLO-2D model is calculated multiplied by the volume expansion coefficient. Finally, the parameters and rate were input of the FLO-2D model (Table 4), and the calculation results were more reliable and realistic without human intervention during the whole process.

Table 4. Basic characteristics and FLO-2D simulation parameters of Quanjia Bay.

Projects	Frequency of Rainstorms			Simulation Parameters	Value
	P = 5%	P = 2%	P = 1%		
Watershed area/(km^2)		1.26		Calculation grid/(m)	5 × 5
Total material sources/(10^4 m^3)		242.38		Manning roughness coefficient	0.15/Residential district
Debris flow capacity/(t/m^3)	1.77	1.89	1.97		0.05/Road
Debris flow peak/(m^3/s)	6.49	10.38	12.97		0.22/Cultivated land
Sediment correction factor	0.89	1.17	1.44		0.2/Bare ground
Sediment blockage factor		3.5			0.8/Woodland
Debris flow discharge/(m^3/s)	42.94	78.8	110.82	laminar flow friction factor K	2280
Volumetric concentration	0.47	0.54	0.59	α_1	0.811
Debris flow amplification factor	1.89	2.17	2.44	α_2	0.00462
Simulation flow/(m^3/s)	81.21	170.93	270.45	β_1	13.72
Simulation time/(h)	0.3	0.8	1.5	β_2	11.24
Simulation accuracy/(%)	81.38	75.53	86.74	Sediment specific gravity/(t/m^3)	2.65

The simulation of FLO-2D results (Figure 5) indicates that the areas of the very high-risk zone, high-risk zone, medium-risk zone and low-risk zone under 100-year (1% frequency) precipitation are 4.45 × 10^4 m^2, 1.36 × 10^4 m^2, 3.26 × 10^4 m^2, 4.35 × 10^4 m^2. The areas of the very high-risk zone, high-risk zone, medium-risk zone and low-risk zone under 50-year (2% frequency) precipitation are 2.80 × 10^4 m^2, 1.10 × 10^4 m^2, 1.96 × 10^4 m^2, and 2.95 × 10^4 m^2. The areas of very high-risk zone, high-risk zone, medium-risk zone and low-risk zone under 20-year (5% frequency) precipitation are 1.79 × 10^4 m^2, 0.42 × 10^4 m^2, 0.98 × 10^4 m^2, 2.04 × 10^4 m^2.

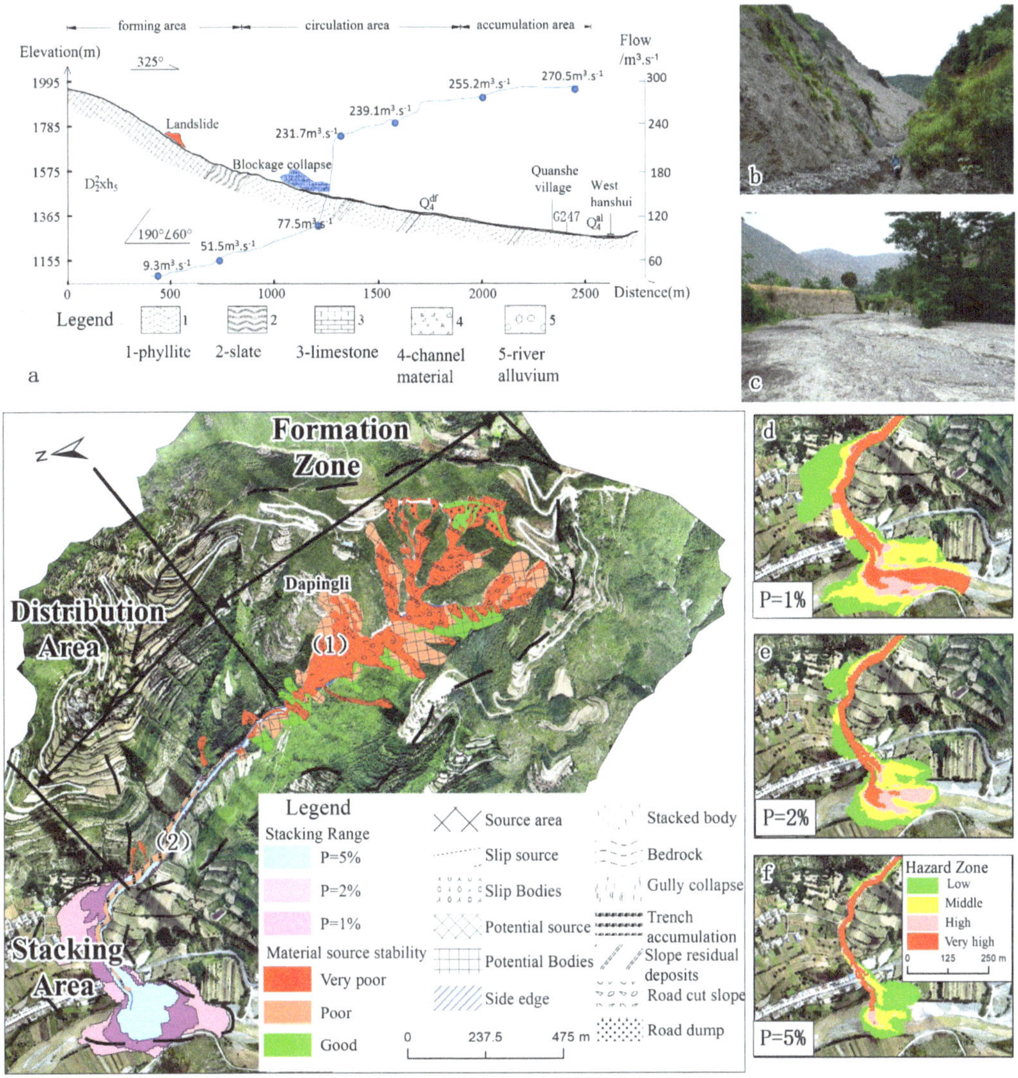

Figure 5. Simulation results of debris flow risk in Jianyuwan under different precipitation conditions. (**a**). Longitudinal profile of the main channel. (**b**). Blockage of the main channel downstream of formation area. (**c**). The situation of the main channel downstream of the circulation area. (**d**). A total of 100 hazard zones of debris flow under the condition of one rainfall. (**e**). A total of 50 risk zones of debris flow under one rainfall condition. (**f**). A total of 20 hazard zones of debris flow in case of rainfall.

4.1.3. Typical Landslide Evaluation Demonstration

The landslide in Panping Village is located approximately 2 km northwest of Longlin Town, on the left bank of the Datang River, a right-bank tributary of the West Hanshui. The landslide is generally tongue shaped on the plane, with obvious rear edge circle chair-like terrain, there are two secondary slides composed of approximately 250 m in length and 180 m in width. The landslide occurs at an elevation of 1320~1481 m, with a relative height difference of approximately 161 m, an average thickness of approximately 25 m, a volume is approximately 112.5 × 104 m^2, a main slide direction is 154°, and a total slope is 27°. The field investigation shows that the disaster mode of the landslide is a small avalanche slip collapse overburden damage under the unloading effect of the back wall of the old landslide.

In this paper, the base map data used for numerical simulation is the DEM with an accuracy of 0.5 m resolution obtained by UAV mapping in 2020. Based on the above proposed 2D calculation model of the landslide process based on the finite volume method, the river-Flow 2D landslide motion simulation system is used to obtain the sliding velocity, sliding distance, and thickness of the slide during the motion. In this example, the soil friction model adopts the Bingham standard friction resistance model, which is widely used in the calculation and simulation of the sliding distance, especially for the debris-fluid landslides with high water content of slide material, which can obtain more ideal calculation results, and the calculation formula is shown in Equation (11). In this paper, the parameters in Table 5 are used for the simulation calculation, and the simulation test and movement process analysis are carried out for the landslide of Panping Village under different precipitation conditions.

Table 5. Model calculation parameters under different precipitation conditions.

Projects	P = 5%	P = 2%	P = 1%
Internal friction angle $\theta b/(°)$	14.4	12.96	11.6
Slip density $\rho/(\text{kg/m}^3)$	20.2	23.23	25.05
Volumetric concentration C_V	0.618	0.802	0.912
Slip body yield stress $\tau y/(\text{MPa})$	0.886	1.422	1.886
Slip viscous stress $\tau\mu/(\text{MPa})$	0.141	0.212	0.270

In the landslide-debris flow movement stage, for the building, the maximum thickness of its location during the movement of the slide is one of the direct factors affecting its deformation and damage condition, so the thickness of the landslide is chosen as an important index for evaluating the landslide hazard. According to the previous studies and the actual situation of the landslide, the landslide hazard is divided into four levels according to the thickness of the movement accumulation, low hazard zone when $H \leq 1$ m, medium hazard zone when $1 < H \leq 3$ m, high hazard zone when $3 < H \leq 5$ m, and very high hazard zone when $H > 5$ m. The simulation results (Figure 6) show that the areas of very high hazard zone, high hazard zone, medium hazard zone, and low hazard zone under the 100-year (1% frequency) precipitation condition of the Panping Village landslide are 1.94×10^4 m^2, 1.62×10^4 m^2, 1.79×10^4 m^2, 1.02×10^4 m^2. The areas of very high hazard zone, high hazard zone, medium hazard zone, and low hazard zone under the 50-year (2% frequency) precipitation condition are 1.36×10^4 m^2, 1.24×10^4 m^2, 1.17×10^4 m^2, 1.42×10^4 m^2. The areas of very high hazard zone, high hazard zone, medium hazard zone, and low hazard zone under the 20-year (5% frequency) precipitation condition are 0.54×10^4 m^2, 1.11×10^4 m^2, 1.07×10^4 m^2, 1.83×10^4 m^2.

According to the above-mentioned risk evaluation process of the urban geohazard chain, geological hazards such as landslides and in the first-class slope zone on both sides of the river valley are analyzed one by one under different precipitation conditions, and the hazard area is predicted. Finally, according to the three-level superposition of slope destabilization, landslide damage and mudslide occurrence, the hazard blocks or strip slope units composed of grid cells in the study area are reasonably delineated to form a geological hazard zoning map of the watershed. The simulation results (Figure 7) show that the areas of very high hazard zone, high hazard zone, medium hazard zone and low hazard zone under 100-year (1% frequency) precipitation in Longlin Town are 2.36 km^2, 4.64 km^2, 13.97 km^2 and 17.43 km^2, respectively. The areas of very high hazard zone, high hazard zone, medium hazard zone and low hazard zone under 50-year (2% frequency) precipitation are 1.19 km^2, 1.78 km^2, 7.28 km^2, and 28.14 km^2, respectively. The area of the very high-risk zone, high-risk zone, medium-risk zone and low-risk zone under 20-year (5% frequency) precipitation condition is 0.53 km^2, 0.75 km^2, 4.13 km^2, and 32.99 km^2, respectively.

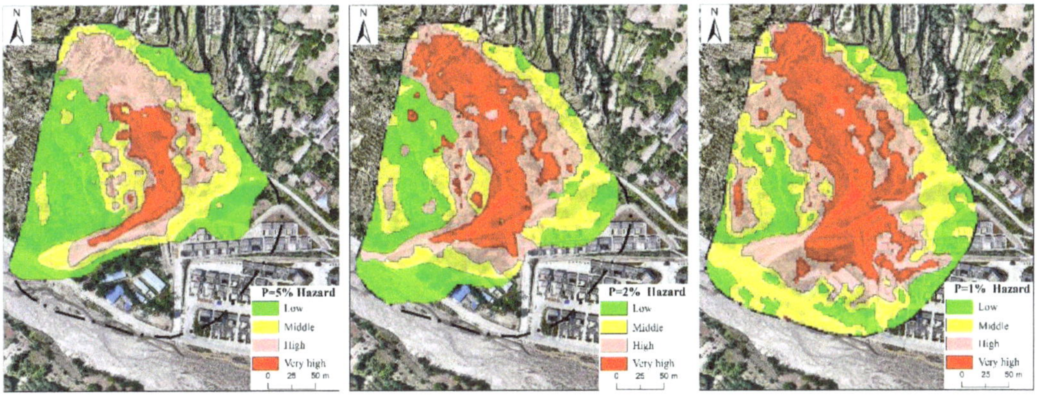

Figure 6. Simulation results of landslide hazard in Panping Village under different precipitation conditions.

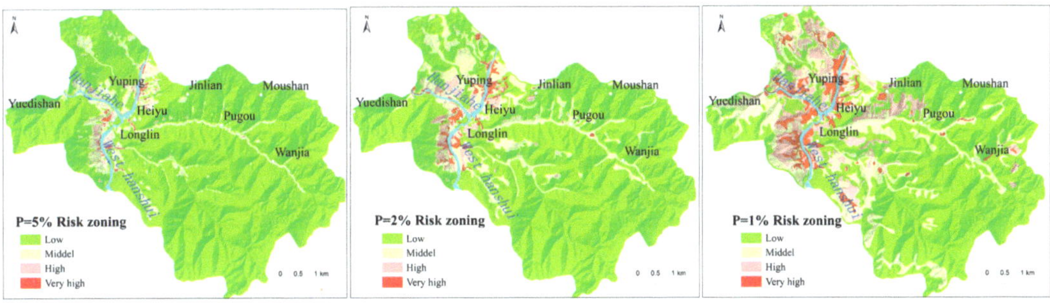

Figure 7. Zoning map of geohazards Longlin town under different conditions.

4.2. Results

4.2.1. Assessment of the Vulnerability of Towns to Geological Hazards

Considering the characteristics of the variety of structures and functions of disaster-bearing bodies in mountainous towns, we extract the types of disaster-bearing bodies automatically used by the hyperspectral curve features and image recognition technology based on the high-resolution images of Gaofen-2 satellite and UAV aerial photography, which improves the identification efficiency of a large number of disaster-bearing bodies. Synthesize on field surveys, interviews, urban planning, and construction information, the information on disaster-bearing bodies is optimized, and a real-time state database of spatial attributes of urban disaster-bearing bodies is constructed, which is summarized into 4 items and 22 categories. Among them, the population includes two categories—population density and population age; the buildings include nine categories—building areas, government administrative districts, schools, hospitals, factories, commercial houses, supermarkets, tourist attractions and temples; the roads includes four categories: national roads, township roads, general roads and bridges; ecological environment includes seven categories: forest land, grassland, cultivated land, garden land, green land, bare land and water.

Geological hazard vulnerability analysis is a comprehensive analysis of the resilience of a hazard-bearing body. The degree of the vulnerability of a hazard-bearing body depends on the sensitivity of the hazard-bearing body to the effects of the geohazard, which is usually expressed by the value (or number) of the hazard-bearing body and its vulnerability index.

The comprehensive value of the hazard-bearing body can be obtained by calculating the average unit value of the hazard-bearing body and the actual number of affected bodies. The vulnerability of a hazard-bearing body is a description of how easily which it can be destroyed by a disaster, and it can be expressed as a number between 0 and 1, with the larger the value, the higher destroy degree. The quantitative description of the vulnerability of a hazard-bearing body is complicated and is mainly influenced by the structural strength of the hazard-bearing body itself and the damage form of the geological hazard. For example, if a hazard-bearing body is located at different places about landslides and hazards, it will be damaged in different ways. The buildings and structures in its main flow line will be mainly affected by impact hazards, and in the landslide's edge or the fan's front edge will be affected by siltation hazards.

The buried disaster-bearing body is difficult to be reused mostly, and it is difficult to play the original planning and design effect due to the change in topography even if the structure is intact. The vulnerability calculation method of a disaster-bearing body is illustrated by the example of a house or structure. The vulnerability index of a disaster-bearing body affected by the siltation hazard is the ratio of the siltation thickness of landslide, debris flow and other hazards to the effective height of the disaster-bearing body itself (Equation (12)).

$$C_d = H_d / H_c. \tag{12}$$

where C_d is the vulnerability index of the disaster-bearing body. H_d is the burial depth of the debris-flow siltation (m). H_c is the effective height of the building or structure (m); if $(H_d/H_c) \geq 1$, it means that the building or structure has been completely buried by the landslide, and its value is 1.

The vulnerability of the disaster-bearing body is calculated by combining the results of the simulation calculation of the formation mechanism and movement process of geological hazards such as landslides, and considering the spatial variability of the disaster intensity. Finally, the vulnerability of the disaster-bearing body in Longlin Town is superposition evaluated according to the principle of choice high and forms a zoning map of the vulnerability of the geological hazard-bearing body in the populated area of Longlin Town (Figure 8).

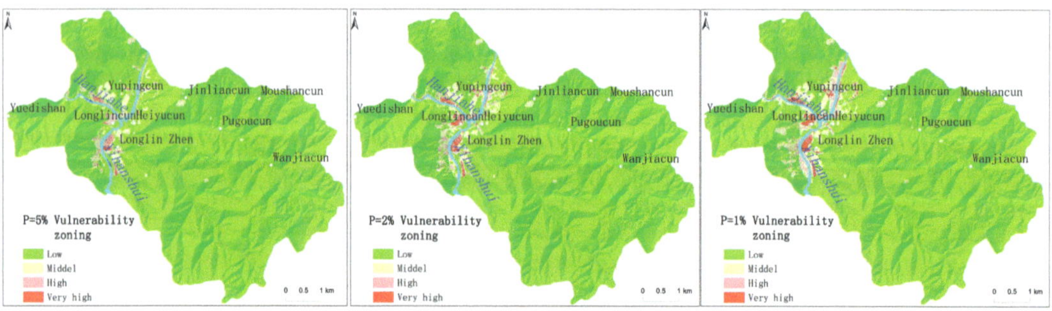

Figure 8. Vulnerability zoning map of geohazards in the Longlin Town Collective Area under different conditions.

4.2.2. Risk Assessment of Urban Geological Hazards

The risk value of each evaluation unit under different precipitation conditions is calculated according to the definition of hazard risk [65–67] based on the results of the geological hazard and vulnerability analysis of Longlin Town.

$$R = H \times V \times P_i \tag{13}$$

where R is the value of the riskiness index of the evaluation unit. H is the value of the hazard index of the evaluation unit. V is the value of the vulnerability index of the evaluation unit.

P_i is the probability of risk occurrence under different precipitation working conditions. Before calculation, all kinds of indices in Equation (13) must be normalized, and the normalization method is as follows:

$$H' = (H - H_{min})/(H_{max} - H_{min})$$

$$V' = (V - V_{min})/(V_{max} - V_{min})$$

where H' is the normalized value of hazard level. H is the value of the hazard index H_{max} and H_{min} are the maximum and minimum hazard values, respectively. V' is the normalized value of vulnerability. V is the normalized value of vulnerability. V_{max} and V_{min} are the maximum and minimum vulnerability values, respectively.

Among them, the P_i calculation method takes the previous geological hazard statistical samples of the study area as an example. A logistic regression statistical model is used to determine the spatial and temporal probability of geological hazard risk in Longlin Town based on the completion of regional geological hazard risk zoning. The P_i is 0.72 for 100-year rainfall conditions, 0.23 for 50-year rainfall conditions, and 0.08 for 50-year rainfall conditions, respectively. After normalizing the risk and vulnerability of geohazards in Longlin Town, the calculation method of Equation (13) is used to calculate the risk probability of geohazards under different precipitation conditions. The comprehensive risk degree of geohazards in Longlin Town is obtained by multiplying the normalized data of hazard of geohazard chains and the normalized data of vulnerability of disaster-bearing bodies. The risk evaluation results classification was conducted used by the method of the natural breakpoint and characteristic points according to the risk characteristics of geological hazards in Longlin Town and the objectives of risk control. The risk degree is divided into four levels: very high risk (0.697~1), high risk (0.538~0.697), medium risk (0.356~0.538), and low risk (0~0.356). The blocks of each risk level were indicated by different spots based on the risk level classification standard, and the raster units of the same risk level were combined to draw the geological hazard risk zoning map of Longlin Town (Figure 9).

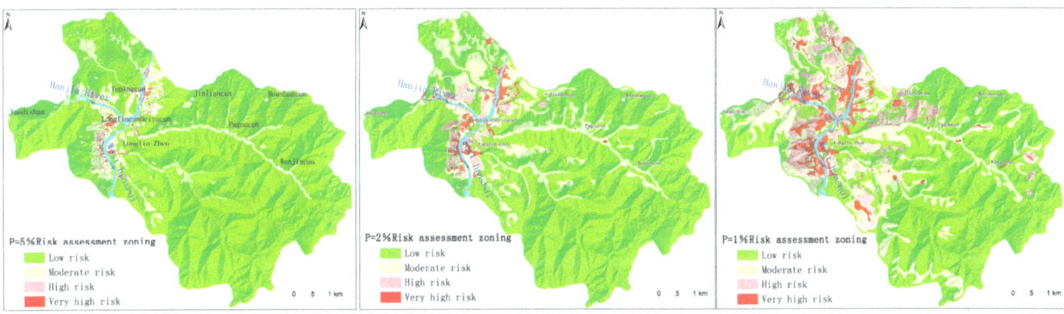

Figure 9. Regional map of geohazards Longlin Town Collective Area under different conditions.

The results show that the area of very high-risk zone, high-risk zone, medium-risk zone, and low-risk zone under 100-year (1% frequency) precipitation conditions are 1.91 km^2, 4.54 km^2, 7.02 km^2, and 24.93 km^2, respectively. The area of the very high-risk zone, high-risk zone, medium-risk zone and low-risk zone under 50-year (2% frequency) precipitation conditions are 0.64 km^2, 1.41 km^2, 5.14 km^2, and 31.21 km^2, respectively. The area of the very high-risk zone, high-risk zone, medium-risk zone and low-risk zone under 20-year (5% frequency) precipitation conditions are 0.15 km^2, 0.64 km^2, 3.32 km^2, and 34.29 km^2, respectively.

Under different rainfall frequencies, 75.23% of the areas always maintain low risk, which is mainly the wasteland or forest where people are rarely found. The area suffered from geological hazards with very few and small scales, and danger and vulnerability and the risk of damage caused by disasters are very low. Among in the 24.38% of the areas, risk level increases with the decrease in rainfall frequency. For example, the risk level of Yuping Village from medium-low risk gradually increases to high–very high risk under the rainfall condition of from one rainfall in 20 years to one rainfall in 100 years. Therefore, regular inspection and professional monitoring and early warning facilities are required to protect people's production and living safety. A total of 0.39% of the areas always remain very high risk due to the wide distribution of geological hazards, large scale, strong destructive power, concentrated distribution of population and property, high hazard and vulnerability, such as Longlin Village, Shuandu Village, and the residential areas on both sides of the Quanjia Bay ditch, which should strengthen risk control by immediately implement of comprehensive disaster prevention projects.

5. Discussion

5.1. Disaster-Forming Pattern Identification Markers

The risk as assessment of urban geological hazards has improved the level of early identification and prediction of risk areas and sources, mitigated the risk of geological hazards at source effectively, and provided a scientific basis for territorial spatial planning and geological hazard prevention and control. In order to better study the distribution characteristics of geological hazards, assess their risk and strengthen the control of geological hazard risks at source in complex mountainous cities, a refined risk assessment was conducted for 19 medium and large geological hazards in Longlin Town, such as the Quanjia Bay debris flow and the Panping Village landslide. There are a total of 71 geological hazards were identified in Longlin Town, the expression of characteristics, such as geological and potential hazard's disaster environment, triggering factors and disaster-bearing bodies, and the typical potential hazard's morphological, deformation and situation, were realized through the construction of the Three Investigations system and formation of the geological hazard knowledge map. It plays an important role in improving the identification, monitoring and early warning ability of geological hazards, solving the problem of where are the hidden hazards effectively and strengthening the foundation of geological hazard risk control.

5.2. Development and Evolution of Disasters under Different Rainfall Scenarios

Influenced by the rainfall conditions, the area of very high and high risk increases as the rainfall level upgrading, e.g., the area of very high risk in the study region is 6.45 km^2 under 100-year rainfall conditions, which are 3.14, and 8.16 multiple more than that in under 50-year and 20-year rainfall conditions, respectively. The area of high risk is 7.0 km^2 under 100-year rainfall conditions, which are 2.36- and 5.47-fold higher than in under 50-year and 20-year rainfall conditions, respectively. Under different rainfall frequencies, 75.23% of the area always remained low risk, 24.38% of the area's risk level increased with decreasing rainfall frequency, and 0.39% of the area always remained very high risk. Our research obtains the mapping of geological hazards and risk regionalization in the study area under different precipitation frequencies. The results show that rainfall can not only scour the loose accumulation of rock and soil bodies on the slope of landslides but also exacerbate the deformation and damage of landslides by the formation of high dynamic water pressure on the potential slip surface separated from water relatively. The area of very high and high risk reaches the biggest under 100-year extraordinary rainstorm conditions, it is indicated that rainfall has a very significant role in the induction of geologic hazard. The geological conditions of the study area also impacted the occurrence of geological disasters. The region is characterized by active neotectonics movement, discordant valley landforms by the denudation and cutting strongly, complicated and broken rock structure, and developed weak rock. The formation of soft and fluid plastic soften belt on the contact surface between the soil and rock due to the strength of the soil and the lower part of the soft rock is greatly reduced, which are induced landslides by reducing the stability of the slope.

5.3. Suggestions for Geological Hazard Risk Control

(1) Geological hazard risk control is a way to minimize the risk and possible loss of geological hazards by evaluating the possible geological hazard risks and proposing targeted risk control measures based on the systematic understanding of the geological hazard risks formation process and mechanism, such as the cause, mechanical mechanism, motion law and disaster formation mechanism, and the physical characteristics of different types of geological hazards and their disaster formation characteristics. The general idea and specific measures for mitigating the risk of geological hazards in the Longlin Town catchment area are proposed to provide technical support for disaster prevention and mitigation and land use spatial planning control.

(2) Comprehensively analyze the engineering geological conditions, engineering technical difficulty, and engineering cost of the risk section, and carry out engineering treatment of high-risk geological hazard sites in a hierarchical and targeted manner to reduce geological hazard risks. According to the development law, movement characteristics and disaster formation mode of geological hazard, based on the accurate judgment of geological hazards and threat scope, in accordance with the principle of comprehensive analysis and differentiation of priorities, the staff will implement step-by-step work such as engineering treatment and risk-avoidance and relocation. Finally, taking into account the return on investment and the conditions of implementation, appropriate measures are selected from among the available mitigation approaches and technologies for effective disaster risk management. Under different rainfall frequencies, taking the rainfall condition of one in 50 years as an example, a risk control chart (Figure 10) is established with a combination of point and polygon with an Area Grid-based Double Control mechanism to realize the organic combination of prevention and control of hidden hazard points and risk source areas and to support and guide the dynamic risk control of geological hazards. After the comprehensive risk control measures, the area of high and very high risk areas under 50-year rainfall conditions is reduced by 39.41%, among which the area of very high risk areas can be reduced by 87.81%, the investment efficiency ratio of comprehensive risk control is 80.78%, the risk reduction ratio of comprehensive risk control is 9.78%, and the risk reduction effect of risk areas is good, and the risk reduction rate of comprehensive risk control is 92.11%, and the risk reduction is obvious.

(3) It can reduce the probability of encountering disaster-bearing bodies and disaster events as well as the value of geological hazard losses according to a series of works, which were including strengthen the combination of monitoring and early warning by general and specialization, carrying out popular science propaganda and technical training continuously, improve the awareness level of the public, and standardize production and living activities.

In the populated areas along the banks of West Hanshui and Hanjia Rivers, with an area of approximately 3.55 km^2 in the catchment area, we will improve the disaster prevention knowledge and awareness of the residents in the affected areas through policies, propaganda training and social management. Carrying out some work enhanced the ability of group measurement and monitoring and emergency avoidance, which are strengthening propaganda and training, professional guidance, inspection and control, and emergency drills in flood and key areas. Further, the level of disaster prevention and mitigation can be improved by summarizing the experience and lessons learned in disaster prevention and mitigation and revising the behaviors, habits, and guidelines for disaster prevention and mitigation continuously.

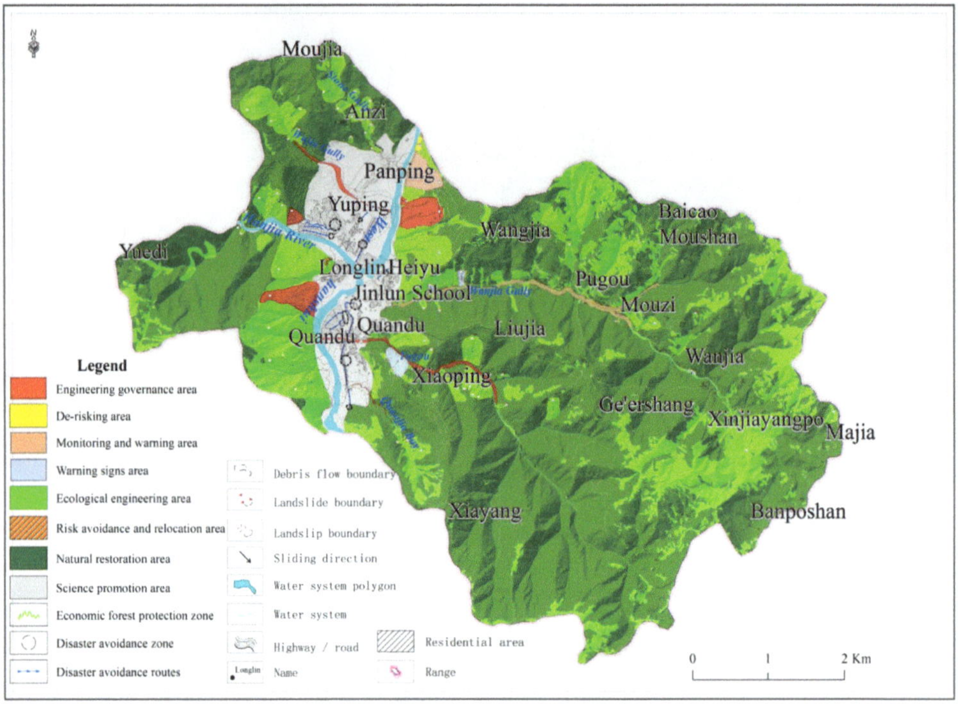

Figure 10. Recommended map of Comprehensive Risk Control of Geohazard in Longlin Town (20a).

6. Conclusions

(1) This paper analyzes and studies the dynamic process and risk management and control objectives of geological hazards, and puts forward the ideas and methods of geological hazard risk assessment and control in complex mountainous cities and towns, based on the identification of geological hazard risk source, research on geological hazard formation mode, geological hazard risk analysis, evaluation of vulnerability of potential disaster-bearing bodies, and proposals of geological hazard risk assessment and control in the township area. The effective service support for residential area construction and disaster prevention planning of disaster points is a double risk control technology method worthy of popularization and application. After the comprehensive risk control measures, the area of high and very high-risk areas under 50-year rainfall conditions are reduced by 39.41%, and the risk reduction ratio of comprehensive risk control is 9.78%, which is a good risk reduction effect.

(2) Taking Longlin Town, Lixian County, Longnan Mountainous Area, a typical mid-alpine valley landform, as an example, a total of 71 geological hazards were identified in the study area through remote sensing identification and detailed investigation, including 4 landslides, 53 landslides, 7 gully-type debris flows, and 7 slope-type debris flows. Through the analysis of slope structure, rock, and soil mass structure characteristics, properties, and deformation, it is concluded that the main disaster modes of landslides in this area are small-scale landslide collapse and overburden damage under the unloading effect of the back wall of the old landslide, pushover damage under the excavation and erosion effect of the middle front edge of the landslide, and secondary landslide slide damage under the erosion of the side edge of the landslide. The main disaster mode of debris flow is two types, gullying uncovered soil-siltation type damage slope runoff erosion slip type-diffuse flow type damage. On this basis, an intuitive and visual early detection map of landslides is constructed.

(3) The hydrologic-fluid coupling model was used to simulate and analyze the impact range and intensity of hazards under different precipitation conditions in the study area. According to the three-level superposition of slope destabilization, landslide damage and occurrence, the risk assessment of geological hazards in the study area was realized. The results show that 75.23% of the area always maintain low risk under different rainfall frequency, 24.38% of the regional risk level increases with the decrease in rainfall frequency, 0.39% of the areas always maintain extremely high risk, and the research results show that the research areas should immediately implement a comprehensive disaster prevention project to strengthen risk management and control.

The research on urban geological disaster risk prevention and control is still in the initial stage, and in the future, it will further deepen and improve the risk analysis based on the formation mechanism and disaster dynamic process. It is significant for improving the risk control system, responsibility system and technical methods, strengthening the investigation and hidden hazard identification ability of potential severe and major high-level geological disasters, and attacking the multi-scale and multi-hazard risk development and coordinated prevention and control technology.

Author Contributions: Conceptualization, W.D., G.W. and Q.Y.; methodology, G.W. and Q.Y.; software, W.D., G.W. and L.H.; validation, G.W., Q.Y., Y.G. and X.C.; formal analysis, Y.X. and Q.Y.; investigation, G.W., Q.Y. and Y.G.; resources, G.W. and Q.Y.; data curation, W.D., G.W., Q.Y., S.X. and X.Y.; writing—original draft preparation, W.D., G.W. and Q.Y.; writing—review and editing, W.D., G.W. and Q.Y.; supervision, G.W. and Q.Y.; project administration, G.W. and Y.G. All authors have read and agreed to the published version of the manuscript.

Funding: This study was financially supported by the China Geological Survey (Nos. DD20230008, DD20221747, DD20160083) and the National Key Research and Development Program of China (Nos. 2022YFC3003403, 2018YFC0603701).

Data Availability Statement: The data presented in this study are all available in this article.

Acknowledgments: The authors are thankful to the National Aeronautics and Space Administration (NASA) for the ASTER DEM, to the United States Geological Survey (USGS) for providing SRTM DEM, and to the reviewers for their constructive suggestions on improving the manuscript. All authors have agreed to the final version.

Conflicts of Interest: The authors declare that they have no known conflict of interest or personal relationship that could have appeared to influence the work reported in this paper.

References

1. Alam, E.; Ray-Bennett, N.S. Disaster risk governance for district-level landslide risk management in Bangladesh. *Int. J. Disaster Risk Reduct.* **2021**, *59*, 102220. [CrossRef]
2. Damalas, A.; Mettas, C.; Evagorou, E.; Giannecchini, S.; Iasio, C.; Papadopoulos, M.; Konstantinou, A.; Hadjimitsis, D. Development and Implementation of a DECATASTROPHIZE platform and tool for the management of disasters or multiple hazards. *Int. J. Disaster Risk Reduct.* **2018**, *31*, 589–601. [CrossRef]
3. Jaiswal, P.; Westen, C.; Jetten, V. Quantitative landslide hazard assessment along a transportation corridor in southern India. *Eng. Geol.* **2010**, *116*, 236–250. [CrossRef]
4. Shi, J.S.; Zhang, Y.S.; Dong, C.; Wu, S.R. GIS-based landslides hazard zonation of the New Badong County site. *Acta Geosci. Sin.* **2005**, *26*, 275–282.
5. Tang, Y.M.; Zhang, M.S.; Li, L.; Xue, Q. Discrimination to the landslide susceptibility, hazard and risk assessmennt. *Hydrogeol. Eng. Geol.* **2011**, *38*, 125–129.
6. Wu, S.R.; Shi, J.S.; Zhang, C.S.; Wang, T. Preliminary discussion on technical guideline for geohazard risk assessment. *Geol. Bull. China* **2009**, *28*, 995–1005.
7. Xu, Q.; Zhang, Y.F.; Chen, W. Vulnerability assessment of geohazards in south-west mountainous area Danba County, Sichuan, China as an example. *Geol. Bull. China* **2010**, *29*, 729–738.

8. Zhang, C.S.; He, S.J.; Xin, P.; Sun, W.F.; Tan, C.X.; Wu, S.R.; Wang, T.; Liu, X. Risk evaluation of geological hazard in Weibin District, Bao Ji City, Shanxi Province, China. *Geol. Bull. China* **2009**, *28*, 1054–1063.
9. Guo, F.Y.; Meng, X.M.; Zhang, Y.J.; Li, S.; Xie, Z.T.; Xiong, M.Q.; Guo, P. Geological disaster risk assessment methods for mountainous cities and towns in Gansu Province. *J. Lanzhou Univ. Nat. Sci.* **2014**, *50*, 604–610.
10. Cui, P.; Zou, Q. Theory and method of risk assessment and risk management of debris flows and flash floods. *Prog. Geogr.* **2016**, *35*, 137–147.
11. Zhang, M.S.; Xue, Q.; Jia, J.; Xu, J.W.; Gao, B.; Wang, J.Y. Methods and practices for the investigation and risk assessmengt of Geo-hazards in mountainous towns. *Northwestern Geol.* **2019**, *52*, 125–135.
12. Wang, G.F.; Ye, Z.N.; Li, G.; Tian, Y.T.; Chen, Z.L.; Deng, B. Geological hazard risk assessment of Zhouqu county in Bailong river basin. *J. Catastrophol.* **2019**, *34*, 128–133.
13. Zhou, C.; Chang, M.; Xu, L.; Che, H.X. Risk assessment of typical urban mine geological disasters in Guizhou province. *Geomat. Inf. Sci. Wuhan Univ.* **2020**, *45*, 1782–1791.
14. Huang, B.L.; Yin, Y.P.; Li, B.; Feng, W.L.; Qin, Z.; Zhang, P. Study of risk assessment and mitigation for landslide-induced impulse wave near towns in reservoir areas. *Acta Geol. Sin.* **2021**, *95*, 1949–1961.
15. Xiao, L.; Zhang, Y.; Peng, G. Landslide Susceptibility Assessment Using Integrated Deep Learning Algorithm along the China-Nepal Highway. *Sensors* **2018**, *18*, 4436. [CrossRef]
16. Su, H.P.; Su, S.; Ma, J.Z.; Liu, D.F.; Zhang, P. Risk assessment of single debris flow gully in south Gansu province, China. *Mt. Res.* **2016**, *34*, 337–345.
17. Du, J.; Yin, K.L.; Wang, H.H. Simulation of three-dimensional movement of landslide-debris flow based on finite volume method. *Chin. J. Rock Mech. Eng.* **2015**, *34*, 480–488.
18. Wang, G.F.; Yang, Q.; Chen, Z.L.; Mao, J.R. Risk assessment of debris flow in the Ganjia gully of the Bailongjiang basin. *J. Sediment. Res.* **2020**, *45*, 66–73.
19. Cascini, L. Applicability of landslide susceptibility and hazard zoning at different scales. *Eng. Geol.* **2008**, *102*, 164–177. [CrossRef]
20. Chen, W.; Pourghasemi, H.; Zhao, Z. A GIS-based comparative study of Dempster-Shafer, logistic regression, and artificial neural network models for landslide susceptibility mapping. *Geocarto Int.* **2016**, *32*, 367–385. [CrossRef]
21. van Westen, C.J.; Castellanos, E.; Kuriakose, S.L. Spatial data for landslide susceptibility, hazard, and vulnerability assessment: An overview. *Eng. Geol.* **2008**, *102*, 112–131. [CrossRef]
22. Cui, P.; Peng, J.; Shi, P.; Tang, H.; Ouyang, C.; Zou, Q.; Liu, L.; Li, C.; Lei, Y. Scientific challenges of research on natural hazards and disaster risk. *Geogr. Sustain.* **2021**, *2*, 216–223. [CrossRef]
23. Ma, Z.; Mei, G. Deep learning for geological hazards analysis: Data, models, applications, and opportunities. *Earth-Sci. Rev.* **2021**, *223*, 103858. [CrossRef]
24. Ren, F.; Chen, W.; Han, W. Study on reason and spatial-temporal distribution characteristics of debris flow in Longnan area along G212. *Chin. J. Rock Mech. Eng.* **2008**, *27*, 3237–3243.
25. Chen, C.; Ren, J.; Meng, G.; Yang, P.; Zhang, J.; Su, X. Analysis of modern activity of major faults in northeast margin of Baryan-Har Block. *J. Geod. Geodyn.* **2012**, *3*, 27–30.
26. Yang, X.; Feng, X.; Huang, X.; Song, F.; Li, G.; Chen, X.; Zhang, L.; Huang, W. The Late Quaternary activity characteristics of the Lixian—Luojiabu fault: A discussion on the seismogenic mechanism of the Lixian M8 earthquake in 1654. *Chin. J. Geophys.* **2015**, *58*, 504–519.
27. Han, Z.; Xaing, H.; Ran, Y. Activity analysis of Lixian-Luojiapu fault zone in the east boundary of Tibetan Plateau since the Late-Pleistocene. *Seismol. Geol.* **2001**, *1*, 43–48.
28. Wei, R.; Zeng, Q.; Davies, T.; Yuan, G.; Wang, K.; Xue, X.; Yin, Q. Geohazard cascade and mechanism of large debris flows in Tianmo gully, SE Tibetan Plateau and implications to hazard monitoring. *Eng. Geol.* **2018**, *233*, 172–182. [CrossRef]
29. Bayer, B.; Simoni, A.; Schmidt, D.; Bertello, L. Using advanced InSAR techniques to monitor landslide deformations induced by tunneling in the Northern Apennines, Italy. *Eng. Geol.* **2017**, *226*, 20–32. [CrossRef]
30. Ciampalini, A.; Raspini, F.; Bianchini, S.; Frodella, W.; Bardi, F.; Lagomarsino, D.; Di Traglia, F.; Moretti, S.; Proietti, C.; Pagliara, P.; et al. Remote sensing as tool for development of landslide databases: The case of the Messina Province (Italy) geodatabase. *Geomorphology* **2015**, *249*, 103–118. [CrossRef]
31. Sun, Q.; Zhang, L.; Ding, X.L.; Hu, J.; Li, Z.W.; Zhu, J.J. Slope deformation prior to Zhouqu, China landslide from InSAR time series analysis. *Remote Sens. Environ.* **2015**, *156*, 45–57. [CrossRef]
32. Xu, C.; Xu, X. Comment on "Spatial distribution analysis of landslides triggered by 2008.5.12 Wenchuan Earthquake, China" by Shengwen Qi, Qiang Xu, Hengxing Lan, Bing Zhang, Jianyou Liu [Engineering Geology 116 (2010) 95–108]. *Eng. Geol.* **2012**, *133–134*, 40–42. [CrossRef]
33. Zhao, C.; Liu, C.; Zhang, Q.; Lu, Z.; Yang, C. Deformation of Linfen-Yuncheng Basin (China) and its mechanisms revealed by Π-RATE InSAR technique. *Remote Sens. Environ.* **2018**, *218*, 221–230. [CrossRef]
34. Tsunetaka, H.; Hotta, N.; Imaizumi, F.; Hayakawa, Y.S.; Masui, T. Variation in rainfall patterns triggering debris flow in the initiation zone of the Ichino-sawa torrent, Ohya landslide, Japan. *Geomorphology* **2021**, *375*, 107529. [CrossRef]
35. Ba, R.; Deng, Q.; Liu, Y.; Yang, R.; Zhang, H. Multi-hazard disaster scenario method and emergency management for urban resilience by integrating experiment–simulation–field data. *J. Saf. Sci. Resil.* **2021**, *2*, 77–89. [CrossRef]

36. Huang, Y.; Dai, Z. Large deformation and failure simulations for geo-disasters using smoothed particle hydrodynamics method. *Eng. Geol.* **2014**, *168*, 86–97. [CrossRef]
37. Małkowski, P.; Niedbalski, Z. A comprehensive geomechanical method for the assessment of rockburst hazards in underground mining. *Int. J. Min. Sci. Technol.* **2020**, *30*, 345–355. [CrossRef]
38. Mohammadi, S.; Taiebat, H. Finite element simulation of an Excavation-triggered landslide Using large deformation theory. *Eng. Geol.* **2016**, *205*, 67–72. [CrossRef]
39. Pistolesi, M.; Cioni, R.; Rosi, M.; Aguilera, E. Lahar hazard assessment in the southern drainage system of Cotopaxi volcano, Ecuador: Results from multiscale lahar simulations. *Geomorphology* **2014**, *207*, 51–63. [CrossRef]
40. Prieto, J.A.; Journeay, M.; Acevedo, A.B.; Arbelaez, J.D.; Ulmi, M. Development of structural debris flow fragility curves (debris flow buildings resistance) using momentum flux rate as a hazard parameter. *Eng. Geol.* **2018**, *239*, 144–157. [CrossRef]
41. Caviedes-Voullième, D.; García-Navarro, P.; Murillo, J. Influence of mesh structure on 2D full shallow water equations and SCS Curve Number simulation of rainfall/runoff events. *J. Hydrol.* **2012**, *448–449*, 39–59. [CrossRef]
42. Naef, D.; Rickenmann, D.; Rutschmann, P.; McArdell, B.W. Comparison of flow resistance relations for debris flows using a one-dimensional finite element simulation model. *Nat. Hazards Earth Syst. Sci.* **2006**, *6*, 155–165. [CrossRef]
43. Siviglia, A.; Stecca, G.; Vanzo, D.; Zolezzi, G.; Toro, E.F.; Tubino, M. Numerical modelling of two-dimensional morphodynamics with applications to river bars and bifurcations. *Adv. Water Resour.* **2013**, *52*, 243–260. [CrossRef]
44. Septian, A.; Llano-Serna, M.A.; Ruest, M.R.; Williams, D.J. Three-dimensional Kinematic Analysis of Bingham Canyon Mine Pit Wall Slides. *Procedia Eng.* **2017**, *175*, 86–93. [CrossRef]
45. Kumar, V.; Gupta, V.; Jamir, I.; Chattoraj, S.L. Evaluation of potential landslide damming: Case study of Urni landslide, Kinnaur, Satluj valley, India. *Geosci. Front.* **2019**, *10*, 753–767. [CrossRef]
46. Salvatici, T.; Di Roberto, A.; Di Traglia, F.; Bisson, M.; Morelli, S.; Fidolini, F.; Bertagnini, A.; Pompilio, M.; Hungr, O.; Casagli, N. From hot rocks to glowing avalanches: Numerical modelling of gravity-induced pyroclastic density currents and hazard maps at the Stromboli volcano (Italy). *Geomorphology* **2016**, *273*, 93–106. [CrossRef]
47. Zhang, W.; Ji, J.; Gao, Y.; Li, X.; Zhang, C. Spatial variability effect of internal friction angle on the post-failure behavior of landslides using a random and non-Newtonian fluid based SPH method. *Geosci. Front.* **2020**, *11*, 1107–1121. [CrossRef]
48. Chigira, M.; Duan, F.; Yagi, H.; Furuya, T. Using an airborne laser scanner for the identification of shallow landslides and susceptibility assessment in an area of ignimbrite overlain by permeable pyroclastics. *Landslides* **2004**, *1*, 203–209. [CrossRef]
49. Ciampalini, A.; Raspini, F.; Frodella, W.; Bardi, F.; Bianchini, S.; Moretti, S. The effectiveness of high-resolution LiDAR data combined with PSInSAR data in landslide study. *Landslides* **2015**, *13*, 399–410. [CrossRef]
50. Horton, P.; Jaboyedoff, M.; Rudaz, B.; Zimmermann, M. Flow-R, a model for susceptibility mapping of debris flows and other gravitational hazards at a regional scale. *Nat. Hazards Earth Syst. Sci.* **2013**, *13*, 869–885. [CrossRef]
51. Miao, S.; Zhu, Q.; Zhang, B.; Ding, Y.; Zhang, J.; Zhu, J.; Zhou, Y.; He, H.; Yang, W.; Chen, L. Knowledge-guided consistent correlation analysis of multimode landslide monitoring data. *Int. J. Geogr. Inf. Sci.* **2017**, *31*, 2255–2271. [CrossRef]
52. Bui, D.T.; Tuan, T.A.; Klempe, H.; Pradhan, B.; Revhaug, I. Spatial prediction models for shallow landslide hazards: A comparative assessment of the efficacy of support vector machines, artificial neural networks, kernel logistic regression, and logistic model tree. *Landslides* **2015**, *13*, 361–378.
53. Gorum, T. Landslide recognition and mapping in a mixed forest environment from airborne LiDAR data. *Eng. Geol.* **2019**, *258*, 105155. [CrossRef]
54. Lan, H.; Martin, D.; Zhou, C.; Lim, C. Rockfall hazard analysis using LiDAR and spatial modeling. *Geomorphology* **2010**, *118*, 213–223. [CrossRef]
55. Kromer, R.A.; Hutchinson, D.J.; Lato, M.J.; Gauthier, D.; Edwards, T. Identifying rock slope failure precursors using LiDAR for transportation corridor hazard management. *Eng. Geol.* **2015**, *195*, 93–103. [CrossRef]
56. Skempton, A.W.; Delory, F.A. Stability of natural slopes in London clay. In *Selected Papers on Soil Mechanics*; Thomas Telford Publishing: London, UK, 1984; pp. 70–73.
57. Wang, G.F.; Chen, Z.L.; Mao, J.R.; Tian, Y.T.; Gao, Y.L.; Deng, B. Debris flow risk assessment in Bailong river basin under different engineering scenarios and rainfall frequency conditions. *J. Shandong Univ. Sci. Technol. Nat. Sci.* **2020**, *39*, 30–40.
58. Erena, S.H.; Worku, H.; De Paola, F. Flood hazard mapping using FLO-2D and local management strategies of Dire Dawa city, Ethiopia. *J. Hydrol. Reg. Stud.* **2018**, *19*, 224–239. [CrossRef]
59. Fang, Q.; Tang, C.; Chen, Z.; Wang, S.; Yang, T. A calculation method for predicting the runout volume of dam-break and non-dam-break debris flows in the Wenchuan earthquake area. *Geomorphology* **2019**, *327*, 201–214. [CrossRef]
60. Caccavale, M.; Matano, F.; Sacchi, M. An integrated approach to earthquake-induced landslide hazard zoning based on probabilistic seismic scenario for Phlegrean Islands (Ischia, Procida and Vivara), Italy. *Geomorphology* **2017**, *295*, 235–259. [CrossRef]
61. Harp, E.L.; Keefer, D.K.; Sato, H.P.; Yagi, H. Landslide inventories: The essential part of seismic landslide hazard analyses. *Eng. Geol.* **2011**, *122*, 9–21. [CrossRef]
62. Neaupane, K.; Piantanakulchai, M. Analytic network process model for landslide hazard zonation. *Eng. Geol.* **2006**, *85*, 281–294. [CrossRef]
63. Van Den Eeckhaut, M.; Hervás, J. State of the art of national landslide databases in Europe and their potential for assessing landslide susceptibility, hazard and risk. *Geomorphology* **2012**, *139–140*, 545–558. [CrossRef]
64. Qiao, J.P.; Yang, Z.J. The K-P method based failure probability calculation on landslides. *Chin. J. Geol. Hazard Control* **2013**, *24*, 1–5.

65. Chen, L.X.; van Westen, C.J.; Haydar, H.; Roxana, L.C.; Thea, T.; Diana, C.R.; Dhruba, P.S. Integrating expert opinion with modelling for quantitative multi-hazard risk assessment in the Eastern Italian Alps. *Geomorphology* **2016**, *273*, 150–167. [CrossRef]
66. Martha, T.R.; van Westen, C.J.; Kerle, N.; Jetten, V.; Kumar, K.V. Landslide hazard and risk assessment using semi-automatically created landslide inventories. *Geomorphology* **2013**, *184*, 139–150. [CrossRef]
67. Rabiya, M. Review of National Multi-Hazard Early Warning System Plan of Pakistan in context with Sendai Framework for Disaster Risk Reduction. *Procedia Eng.* **2018**, *212*, 206–213.

water

MDPI

Article

Spatial Distribution, Migration, and Ecological Risk of Cd in Sediments and Soils Surrounding Sulfide Mines—A Case Study of the Dabaoshan Mine of Guangdong, China

Weikang Sheng [1], Qingye Hou [1,*], Zhongfang Yang [1] and Tao Yu [2]

1 School of Earth Science and Resources, China University of Geosciences, Beijing 100083, China; shengwk95@163.com (W.S.); yangzf@cugb.edu.cn (Z.Y.)
2 School of Science, China University of Geosciences, Beijing 100083, China; yutao@cugb.edu.cn
* Correspondence: qingyehou@cugb.edu.cn

Abstract: Acid mine drainage (AMD) resulting from metal sulfide mining activities can lead to contamination by potentially toxic elements (PTEs) primarily concentrated around the mining area and gradually spreading outward. However, ecological risks do not correspond directly to PTE concentrations, making it challenging to effectively manage the mining environment and accurately prevent potential ecological impacts. In this paper, we analyzed Cd levels in sediments, soils, and corresponding rice grains sampled from four villages near Dabaoshan Mine of Guangdong, China, in 2017. Our results reveal that Cd is the most prominent pollutant element, exhibiting significant enrichment and spatial heterogeneity in both soil and sediments and higher accumulation levels in rice grains compared to other PTEs. Cd concentrations in soil decrease from the tailings pond to the river terrace, with a slight increase after Taiping River joins and flows into the alluvial plain. However, the concentrations in sediments show the opposite trend. The bioconcentration factor (BCF) for Cd in agricultural soil from the river terrace is lower than that from the alluvial plain and the degree of exceeding the maximum permit level (MPL) of Cd in rice grains increases along the river. Mineral transformation and topography are important factors in controlling the geochemical behavior of PTEs. Remediation efforts alter the physicochemical properties of the river, resulting in the release of PTEs during schwertmannite transformation followed by their adsorption by clay minerals. Furthermore, the random forest (RF) analysis highlights that the bioavailability and potential ecological risk of Cd in soils are governed by the occurrence form of Cd in different topographies, mainly controlled by TFe_2O_3, Mn, and CaO in the river terrace and CaO, Al_2O_3/SiO_2, and Mn in the alluvial plain. Therefore, considering the impact of topography on mineral compositions, physicochemical properties, and occurrence form of PTEs in soil and sediments is essential for assessing ecological risk in mining areas.

Keywords: PTEs; sulfide mines; soils; ecological risk; topography

Citation: Sheng, W.; Hou, Q.; Yang, Z.; Yu, T. Spatial Distribution, Migration, and Ecological Risk of Cd in Sediments and Soils Surrounding Sulfide Mines—A Case Study of the Dabaoshan Mine of Guangdong, China. *Water* **2023**, *15*, 2223. https://doi.org/10.3390/w15122223

Academic Editors: Ruiping Liu and Youning Xu

Received: 9 April 2023
Revised: 7 June 2023
Accepted: 11 June 2023
Published: 13 June 2023

1. Introduction

Mining and smelting of metal sulfides result in significant contamination by potentially toxic elements (PTEs) in the surrounding environment, with characteristics of high concentrations of PTEs, multiple contaminants, and large areas with significant ecological risk [1,2]. Soils in mining areas have been contaminated with PTEs globally, posing a major environmental issue in terms of agricultural soil and crop contamination near mining areas [3–7]. Even after mining activities have ceased, PTEs from tailings still continue to be released into the surrounding soil, streams, and groundwater over long periods of time through erosion, weathering, and leaching processes [3].

Dabaoshan Mine is one of China's major large−scale polymetallic sulfide mines [8]. Exploitation of mineral resources has produced a large quantity of tailings containing sulfide minerals. These tailings are continuously oxidized to form acid mine drainage

(AMD) rich in PTEs [9,10]. A large area surrounding the mine has been impacted, posing a threat to the health of local residents [11,12]. The water and soil in 83 villages around Dabaoshan Mine have become contaminated with PTEs, resulting in severe health issues such as cardiovascular, kidney, nervous, blood, and skeletal diseases [13,14], and even esophageal cancer, liver cancer, and other cancers [15,16]. To avoid further contamination, mining activities were halted in 2016. In addition, in response to government policies, mining companies have conducted a series of restoration efforts for AMD treatment, including building mud impoundments to restrict the movement of waste and soil [17], constructing sewage treatment plants [18], and adding lime to the rivers near the mine [19].

PTEs can enter the human body via various exposure pathways from contaminated soil, leading to adverse health effects [20]. Although PTE contamination is primarily concentrated around the mining area and gradually spreads outward, extensive research has demonstrated that the areas with the highest human health risks do not align with the areas exhibiting the highest PTE concentrations in soil [15,21–25]. In the Dabaoshan Mine area, concentrations of PTEs in soils and suspended particles gradually decrease as distance increases from the headwaters of the river [22,23]. Furthermore, PTEs are characterized by high potential bioavailability owing to the high non−residual fractions of PTEs in the sediments and soil [24], and the average concentrations of Cd and Pb in rice grains grown near the mining area exceed the maximum permit levels (MPLs) [25]. Shangba Village is the highest risk area, with Cd being the main pollutant contributing to human health risks, while the risks near the tailings area are within acceptable ranges according to human health risk assessments [21,24]. Areas farther away from the mining area exhibit higher toxicity and greater ecological risks, indicating no obvious correspondence between PTEs in soil and crops, and posing challenges for effective pollution control and risk prevention in mining areas. In addition, the distribution and migration of PTEs around the Dabaoshan Mine after restoration efforts have been rarely studied. Therefore, the objectives of this study were to (1) investigate the characteristics, spatial distributions, and migration of Cd in the environment surrounding the mine, and (2) identify the key factors governing the spatial distributions and bioavailability of Cd.

2. Materials and Methods

2.1. Study Area

Dabaoshan Mine is the largest mine in southern China, and is located in Shaoguan City, Guangdong Province, China (24°31′37″ N, 113°42′49″ E) [8,26]. The Dabaoshan Mine spreads from north to south, with higher elevations in the north (800–1200 m) and lower elevations in the south, including low hills and alluvial plains [27]. The region has a subtropical monsoon climate, with an average annual temperature of 20.3 °C and an average annual precipitation of 1782 mm [28]. Dabaoshan Mine is a well−known large−scale multi−metal sulfide mine with low−temperature mineralization. The main mineralization includes limonite in the upper part, copper−sulfur ores in the middle part, and lead−zinc ores in the lower part, along with non−ferrous metals such as tungsten, bismuth, molybdenum, gold, and silver [29,30].

Since the 1970s, large−scale mining activities have been carried out in the area, resulting in the generation of large amounts of AMD from the waste stacking, ore beneficiation, and washing processes [31]. AMD is discharged from the water outlets of the Tielong and Caoduikeng tailings ponds and flows into the Hengshi and Fanshui Rivers, which converge 1.6 km north of Liangqiao Village before entering the mainstream Hengshi River. The river flows through several villages, including Liangqiao Village (LQ, ~5 km), Tangxin Village (TX, ~9 km), Yanghe Village (YH, ~12 km), and Shangba Village (SB, ~15 km), before it joins the Taiping River, which is not directly affected by mining activities. Shangba Village is known as an endemic cancer village due to its high mortality rate [25]. For this study, we defined the LQ–SB section of the Hengshi River as the upstream area of the Hengshi River. The LQ–TX area has mainly hills and terraces accompanied by high river flow velocity due to a drop in elevation and a narrow valley, while the YH–SB area has mainly alluvial plains, where the riverbed gradually widens and the slope is very gentle [32]. The geological map is shown in Figure S1. Granite outcrops are present and the surface rocks are highly weathered. The study area is covered with quaternary alluvium along Hengshi River and the predominant soil type is laterite [13].

2.2. Sampling

The sampling sites are shown in Figure 1. Sampling was carried out in July 2017 during the rainy season. Six natural forest soil samples were collected in the upstream area of the Hengshi River, including one from TX, two from YH, and three from SB. The sampling depth was 0–20 cm and the sampling density was 4 points/km^2. Four to six subsamples were collected according to the "S" or "X" sampling method. After removing roots and other debris, they were thoroughly mixed and divided into four parts with 1.0–1.5 kg per subsample, which were then placed in sample bags. After drying in a cool place and removing debris, the samples were smashed with a rubber hammer, ground, and sieved through a 10−mesh nylon sieve, and mixed well; each sample weighed >100 g. The samples were then placed in clean polyethylene bags and sent to the laboratory for analysis.

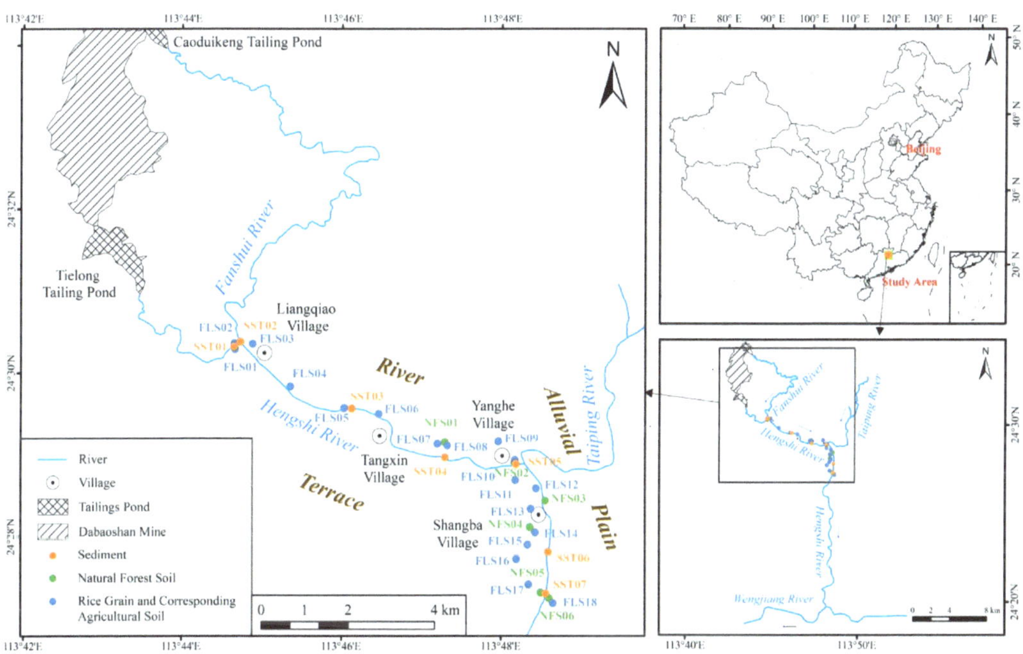

Figure 1. Sampling sites from the upper Hengshi River through the Dabaoshan Mine.

During the rice harvesting period, 18 rice grains and corresponding agricultural soil samples were collected, including 4 from LQ, 4 from TX, 5 from YH, and 5 from SB. Between 3 and 5 soil subsamples were collected at each sampling point and 5–20 plants were collected at each subsample point. The soil and plant subsamples were then mixed into composite samples. The whole rice plant was pulled out of the ground and the ears of rice were cut and placed in clean mesh bags. The root soil was shaken into a clean cotton bag. The ears of rice were air dried in a cool, ventilated, clean, and dust−free place. The panicle was then threshed, removing impurities and husks, ground with an agate mortar to 60 mesh (<0.25 mm), mixed well, and used for further chemical analysis. Plant residues, such as straw and other non−soil components, were removed from the soil samples, and then the samples were air dried. The soil was regularly turned over and beaten during this process. All soil components were sieved through a 10−mesh nylon sieve and mixed well for further chemical analysis.

Seven sediment samples were collected from the Hengshi River. During the collection process, sand was avoided as much as possible, and surface sediments were collected from the 0–5 cm horizon, sealed in polyethylene plastic bags, and numbered. The pH of the river water at the sampling sites was measured and recorded with a handheld pH meter. After air drying, the river sediment samples were sieved through a 10−mesh nylon sieve and ground with a pollution−free ball mill to 200 mesh for analysis.

2.3. Chemical Analysis

Sample analysis was conducted by the Hefei Mineral Resources Supervision and Testing Center, the Ministry of Land and Resources, People's Republic of China. The oxides content was measured directly by wavelength dispersive X−ray fluorescence spectrometry (PANALITICA AXIOS PW4400, Almelo, The Netherlands). Soil samples were dissolved using a mixed acid solution (5 mL HCl, 3 mL HNO_3, 7 mL HF, and 0.25 mL $HClO_4$) and then the samples were dissolved using aqua regia and made up to a volume of 25 mL using high−purity water. Soil samples weighing 5 g were extracted with 50 mL of 0.01 M $CaCl_2$ solution. After centrifugation for 10 min, the supernatant was filtered. Rice grains were washed with deionized water, dried at 105 °C for 5 min, and dried at 70 °C to a constant weight. The plant samples were ground using a stainless−steel grinder and then dissolved using a mixed acid solution (HNO_3:$HClO_4$ = 4:1) until a transparent and clear solution was obtained. Concentrations of Cd, Pb, and Zn in the above solutions were determined using inductively coupled plasma mass spectrometry (ICP−MS) (X−SERIES II, Thermo Fisher Scientific, Waltham, MA, USA) [33]. For soil pH measurement, 10 g of an air−dried soil sample passed through a 2 mm sieve was weighed precisely to 0.01 g and added to a 50 mL high−form beaker. Then, 25 mL of decarbonated water was added (sample to liquid ratio 1:2.5) and the mixture was stirred for 1 min to disperse the sample particles. After standing for 30 min, the soil pH was measured. The cation exchange capacity (CEC) was measured using the $BaCl_2$−H_2SO_4 method. Soil organic matter (SOM) was measured by wet oxidation in an acid dichromate solution, followed by back titration of the remaining dichromate using a ferrous ammonium sulfate standard solution. National standard materials, GBW07405/GBW07458, GBW07304a, and GBW10010 were used to control the data accuracy for soil, sediment, and rice, respectively, and coded samples were added to control the precision of the data. The detailed quality control analysis of data is shown in Supporting Information, and all analytical data met the quality requirements.

2.4. Data Analysis

Basic descriptive statistics were derived to summarize the concentrations of major elements and Cd, Pb, and Zn in the study area. Descriptive statistics and Pearson correlations were analyzed using SPSS 26 (IBM SPSS Statistics). Pearson correlation analysis estimates the linear relationship between two variables. The analysis renders correlation coefficients ranging from −1 to 1, where −1 indicates a perfect negative linear relationship, 0 indicates no correlation, and 1 indicates a perfect positive linear relationship.

Linear regression fit analysis was performed using Origin 2020 (Origin lab Corporation). The cumulative probability distributions were fitted by a logistic function as follows:

$$y = 1/(1 + \exp((a - x)/b)$$

where y is the probability of a given value of x, and a and b are constants.

Random forest analysis was performed using R software (version 4.2.2). Random forest (RF) is a machine learning algorithm that can be used to assess the relative importance of independent variables on a dependent variable [34]. As part of the bootstrap resampling procedure, n samples were randomly selected from the training set to construct decision trees. Meanwhile, a random selection of k ($k \leq$ total number of features) candidate features was performed, and the optimal attribute was selected as the split node [35].

Geographical distribution maps were generated using ESRI−ArcGIS geospatial software (version 10.8).

3. Results

3.1. Concentrations of PTEs in Soils and Sediments

Cd, Pb, and Zn concentrations in natural forest soils, agricultural soils, and sediments in the upper Hengshi River are shown in Figure 2. The corresponding oxides and physicochemical properties are shown in Tables S3–S5. The average concentrations of Cd, Pb, and Zn in sediments are 10.31 mg/kg, 455.01 mg/kg, and 1943.97 mg/kg, respectively. The average concentrations of Cd, Pb, and Zn in natural forest soils are 0.73 mg/kg, 293.13 mg/kg, and 241.90 mg/kg, respectively. The average concentrations of Cd, Pb, and Zn in agricultural soils are 0.50 mg/kg, 96.42 mg/kg, and 179.99 mg/kg, respectively. Using the risk screening values (RSVs) and risk intervention values (RIVs) for Cd, Pb, and Zn shown in Table S1 as standards [36], 67% and 53% of the natural forest soils and agricultural soils exceed the Cd risk screening values (RSVs), respectively. A total of 17% of the natural forest soils exceed the Pb risk intervention values (RIVs), while 33% of the natural forest soils exceed the RSV of Pb. A total of 43% of the agricultural soils exceed the RSV of Pb, and 33% and 47% of the natural forest soils and agricultural soils exceed the RSV of Zn, respectively. Cd is the most serious pollutant in the soils around the Dabaoshan Mine, followed by Pb and Zn. Compared with natural forest soils, the degree of PTE pollution in agricultural soils is lower.

The potential ecological risks of PTEs in sediments were assessed using the threshold effect level (TEL) and probable effect level (PEL) recommended by Sediment Quality Guidelines (SQGs) [37]. When PTEs concentrations are below the TEL, they are considered to pose no or low ecological risk. PTEs concentrations between the TEL and PEL indicate a moderate ecological risk, while concentrations surpassing the PEL indicate a high ecological risk. TELs and PELs for Cd, Pb, and Zn are given in Table S2. Concentrations of Cd, Pb, and Zn in sediments of the upper Hengshi River (Figure 2) are higher than their PELs, indicating a high ecological risk with potential toxic effects on aquatic organisms.

Spatial variations in PTEs concentrations in soils and sediments are shown in Figure 3. Contrary to previous studies, the concentrations of PTEs do not exhibit a consistent decrease [22,23]. Concentrations of Cd, Pb, and Zn in agricultural soils gradually decrease from 1.08 to 0.19 mg/kg from LQ near the mining area to TX, then show a significant positive peak near YH and decrease near SB. Natural forest soils show an overall trend of gradually increasing near TX, then sharply decreasing near YH. Concentrations of Cd, Pb, and Zn in sediments increase from 5.29 to 20.22 mg/kg from LQ to TX, then sharply decrease near YH and gradually increase again near SB.

3.2. Concentrations of PTEs in Rice Grains

Cd concentrations in rice grains in the upper Hengshi River range from 0.014 mg/kg to 2.69 mg/kg with an average value of 1.03 mg/kg, with 95.24% exceeding maximum permit levels (MPLs) (0.20 mg/kg) for rice grains [38]. Concentrations of Pb and Zn in rice grains range from 0.02 to 0.39 mg/kg, and from 15.6 to 23.9 mg/kg, with average values of 0.12 mg/kg and 18.99 mg/kg, respectively. There are no rice grain samples in which concentrations of Pb and Zn exceed MPLs (Table 1). The degree of exceeding the MPL of Cd in rice grains from river terrain ranges from −93% to 290%, with a median value 238%. The degree of exceeding the MPL of Cd in rice grains from alluvial plains ranges from 258% to 1245%, with a median value 573%.

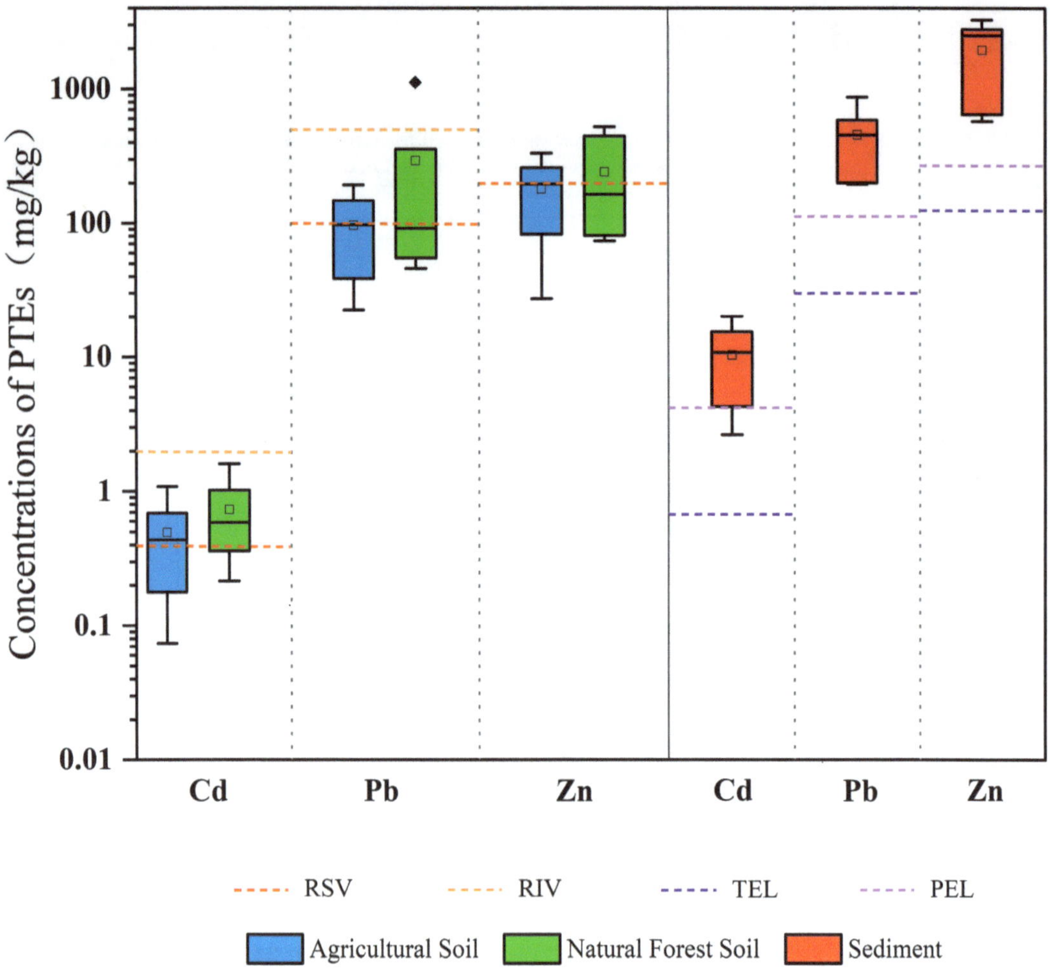

Figure 2. The boxplots for the concentrations of Cd, Pb, and Zn in agricultural soils, natural forest soils, and sediments in the study area (RSV: risk screening value, RIV: risk intervention value, TEL: threshold effect level, PEL: probable effect level).

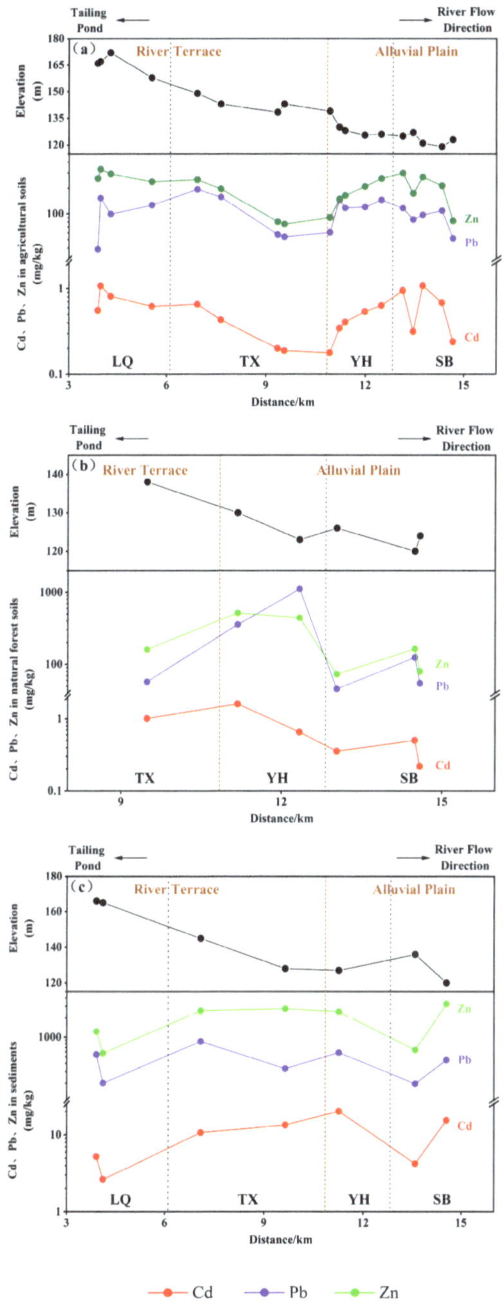

Figure 3. The spatial variation in Cd, Pb, and Zn content in different media and elevations in the study area, including agricultural soils (**a**), natural forest soils (**b**), and sediments (**c**). (Brown dash line: the boundary of topography; black dash line: the boundary of villages. LQ: Liangqiao Village, TX: Tangxin Village, YH: Yanghe Village, SB: Shangba Village).

Table 1. Statistical summaries of Cd, Pb, and Zn concentrations in rice grains from the study area (mg/kg).

	Min	Max	Mean	Median	SD	Coefficients Variation	MPL *	Exceedance Rate (%)
Cd	0.014	2.69	1.03	0.76	0.60	0.59	0.20	95.24%
Pb	0.02	0.39	0.12	0.08	0.08	0.72	0.40	0
Zn	15.60	23.90	18.99	18.64	1.75	0.09	50.00	0

Note(s): * MPL: The Chinese maximum permit levels (MPLs) for rice grains [38].

The concentration of Cd in rice grains and corresponding agricultural soils is shown in Figure 4. Except for FLS08, the concentration of Cd in rice grains gradually increases from the tailings pond to YH, reaches the maximum value near YH, and then gradually decreases. The variation characteristic of Pb in rice grains is similar to that of Cd. There is no obvious positive correlation between concentrations of PTEs in rice grains and in corresponding agricultural soils.

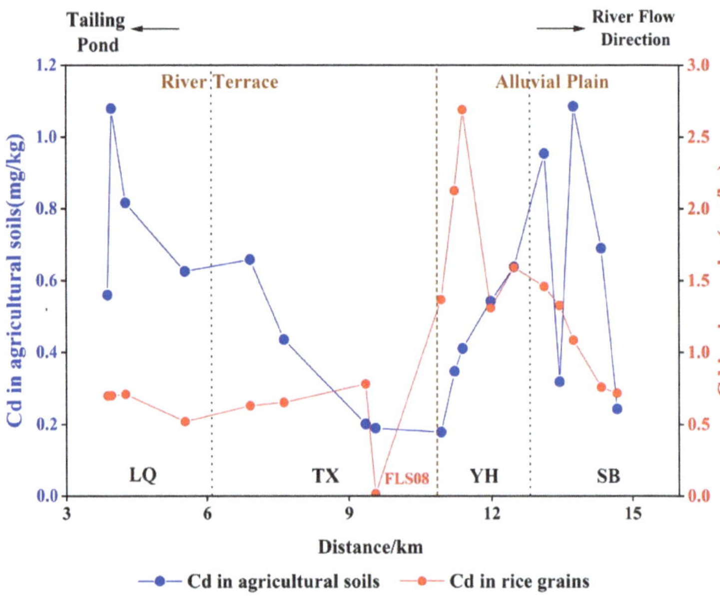

Figure 4. The spatial variation in Cd concentrations in rice grains and corresponding agricultural soils from the study area. (Brown dash line: the boundary of topography; black dash line: the boundary of village. LQ: Liangqiao Village, TX: Tangxin Village, YH: Yanghe Village, SB: Shangba Village).

4. Discussions

According to our results, although Cd, Pb, and Zn in sediments posed high ecological risk, Cd and Pb concentrations in soils were both lower than the risk intervention value, while in less than half of the soil, Zn exceeded the risk screening value, indicating there is no strict control area in Dabaoshan Mining area. However, the Cd concentration in 95.24% of rice grains exceeded the MPL, while concentrations of Pb and Zn in rice grain were found to be safe. These results demonstrate that, under a series of restoration efforts, the ecological risks posed by Pb and Zn have been gradually reduced and Cd contamination is the most severe in soils and rice grains around Dabaoshan Mine. Furthermore, the coefficients of variation in Cd, Pb, and Zn concentrations in rice grains (Table 1) revealed that Zn presented no significant spatial variation and Cd and Pb have similar variation characteristics. Therefore, in the following discussions, we focus on the factors affecting the Cd geochemical behavior in sediments, soil, and the soil–rice system.

4.1. Factors Affecting Cd Geochemical Behavior in Sediments

Cd concentrations in sediments increase from the tailings pond to the river terrace approximately 10 km away, then slightly decrease after Taiping River joins and flows into a small alluvial plain (Figure 3c). Cd is influenced by multiple geochemical processes during its migration and deposition in the river, including adsorption–desorption, hydraulic transport, and co−precipitation. Large amounts of AMD with low pH and high concentrations of PTE ions were produced by natural weathering and mining activities at Dabaoshan Mine, which changed the basic physicochemical properties of water and sediments in the Hengshi River and significantly affected the geochemical behavior of Cd in the river.

An important driving force for migration of PTEs in the river is hydraulic conditions, which are directly related to the elevation change of the river caused by topography. As the Hengshi River flows downstream along the valley, there is a significant decrease in river elevation, from 175 to 128 m. The river channel is narrow and river flow velocity is high in the section of the river terrace from LQ to TX. The riverbed gradually widens and the slope becomes very gentle in the alluvial plain section from YH to SB.

Cd in both dissolved and suspended forms originating from the tailings pond is not easily deposited in the river terrace and can accumulate in the alluvial plain, as supported by previous studies (Figure 3) [17,39,40].

The pH of water and the presence of secondary minerals in the river are also important factors controlling the migration of Cd in the upper Hengshi River. Figure 5 shows the changes in water pH under restoration efforts in the study area. In 2012, the water in Tielong tailings pond had an extremely low pH of approximately 2.39–2.81, primarily due to mine waste and runoff from the surrounding hills [41]. Since the Hengshi River flows through a very narrow valley with limited surface runoff, there is a lack of dilution in the river terrace. Cd in the river water migrated downstream in the form of divalent free ions or complexed compounds [26]. During the time of our sampling, the pH of the river water increased due to restoration measures, such as adding lime to the river, and the dilution effect of the Taiping River which is not contaminated by mining activities. However, the presence of various secondary minerals also plays a vital role in adsorption and migration of Cd in the river, which are susceptible to variation in pH. Schwertmannite is a hydroxy sulfate mineral that commonly exists in AMD, with a chemical formula of $Fe_8O_8(OH)_{8-2x}(SO_4)_x$ $1 \leq x \leq 1.75$ [42]. Schwertmannite exhibits a very high affinity for PTEs and has been observed in Hengshi River sediments [43,44]. Cd may co−precipitate and enter the solid phase during formation of schwertmannite and can be incorporated into the mineral lattice under very acidic conditions. However, as the pH of river water gradually increases, schwertmannite will gradually transform into more stable minerals such as goethite and ferrihydrite, causing Cd to be re−released predominantly in ionic form and absorbed in clay minerals [41,43].

4.2. Factors Affecting Cd Geochemical Behavior in Soil

The concentrations and distributions of Cd, Pb, and Zn in natural forest soils and agricultural soils show minimal differences (Figure 2). Cd, Pb, and Zn concentrations in agricultural soils are generally lower than those in natural forest soils within the same village, indicating that agricultural activities have not exacerbated Cd, Pb, and Zn pollution in the soils. The estimated contribution of fertilizer to soil Cd is only 1.21% based on input fluxes of Cd to agricultural soil near the Dabaoshan Mine [32]. Instead, crop harvesting may remove PTEs from soil. Therefore, PTEs in soils around Dabaoshan Mine are less influenced by anthropogenic activities, and AMD resulting from mining activities is likely the main source [45].

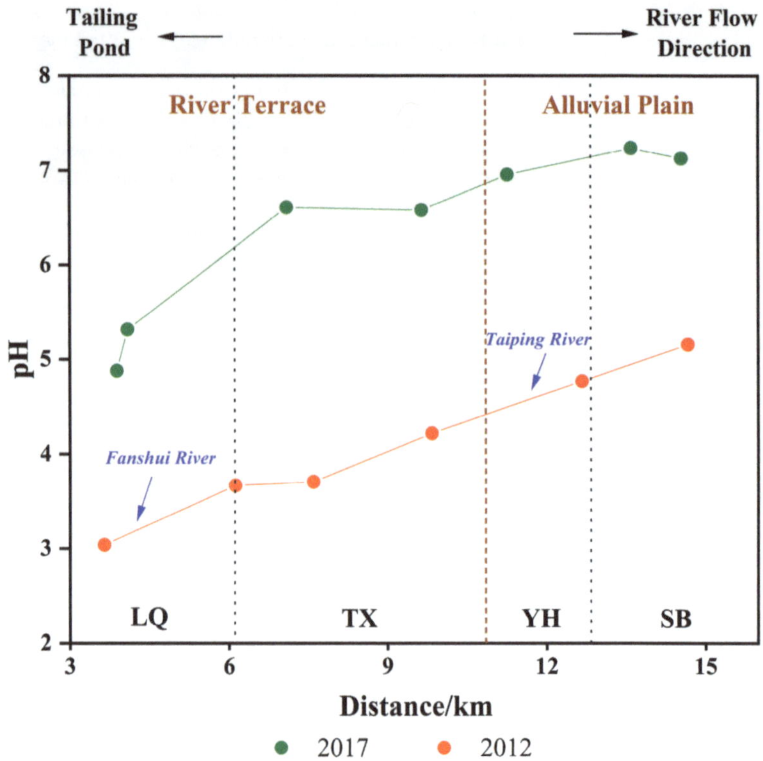

Figure 5. The spatial variation in river pH in 2012 [41] and in 2017 corresponding to sediments in the study area (brown dash line: the boundary of topography; black dash line: the boundary of village. LQ: Liangqiao Village, TX: Tangxin Village, YH: Yanghe Village, SB: Shangba Village).

Cd concentrations in soil showed the opposite trend to that in sediments (Figure 3a,b). The concentrations of Cd in soils are strongly affected by physicochemical properties of soil, including clay minerals and Fe−Mn oxides [46,47]. The Al_2O_3/SiO_2 ratio is an indicator of soil maturity, where higher values indicate finer soil texture and higher clay particle content [48]. The spatial variations in oxides, Al_2O_3/SiO_2, CEC, and pH in agricultural soils are shown in Figure 6.

In the river terrace, Al_2O_3/SiO_2 of agricultural soils shows a gradual decrease in the downstream direction, along with decreasing concentrations of oxides, CEC, and pH, which exhibit similar spatial distributions as those of Cd in agricultural soils (Figure 6). However, in the alluvial plain, Al_2O_3/SiO_2 of the agricultural soils gradually increases, while the concentration of TFe_2O_3 gradually decreases. The spatial patterns of these indexes in the agricultural soils do not align with spatial distributions of Cd concentrations, indicating that Cd concentrations in agricultural soils in the river terrace and alluvial plain may be affected by different factors.

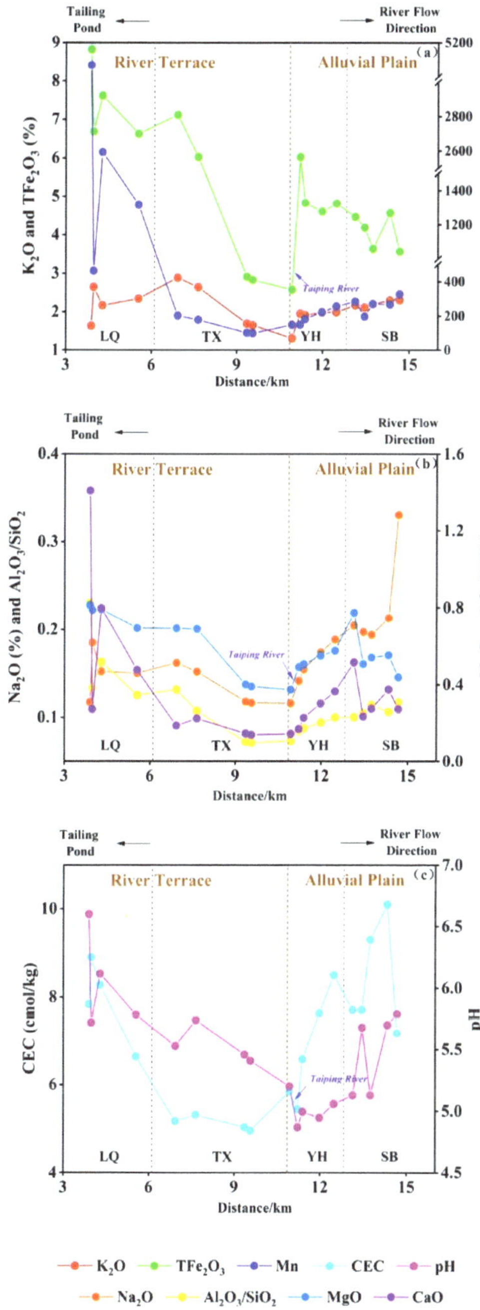

Figure 6. The spatial variation in oxides, Al_2O_3/SiO_2, CEC, and pH in agricultural soils from the study area ((**a**): K_2O, TFe_2O_3 and Mn; (**b**): Na_2O, Al_2O_3, MgO and CaO; (**c**): CEC and pH. Brown dash line: the boundary of topography; black dash line: the boundary of village. LQ: Liangqiao Village, TX: Tangxin Village, YH: Yanghe Village, SB: Shangba Village).

Random forest (RF) is a machine learning algorithm that can be used to assess the relative importance of soil physicochemical properties on Cd concentrations in soil [34]. In river terrace soils, TFe_2O_3, Mn, and CaO were identified as the most important factors affecting Cd concentration, with importance scores of 10.15%, 10.40%, and 9.72%, respectively (Figure 7). The steep slopes in the riverbed promote oxidation of the river water and the formation of Fe−Mn oxides. However, these conditions hinder sedimentation, resulting in colloids and oxides being transported downstream and deposited in more favorable locations [49]. The presence of CaO in the soil can reduce the solubility of Cd ions and decrease migration of Cd in the soil [50]. In the alluvial plain, the primary factors affecting Cd concentration in the agricultural soils are CaO, Al_2O_3/SiO_2, and Mn, with importance scores of 8.56%, 7.53%, and 7.22%, respectively. The score of TFe_2O_3 is only 1.90%, indicating that the influence of Fe oxides on Cd concentrations in the alluvial plain is relatively small. Decreased water flow velocity in the alluvial plain results in the preferential deposition of sand and silt, while fine−grained clay minerals migrate further, leading to a gradual increase in Al_2O_3/SiO_2 in agricultural soils, providing more adsorption sites for Cd [51]. The inflow of the Taiping River may be an important factor causing a gradual decrease in TFe_2O_3 concentration of the agricultural soils in the alluvial plain (Figure 6a). The Taiping River is not impacted by mining activities, and has the $Ca-HCO_3^-$ water quality type and a pH of about 7.09 [8]. The water pH of the upper Hengshi River gradually increases because of the dilution by the Taiping River (Figure 5). At the same time, schwertmannite minerals are metastable. When the water pH increases, SO_4^{2-} is gradually released from the schwertmannite mineral structure and is easily transformed into stable crystalline Fe oxide minerals (ferrihydrite, goethite, etc.) in the river, releasing a large amount of Cd [52,53]. This released Cd does not enter the structure of newly formed Fe oxide minerals and instead is adsorbed onto the surface of clay minerals in the form of ions [54].

4.3. Factors Influencing Migration of Cd in the Soil−Rice System

The ability of plants to absorb Cd from soil depends on bioavailability, which is influenced by various parameters such as pH, Eh, SOM, and Ca and Fe concentrations. The bioavailable concentration of Cd in soil is defined as the amount that is extractable by 0.01M $CaCl_2$, representing the most labile fraction [55]. The bioconcentration factor (BCF) represents the ability of rice grains to accumulate Cd from the agricultural soils, and is calculated using the following formula:

$$BCF_{Cd} = C_{grain}/C_{soil}$$

where C_{grain} is the Cd concentration in rice grains and C_{soil} is the Cd concentration in the corresponding agricultural soil.

The BCF for Cd in agricultural soil from the river terrace is lower than that from the alluvial plain and the degree of exceeding the MPL of Cd in rice grains increases from 238% to 573% along the river. The cumulative probability distribution curves for the proportion of $CaCl_2$−extracted Cd in soil and BCF_{Cd} in rice grains were fitted using a logistic function (Figure 8). Variations in BCF_{Cd} and proportion of $CaCl_2$−extracted Cd in soil are large, about 1–2 orders of magnitude, indicating significant spatial variations in the bioavailability of Cd in the study area. The mean values of BCF_{Cd} and proportion of $CaCl_2$−extracted Cd in the river terrace are lower than those in the alluvial plain, which indicates that Cd in agricultural soils posed greater potential ecological risk.

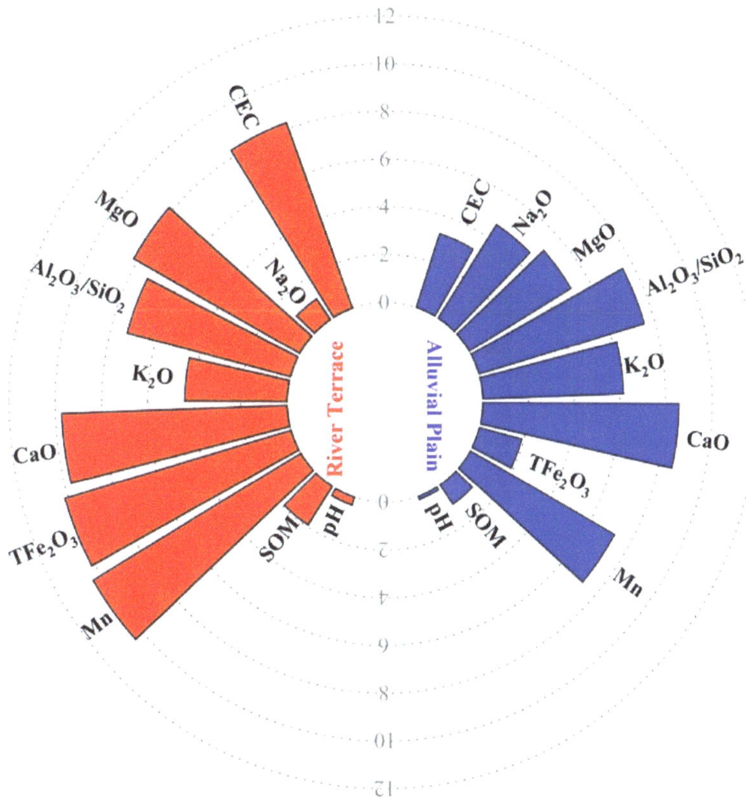

Figure 7. Importance scores of soil physicochemical properties influencing Cd concentration in soils from the river terrace and alluvial plain based on random forest (RF) analyses (unit: %).

Pearson correlation coefficients were calculated after logarithmic transformation (excluding pH) between BCF_{Cd} and soil physicochemical properties to better understand the factors affecting the bioavailability of Cd in agricultural soil (Table S6, Figure 9). In this diagram, the closer the dot to the center, the higher the negative correlation between each variable and Cd. In addition, the mazarine dot and azury dot represent significant correlations at $p < 0.01$ and $p < 0.05$, respectively. The BCF_{Cd} was found to have a significant negative correlation with Al_2O_3/SiO_2, K_2O, CaO, and MgO in agricultural soil. Soil clay minerals provide the main adsorption sites for Cd. The lower Al_2O_3/SiO_2 facilitates Cd desorption reactions, greatly increasing the bioavailability of Cd due to deposition sorting [41]. A decrease in clay mineral content also leads to lower CEC in soil, reducing exchangeable sites for PTE ions and increasing the activity of Cd [56].

An increase in the Ca^{2+} concentration in soil solution can reduce the solubility of Cd^{2+}, thereby reducing the bioavailability of Cd [57]. The high concentrations of Ca^{2+} and Mg^{2+} in the Taiping River result in precipitation of gypsum, which reduces the concentration of Ca^{2+} in soil solution [43].

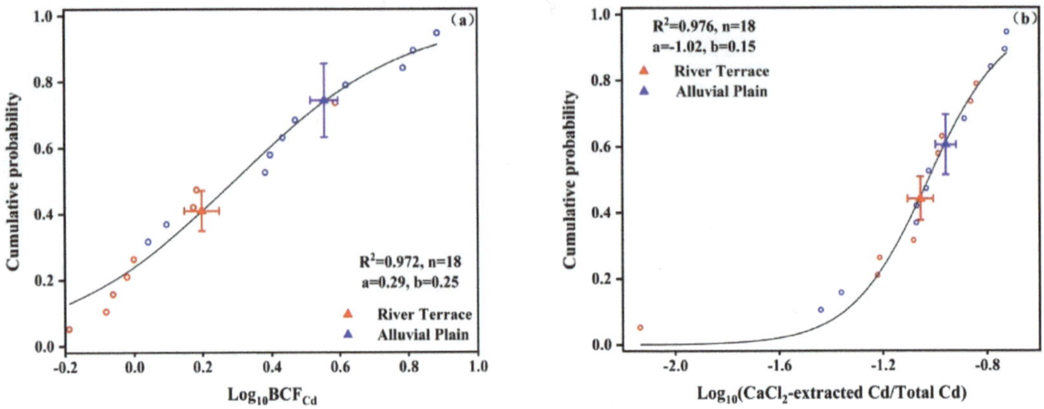

Figure 8. Cumulative probability distributions for: (**a**) BCF_{Cd} (bioconcentration factor), (**b**) proportion of $CaCl_2$−extracted Cd ($CaCl_2$−extracted Cd/total soil Cd). Red hollow dots: soil from river terrace; Blue hollow dots: soil from alluvial plain.

Fe−Mn oxide can promote adsorption of Cd^{2+} in soil solution, thereby reducing its bioavailability [58]. As previously described, Cd ions can co−precipitate with iron oxides such as schwertmannite and be re−released under phase transformations of these metastable minerals, increasing their activity. Meanwhile, reducing conditions caused by agricultural irrigation promote the reduction and dissolution of Fe−Mn oxides, enhancing the mobility and bioavailability of PTEs in the soil.

pH and SOM in the study area did not have a significant impact on the concentration and bioavailability of Cd in soil (Figures 7 and 8). On the one hand, pH can affect dynamic adsorption of Fe−Mn oxides on the surface of clay minerals. Ponthieu et al. [59] showed that when pH < 5.5 (such as on the YH–SB alluvial plain), Cd does not significantly adsorb onto amorphous iron oxides or goethite. Under very low pH conditions, negatively charged sites on the surface of clay minerals can be completely protonated, preventing binding of Cd to these sites [13]. On the other hand, binding of Cd ions to SOM is also controlled by pH. Previous research has shown that a pH change of at least 1–2 units is necessary to observe any effect of SOM on PTEs ions [60,61]. The relatively narrow pH range in the study area ($\Delta pH < 1.5$) makes it challenging to accurately assess the impact of SOM on PTEs ions such as Cd.

Therefore, the decreases in clay mineral content due to deposition sorting and phase transformations of schwertmannite are indeed important factors that increase the bioavailability and ecological risk of Cd in alluvial plains. The resulting high Cd concentration in rice grains has led to an increase in the incidence of cancer.

4.4. *Suggestions for Prevention and Control of PTE Contamination near Sulfide Mines*

Properly managing sulfide mines with a science−based approach is crucial for the adequate treatment of environmental pollution and the effective prevention of ecological risk.

A series of restoration plans have been implemented in the areas affected by Dabaoshan Mine to reduce the impact of the mining activities on downstream ecosystems, including the construction of a sewage treatment plant [18], mud impoundment [17], and adding lime to the rivers near the mine [19]. Construction of the sewage treatment plant and mud impoundment has effectively improved the water quality of the Hengshi River in the dry season. However, the mud impoundment has gradually become full over time and AMD overflows during the rainy season, and the reactivation of PTEs in sediment has increased PTEs concentrations in the Hengshi River [30]. Although the addition of lime promotes the precipitation and adsorption of PTEs in river water, reducing their mobility [19], the increase in water pH may trigger the phase transformation of schwertmannite, releasing PTEs and increasing ecological risks.

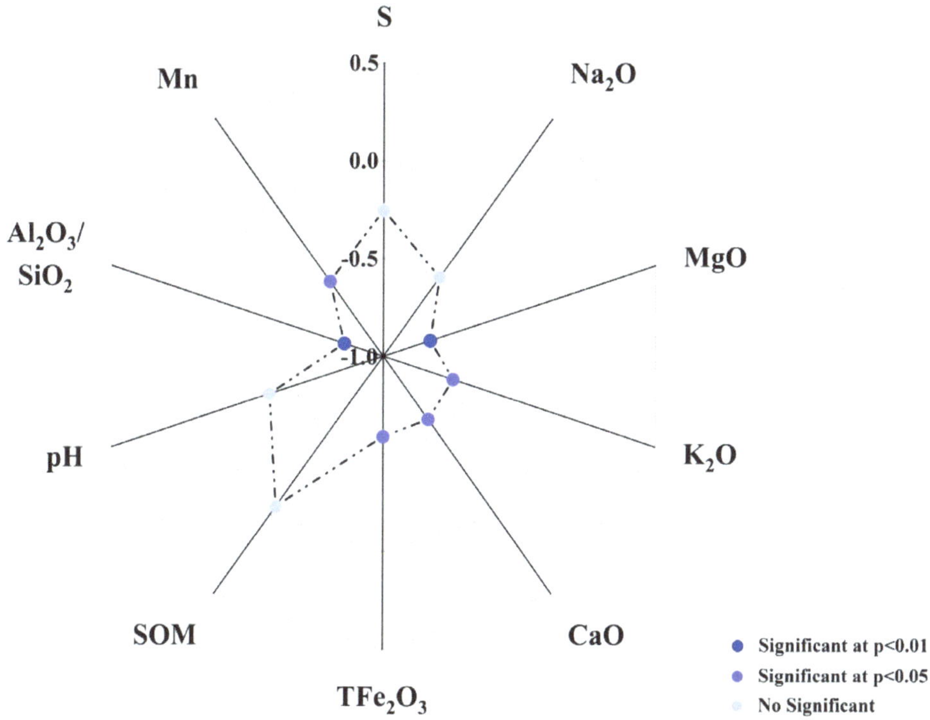

Figure 9. Correlation between the bioconcentration factor of Cd (BCF_{Cd}) and soil properties in agricultural soils from the study area.

Our study reveals significant differences in the concentrations and bioavailability of Cd in soils between river terraces and alluvial plains, along with their influencing factors. Therefore, classifying the surrounding environment of the sulfide mine according to topography can more precisely treat mining pollution, improve ecological risk prevention, and reduce PTE contamination of agricultural soil. In river terraces, it is recommended to reduce PTE loads through natural attenuation, relying on the environmental resilience, and to continuously monitor PTE concentrations in various environmental media [62]. In alluvial plains, plant restoration can be employed by introducing aquatic plants (such as reeds and cattails) in low−flow areas to stabilize sediment and absorb PTEs. Additionally, dredging can be used to remove contaminated sediment to avoid secondary release caused by changes in physiochemical conditions.

5. Conclusions

This study focused on the spatial distributions of Cd, Pb, and Zn in sediments and soils of Dabaoshan Mine, as well as the associated ecological risks after restoration efforts. The following conclusions were reached:

(1) Cd is the most prominent pollutant element, exhibiting significant enrichment and spatial heterogeneity in both soil and sediments, and higher accumulation levels in rice grains, in the Dabaoshan Mine area.

(2) The spatial distributions of Cd in sediments and soils are governed by topography, water velocity, channel width, change in physicochemical properties, and mineral composition of the upper Hengshi River.
(3) Areas posing the highest human health risks do not align with areas exhibiting the highest PTE concentrations in soil. Restoration efforts aimed at mitigating pollution may inadvertently alter the physicochemical properties of the river, leading to the transformation of schwertmannite, which affects the occurrence form of Cd and ultimately increases the potential ecological risk of Cd in soils within the alluvial plain.
(4) Although a series of restoration plans have been implemented, the Cd contamination problem in rice grains still persists. Taking topography into consideration when assessing the ecological risk of PTEs can enable effective pollution control and accurate prevention of potential ecological risks in the mining area.

Supplementary Materials: The following supporting information can be downloaded at: https://www.mdpi.com/article/10.3390/w15122223/s1. Figure S1 The geological map of study area (modified by National Geological Archive). Table S1. The Risk Screening Values (RSVs) and Risk Intervention Values (RIVs) for Cd, Pb and Zn in "Soil Environment Quality: Risk Control Standard for Soil Contamination of Agricultural Land" (GB15618-2018). Table S2. The Threshold Effect Level (TEL) and Probable Effect Level (PEL) of Cd, Pb, and Zn in Sediment Quality Guidelines (SQGs). Table S3. The concentrations of Cd, Pb, Zn, oxides and physicochemical properties of natural forest soil sampled in study area. Table S4. The concentrations of Rice grain and concentrations of Cd, Pb, Zn, oxides and physicochemical properties of corresponding agricultural soils sampled in study area. Table S5. The concentrations of Cd, Pb, Zn, oxides and physicochemical properties of sediments sampled in study area. Table S6. Pearson correlation coefficients between the bioconcentration factor of Cd (BCFCd) and soil properties in agricultural soils from the study area.

Author Contributions: W.S.: Methodology, Data curation, Formal analysis, Writing—original draft; Q.H.: Conceptualization, Methodology, Funding acquisition, Writing—Review and Editing; Z.Y.: Project administration, Supervision, Writing—Review and Editing; T.Y.: Resources. All authors have read and agreed to the published version of the manuscript.

Funding: This research was financially supported by the Natural Science Foundation of China (Project No. 41773019), the National Key R&D Program of China (Project No. 2017YFD0800304) and Guangdong Geological Exploration and Urban Geology Project (Grant No. 2023-25).

Institutional Review Board Statement: Not needed or available.

Informed Consent Statement: Not available or needed for this study.

Data Availability Statement: Data used in this study are available from the first authors upon reasonable request.

Conflicts of Interest: The authors declare no conflict of interest.

References

1. Mahar, A.; Wang, P.; Ali, A.; Awasthi, M.K.; Lahori, A.H.; Wang, Q.; Li, R.; Zhang, Z. Challenges and opportunities in the phytoremediation of heavy metals contaminated soils: A review. *Ecotoxicol. Environ. Saf.* **2016**, *126*, 111–121. [CrossRef] [PubMed]
2. Martin, C.W. Trace metal concentrations along tributary streams of historically mined areas, Lower Lahn and Dill River basins, central Germany. *CATENA* **2018**, *174*, 174–183. [CrossRef]
3. Damian, G.E.; Micle, V.; Sur, I.M.; Băbău, A.M.C. From Environmental Ethics to Sustainable Decision-Making: Assessment of Potential Ecological Risk in Soils Around Abandoned Mining Areas-Case Study "Larga de Sus mine" (Romania). *J. Agric. Environ. Ethic* **2019**, *32*, 27–49. [CrossRef]
4. Huang, S.; Li, Q.; Yang, Y.; Yuan, C.; You, P. Risk Assessment of Heavy Metals in Soils of a Lead-Zinc Mining Area in Hunan Province (China). *J. Chem. Chem. Eng.* **2017**, *66*, 173–178. [CrossRef]
5. Pearson, J.; Ipsen, J.; Sutherland, S.; Wegerson, K.; Onello, E. Risks and costs to human health of sulfide-ore mining near the Boundary Waters Canoe Area Wilderness. *Hum. Ecol. Risk Assess. Int. J.* **2019**, *26*, 1329–1340. [CrossRef]
6. Pietro, M.; Gerardo, B.; Giulia, S.; Monica, S.; Valentina, M.; Christian, M.; Paola, S. Geoheritage Values and Environmental Issues of Derelict Mines: Examples from the Sulfide Mines of Gromolo and Petronio Valley (Eastern Liguria, Italy). *Minerals* **2018**, *8*, 229.
7. Yang, S.; Ge, W.-Y.; Chen, H.-H.; Xu, W.-L. Investigation of soil and groundwater environment in urban area during post-industrial era: A case study of brownfield in Zhenjiang, Jiangsu Province, China. *China Geol.* **2019**, *2*, 501–511. [CrossRef]

8. Zhao, H.; Xia, B.; Qin, J.; Zhang, J. Hydrogeochemical and mineralogical characteristics related to heavy metal attenuation in a stream polluted by acid mine drainage: A case study in Dabaoshan Mine, China. *J. Environ. Sci.* **2012**, *24*, 979–989. [CrossRef]
9. Wang, M.; Song, H.; Chen, W.-Q.; Lu, C.; Hu, Q.; Ren, Z.; Yang, Y.; Xu, Y.; Zhong, A.; Ling, W. Cancer mortality in a Chinese population surrounding a multi-metal sulphide mine in Guangdong province: An ecologic study. *BMC Public Health* **2011**, *11*, 319. [CrossRef]
10. Gao, K.; Jiang, M.; Guo, C.; Zeng, Y.; Fan, C.; Zhang, J.; Reinfelder, J.; Huang, W.; Lu, G.; Dang, Z. Reductive dissolution of jarosite by a sulfate reducing bacterial community: Secondary mineralization and microflora development. *Sci. Total Environ.* **2019**, *690*, 1100–1109. [CrossRef]
11. Zhou, J.-M.; Dang, Z.; Cai, M.-F.; Liu, C.-Q. Soil Heavy Metal Pollution around the Dabaoshan Mine, Guangdong Province, China. *Pedosphere* **2007**, *17*, 588–594. [CrossRef]
12. Chan, W.S.; Routh, J.; Luo, C.; Dario, M.; Miao, Y.; Luo, D.; Wei, L. Metal accumulations in aquatic organisms and health risks in an acid mine-affected site in South China. *Environ. Geochem. Health* **2021**, *43*, 4415–4440. [CrossRef]
13. Li, Y.-T.; Becquer, T.; Dai, J.; Quantin, C.; Benedetti, M.F. Ion activity and distribution of heavy metals in acid mine drainage polluted subtropical soils. *Environ. Pollut.* **2009**, *157*, 1249–1257. [CrossRef] [PubMed]
14. Qu, L.; Xie, Y.; Lu, G.; Yang, C.; Zhou, J.; Yi, X.; Dang, Z. Distribution, fractionation, and contamination assessment of heavy metals in paddy soil related to acid mine drainage. *Paddy Water Environ.* **2017**, *15*, 553–562. [CrossRef]
15. Zhuang, P.; McBride, M.B.; Xia, H.; Li, N.; Li, Z. Health risk from heavy metals via consumption of food crops in the vicinity of Dabaoshan mine, South China. *Sci. Total Environ.* **2009**, *407*, 1551–1561. [CrossRef] [PubMed]
16. Liao, J.; Wen, Z.; Ru, X.; Chen, J.; Wu, H.; Wei, C. Distribution and migration of heavy metals in soil and crops affected by acid mine drainage: Public health implications in Guangdong Province, China. *Ecotoxicol. Environ. Saf.* **2016**, *124*, 460–469. [CrossRef]
17. Chen, M.; Lu, G.; Wu, J.; Yang, C.; Niu, X.; Tao, X.; Shi, Z.; Yi, X.; Dang, Z. Migration and fate of metallic elements in a waste mud impoundment and affected river downstream: A case study in Dabaoshan Mine, South China. *Ecotoxicol. Environ. Saf.* **2018**, *164*, 474–483. [CrossRef]
18. He, Q. *The Migration of Heavy Metals in the Surface Runoff of Dabaoshan Mining Area*; South China University of Technology: Guangzhou, China, 2011.
19. Wei, L.; Liu, Y.; Cai, D.; Li, F.; Luo, D.; Li, C.; Xu, G.; Xiao, T.; Wu, Q.; He, H.; et al. River morphology redistributes potentially toxic elements in acid mine drainage-impacted river sediments: Evidence, causes, and implications. *Catena* **2022**, *214*, 106183. [CrossRef]
20. Hou, D.; O'Connor, D.; Nathanail, P.; Tian, L.; Ma, Y. Integrated GIS and multivariate statistical analysis for regional scale assessment of heavy metal soil contamination: A critical review. *Environ. Pollut.* **2017**, *231*, 1188–1200. [CrossRef]
21. Zhao, H.; Xia, B.; Fan, C.; Zhao, P.; Shen, S. Human health risk from soil heavy metal contamination under different land uses near Dabaoshan Mine, Southern China. *Sci. Total Environ.* **2012**, *417–418*, 45–54. [CrossRef]
22. Su, W.; Xu, Y.; Fan, S.; Fu, S. The distribution regularity and accumulation risk of heavy metals in water and soil along the Hengshi River in the Dabaoshan mining area, Guangdong Province. *Geol. Bull. China* **2014**, *33*, 1231–1238.
23. Huang, J.; Qin, M.; Ma, W.; Yu, J.; Peng, X.; Yang, L. Characteristics analysis and ecological risk assessment of heavy metals contamination in suspend solids in a river affected by acid mine drainage. *Environ. Chem.* **2016**, *35*, 2315–2326.
24. Shu, X.-H.; Zhang, Q.; Lu, G.-N.; Yi, X.-Y.; Dang, Z. Pollution characteristics and assessment of sulfide tailings from the Dabaoshan Mine, China. *Int. Biodeterior. Biodegrad.* **2018**, *128*, 122–128. [CrossRef]
25. Zhuang, P.; Zou, B.; Li, N.Y.; Li, Z.A. Heavy metal contamination in soils and food crops around Dabaoshan mine in Guangdong, China: Implication for human health. *Environ. Geochem. Health* **2009**, *31*, 707–715. [CrossRef] [PubMed]
26. Yang, W.-J.; Ding, K.-B.; Zhang, P.; Qiu, H.; Cloquet, C.; Wen, H.-J.; Morel, J.-L.; Qiu, R.-L.; Tang, Y.-T. Cadmium stable isotope variation in a mountain area impacted by acid mine drainage. *Sci. Total Environ.* **2019**, *646*, 696–703. [CrossRef]
27. Wang, Y.; Dong, R.; Zhou, Y.; Luo, X. Characteristics of groundwater discharge to river and related heavy metal transportation in a mountain mining area of Dabaoshan, Southern China. *Sci. Total Environ.* **2019**, *679*, 346–358. [CrossRef]
28. Liao, J.; Ru, X.; Xie, B.; Zhang, W.; Wu, H.; Wu, C.; Wei, C. Multi-phase distribution and comprehensive ecological risk assessment of heavy metal pollutants in a river affected by acid mine drainage. *Ecotoxicol. Environ. Saf.* **2017**, *141*, 75–84. [CrossRef] [PubMed]
29. Ye, L.; Liu, T.; Yang, Y.; Gao, W.; Pan, Z.; Bao, T. Petrogenesis of bismuth minerals in the Dabaoshan Pb–Zn polymetallic massive sulfide deposit, northern Guangdong Province, China. *J. Asian Earth Sci.* **2014**, *82*, 1–9. [CrossRef]
30. Luo, C.; Routh, J.; Dario, M.; Sarkar, S.; Wei, L.; Luo, D.; Liu, Y. Distribution and mobilization of heavy metals at an acid mine drainage affected region in South China, a post-remediation study. *Sci. Total Environ.* **2020**, *724*, 138122. [CrossRef]
31. Chen, A.; Lin, C.; Lu, W.; Wu, Y.; Ma, Y.; Li, J.; Zhu, L. Well water contaminated by acidic mine water from the Dabaoshan Mine, South China: Chemistry and toxicity. *Chemosphere* **2008**, *70*, 248–255. [CrossRef]
32. Chen, Q.; Ma, M.; You, Y.; Yang, D. Cadmium Input Flux in Farmland Soil Near the Dabaoshan Mining Area of Guangdong Province. *Geol. Miner. Resour. South China* **2020**, *36*, 147–152.
33. Bussan, D.; Harris, A.; Douvris, C. Monitoring of selected trace elements in sediments of heavily industrialized areas in Calcasieu Parish, Louisiana, United States by inductively coupled plasma-optical emission spectroscopy (ICP-OES). *Microchem. J.* **2019**, *144*, 51–55. [CrossRef]
34. Breiman, L. Random forests. *Mach. Learn.* **2001**, *45*, 5–32. [CrossRef]

35. Li, C.; Zhang, C.; Yu, T.; Ma, X.; Yang, Y.; Liu, X.; Hou, Q.; Li, B.; Lin, K.; Yang, Z.; et al. Identification of soil parent materials in naturally high background areas based on machine learning. *Sci. Total Environ.* **2023**, *875*, 162684. [CrossRef]
36. *GB 15618-2018;* Soil Environment Quality-Risk Control Standard for Soil Contamination of Agricultural Land. Standard Press of China: Beijing, China, 2018. (In Chinese)
37. Yu, Z.; Liu, E.; Lin, Q.; Zhang, E.; Yang, F.; Wei, C.; Shen, J. Comprehensive assessment of heavy metal pollution and ecological risk in lake sediment by combining total concentration and chemical partitioning. *Environ. Pollut.* **2020**, *269*, 116212. [CrossRef]
38. *NY 861-2004;* Limits of Eight Elements in Cereals, Legume, Tubes and Its Products. Standard Press of China: Beijing, China, 2005.
39. Zhou, X.; Xia, B. Defining and modeling the soil geochemical background of heavy metals from the Hengshi River watershed (southern China): Integrating EDA, stochastic simulation and magnetic parameters. *J. Hazard. Mater.* **2010**, *180*, 542–551. [CrossRef] [PubMed]
40. Mueller, E.R.; Grams, P.E.; Schmidt, J.C.; Hazel, J.E.; Alexander, J.S.; Kaplinski, M. The influence of controlled floods on fine sediment storage in debris fan-affected canyons of the Colorado River basin. *Geomorphology* **2014**, *226*, 65–75. [CrossRef]
41. Chen, M.; Lu, G.; Guo, C.; Yang, C.; Wu, J.; Huang, W.; Yee, N.; Dang, Z. Sulfate migration in a river affected by acid mine drainage from the Dabaoshan mining area, South China. *Chemosphere* **2015**, *119*, 734–743. [CrossRef]
42. Regenspurg, S.; Brand, A.; Peiffer, S. Formation and stability of schwertmannite in acidic mining lakes. *Geochim. Cosmochim. Acta* **2004**, *68*, 1185–1197. [CrossRef]
43. Xie, Y.; Lu, G.; Yang, C.; Qu, L.; Chen, M.; Guo, C.; Dang, Z. Mineralogical characteristics of sediments and heavy metal mobilization along a river watershed affected by acid mine drainage. *PLoS ONE* **2018**, *13*, e0190010. [CrossRef]
44. Chen, M.; Lu, G.; Wu, J.; Sun, J.; Yang, C.; Xie, Y.; Wang, K.; Deng, F.; Yi, X.; Dang, Z. Acidity and metallic elements release from AMD-affected river sediments: Effect of AMD standstill and dilution. *Environ. Res.* **2020**, *186*, 109490. [CrossRef]
45. Liu, Y.; Gao, T.; Xia, Y.; Wang, Z.; Liu, C.; Li, S.; Wu, Q.; Qi, M.; Lv, Y. Using Zn isotopes to trace Zn sources and migration pathways in paddy soils around mining area. *Environ. Pollut.* **2020**, *267*, 115616. [CrossRef] [PubMed]
46. Yu, X.; Lu, S. Micrometer-scale internal structure and element distribution of Fe-Mn nodules in Quaternary red earth of Eastern China. *J. Soils Sediments* **2015**, *16*, 621–633. [CrossRef]
47. Wu, W.; Qu, S.; Nel, W.; Ji, J. The impact of natural weathering and mining on heavy metal accumulation in the karst areas of the Pearl River Basin, China. *Sci. Total Environ.* **2020**, *734*, 139480. [CrossRef] [PubMed]
48. Makabe, S.; Kakuda, K.-I.; Sasaki, Y.; Ando, T.; Ando, H. Relationship between mineral composition or soil texture and available silicon in alluvial paddy soils on the Shounai Plain, Japan. *Soil Sci. Plant Nutr.* **2009**, *55*, 300–308. [CrossRef]
49. Ma, Y.; Lu, W.; Lin, C. Downstream patterns of bed sediment-borne metals, minerals and organic matter in a stream system receiving acidic mine effluent: A preliminary study. *J. Geochem. Explor.* **2011**, *110*, 98–106. [CrossRef]
50. Guo, C.; Wen, Y.; Yang, Z.; Li, W.; Guan, D.; Ji, J. Factor controlling the bioavailability of soil cadmium in typical karst areas with high geogenic background. *J. Nanjing Univ. Nat. Sci.* **2019**, *55*, 678–687.
51. Guo, L.; Su, N.; Townend, I.; Wang, Z.B.; Zhu, C.; Wang, X.; Zhang, Y.; He, Q. From the headwater to the delta: A synthesis of the basin-scale sediment load regime in the Changjiang River. *Earth-Sci. Rev.* **2019**, *197*, 102900. [CrossRef]
52. Paikaray, S.; Schröder, C.; Peiffer, S. Schwertmannite stability in anoxic Fe(II)-rich aqueous solution. *Geochim. Cosmochim. Acta* **2017**, *217*, 292–305. [CrossRef]
53. Cruz-Hernandez, P.; Carrero, S.; Pérez-López, R.; Fernandez-Martinez, A.; Lindsay, M.; Dejoie, C.; Nieto, J.M. Influence of As(V) on precipitation and transformation of schwertmannite in acid mine drainage-impacted waters. *Eur. J. Miner.* **2019**, *31*, 237–245. [CrossRef]
54. Acero, P.; Ayora, C.; Torrentó, C.; Nieto, J.-M. The behavior of trace elements during schwertmannite precipitation and subsequent transformation into goethite and jarosite. *Geochim. Cosmochim. Acta* **2006**, *70*, 4130–4139. [CrossRef]
55. Houba, V.J.G.; Temminghoff, E.J.M.; Gaikhorst, G.A. Soil analysis procedures using 0.01 M calcium chloride as extraction reagent. *Commun. Soil Sci. Plant Anal.* **2000**, *31*, 1299–1396. [CrossRef]
56. Razzaghi, F.; Arthur, E.; Moosavi, A.A. Evaluating models to estimate cation exchange capacity of calcareous soils. *Geoderma* **2021**, *400*, 115221. [CrossRef]
57. Gray, C.W.; Mclaren, R.G.; Roberts, A.H.C.; Condron, L.M. Solubility, sorption and desorption of native and added cadmium in relation to properties of soils in New Zealand. *Eur. J. Soil Sci.* **1999**, *50*, 127–137. [CrossRef]
58. Wang, J.; Wang, P.-M.; Gu, Y.; Kopittke, P.M.; Zhao, F.-J.; Wang, P. Iron–Manganese (Oxyhydro)oxides, Rather than Oxidation of Sulfides, Determine Mobilization of Cd during Soil Drainage in Paddy Soil Systems. *Environ. Sci. Technol.* **2019**, *53*, 2500–2508. [CrossRef] [PubMed]
59. Ponthieu, M.; Juillot, F.; Hiemstra, T.; van Riemsdijk, W.; Benedetti, M. Metal ion binding to iron oxides. *Geochim. Cosmochim. Acta* **2006**, *70*, 2679–2698. [CrossRef]
60. Kinniburgh, D.G.; van Riemsdijk, W.H.; Koopal, L.K.; Borkovec, M.; Benedetti, M.F.; Avena, M.J. Ion binding to natural organic matter: Competition, heterogeneity, stoichiometry and thermodynamic consistency. *Colloids Surf. A* **1999**, *151*, 147–166. [CrossRef]

61. Milne, C.J.; Kinniburgh, D.G.; van Riemsdijk, W.H.; Tipping, E. Generic NICA–Donnan Model Parameters for Metal-Ion Binding by Humic Substances. *Environ. Sci. Technol.* **2003**, *37*, 958–971. [CrossRef]
62. Dore, E.; Fancello, D.; Rigonat, N.; Medas, D.; Cidu, R.; Da Pelo, S.; Frau, F.; Lattanzi, P.; Marras, P.A.; Meneghini, C.; et al. Natural attenuation can lead to environmental resilience in mine environment. *Appl. Geochem.* **2020**, *117*, 104597. [CrossRef]

water

Article

Acid Mine Drainage Discrimination Using Very High Resolution Imagery Obtained by Unmanned Aerial Vehicle in a Stone Coal Mining Area

Xiaomei Kou [1,2], Dianchao Han [1], Yongxiang Cao [1,2,*], Haixing Shang [1,3], Houfeng Li [4], Xin Zhang [5] and Min Yang [5,*]

1 Northwest Engineering Corporation Limited, Power China Group, Xi'an 710065, China; kouxm0427@nwh.cn (X.K.); handianch@nwh.cn (D.H.); shang_hx@nwh.cn (H.S.)
2 Shaanxi Union Research Center of University and Enterprise for River and Lake Ecosystems Protection and Restoration, Xi'an 710065, China
3 Xi'an Key Laboratory of Clean Energy Digital Technology, Xi'an 710065, China
4 Shaanxi Province Water Conservancy Development Survey and Hanjiang to Weihe River Water Diversion Project Coordination Center, Xi'an 710004, China; lihoufeng@hwrvwd.cn
5 School of Resources Engineering, Xi'an University of Architecture and Technology, Xi'an 710055, China
* Correspondence: caoyongx@nwh.cn (Y.C.); ymin@xauat.edu.cn (M.Y.); Tel.: +86-029-898-10109 (Y.C.)

Abstract: Mining of mineral resources exposes various minerals to oxidizing environments, especially sulfide minerals, which are decomposed by water after oxidation and make the water in the mine area acidic. Acid mine drainage (AMD) from mining can pollute surrounding rivers and lakes, causing serious ecological problems. Compared with traditional field surveys, unmanned aerial vehicle (UAV) technology has advantages in terms of real-time imagery, security, and image accuracy. UAV technology can compensate for the shortcomings of traditional technology in mine environmental surveys and effectively improve the implementat ion efficiency of the work. UAV technology has gradually become one of the important ways of mine environmental monitoring. In this study, a UAV aerial photography system equipped with a Red, Green, Blue (RGB) camera collected very-high-resolution images of the stone coal mining area in Ziyang County, northwest China, and classified the very-high-resolution images by support vector machine (SVM), random forest (RF), and U-Net methods, and detected the distribution of five types of land cover, including AMD, roof, water, vegetation, and bare land. Finally, the accuracy of the recognition results was evaluated based on the land-cover map using the confusion matrix. The recognition accuracy of AMD using the U-Net method is significantly better than that of SVM and RF traditional machine-learning methods. The results showed that a UAV aerial photography system equipped with an RGB camera and the depth neural network algorithm could be combined for the competent detection of mine environmental problems.

Keywords: acid mine drainage; mining environmental survey; unmanned aerial vehicle; support vector machine; random forest; U-Net

Citation: Kou, X.; Han, D.; Cao, Y.; Shang, H.; Li, H.; Zhang, X.; Yang, M. Acid Mine Drainage Discrimination Using Very High Resolution Imagery Obtained by Unmanned Aerial Vehicle in a Stone Coal Mining Area. *Water* **2023**, *15*, 1613. https://doi.org/10.3390/w15081613

Academic Editor: Liliana Lefticariu

Received: 28 March 2023
Revised: 11 April 2023
Accepted: 18 April 2023
Published: 20 April 2023

1. Introduction

Acid mine drainage (AMD) is an environmental phenomenon caused by mining, which may be related to mining activities that artificially expose a large amount of sulfide minerals to the weathering conditions of the earth's surface and form low-pH water under the chemical reaction process involving oxygen, water, and microorganisms. AMD has been defined in previous studies as the process of oxidation of sulfide minerals to form low-pH drainage water [1]. This acidic water with toxic elements could significantly influence the health of natural ecosystems and the survival of aquatic organisms. The main areas affected by AMD are generally rivers, lakes, estuaries, and coastal waters. The formation of AMD always occurs during a long period taking from years to decades, and its environmental impact could last for centuries [2]. Therefore, this environmental problem requires long-term monitoring and sustained remediation. In order to monitor the spatial distribution and intensity of AMD pollution, some efforts have been made, mainly with systematic sampling and laboratory analysis of surface water and sediments. Then, interpolation results are scattered to draw pollution distribution and evaluation maps [3,4]. However, this method is suitable for scientific research and detailed investigation, which require high test results. It is time-consuming, expensive, and covers limited spatial scale for the daily monitoring work of local environmental regulatory authorities.

In many metallic and coal mines of the United States, China, and Russia, the oxidative breakdown of iron-containing sulfide minerals releases ferrous iron and the energy that microbes depend on for survival [5–7]. The ferrous iron continues to be oxidized to ferric iron and produces more sulfuric acid [8,9]. The solubility of ferric iron at near-neutral pH is low, at which point iron hydroxide solids are rapidly formed. Iron hydroxide precipitation can be regarded as evidence for the presence of AMD [2]. It also has the characteristic of adsorbing toxic metals from AMD. The alteration of ferrous iron can produce a series of secondary minerals such as goethite, hematite, and jarosite, which usually appear orange [10]. The pH value plays a crucial role in secondary mineral precipitation, where goethite forms at a pH of between 2 and 12, and hematite forms at a pH of between 7 and 9 [11,12]. The jarosite usually produces at a low pH value (between 1 and 3) [13]. A group of secondary mineral products appears orange to red in color, according to their high concentrations of ferric iron. This characteristic has been previously known as a gossan, a significant remote-sensing detection marker for ore exploration [14–17].

Previous studies have proved that remote-sensing technology facilitates many environmental monitoring efforts [18–20]. Multispectral and hyperspectral sensors have been widely used due to the obvious reflectance variation of minerals such as hematite and goethite in the visible to short-wave infrared region of the electromagnetic spectrum [21,22]. These studies are based on a wide range of spatial scales, which are determined by data-collection platforms including satellite remote-sensing technology for water source protection [23,24]. The indirect estimation of pH values of mine waste [25–28] and iron oxide precipitation on the riverbed mainly used aerial remote sensing [29]. Other studies have collected AMD minerals, performed spectroscopic tests indoors, and established a reference spectroscopic library of iron minerals [30,31]. A multi-rotor UAV carrying a hyperspectral sensor supplies images with both high spatial and high spectral resolution that is better than that of most satellite data [32]. However, hyperspectral and multispectral sensors are not suitable for routine monitoring due to their expensiveness and scarcity. At present, a multi-rotor UAV system with a Red, Green, Blue (RGB) color digital camera has been rapidly developed due to its low price and flexible application direction, and is widely used in land resources and in geological, environmental, and other industrial mapping. The application research of a UAV system with an RGB color digital camera in acid drainage investigation in mines has not been comprehensively discussed yet.

The purpose of this study is to collect aerial remote-sensing data from stone coal mines in Qinling Mountains by using a UAV system with an RGB color digital camera. Three methods, including support vector machine (SVM), random forest (RF), and U-Net methods, were used for acid mine drainage recognition and surface classification mapping. The results of the three methods are comparatively discussed and an efficient, economical method for acid mine drainage monitoring was developed. This study was proposed based on actual mine environmental investigation work, and provided a simple, inexpensive, and efficient environmental problem detection approach for the environmental authorities of local governments and mining companies.

2. Materials and Methods

2.1. Study Area

The study area is located in the stone coal mining area of Qinling Mountains in Ziyang County, Shaanxi Province. The main ore-bearing strata in the mining area are the Upper

Ordovician Lower Silurian Banjiuguan Formation, and the main lithology is a set of black carbonaceous slate with pyrite-rich trachyte [33]. The UAV imagery covered a relatively larger mining company named the Minghua Mining Industry Co., Ltd. The Minghua Mining Industry Co., Ltd. was established in the year of 2010 and produces 30,000 to 50,000 tons of ore annually. This company was shut down by the local government in 2017 due to its serious environmental problems. After nearly 8 years of exploitation in the mining area, a large amount of gangue piles had been stacked on the slope and in the valley, leaving a huge open pit. Under the effect of rainfall infiltration, oxidation occurs and a large amount of acid mine drainage (AMD) was directly discharged into the Xiaomi stream downstream from the mining area, forming an orange-colored river bed several kilometers long (Figures 1 and 2). Xiaomi stream flows downward and inward to the Hanjiang River, bringing huge contamination danger to the regional water environment. At the same time, the Hanjiang River is an important water source for the South to North Water Transfer Project. The acid drainage of the stone coal mines endangers the water source safety of the water transfer project.

Figure 1. (**a**) Location of the stone coal mining area; (**b**) UAV image of the study area; (**c**) the Xiaomi stream polluted by AMD.

(**a**) (**b**)

Figure 2. (**a**) The AMD flows out of the mine portal; (**b**) the Xiaomi stream polluted by AMD.

2.2. *The UAV Aerial Photography System*

The Feima D2000 UAV system (Shenzhen Feima Robotics Technology Co., Ltd., Shenzhen, China), which was used for image acquisition in this study, is a light, long-endurance, multi-rotor UAV system that can simultaneously meet the requirements of high-precision mapping, remote sensing, and video shooting (Figure 3a). An aerial survey module, tilt module, visible video module, thermal infrared video module, and thermal infrared remote-sensing module could optionally be carried and could obtain diversified data. The standard takeoff weight of the system is 2.8 kg, the standard load is 200 g, and the endurance is 74 min [34]. The whole system can be integrated into a work box after modular decomposition, which is convenient for carrying and transportation. Feima D2000 is equipped with high-precision differential GNSS card office, and the network RTK, PPK, and their fusion solution services are provided as standard [35]. This system is competent for 1:500 mapping without control points, supports high-precision POS-assisted aerial triangulation, and realizes phase-control-free application. The image size of the SONY a-6000 digital camera (Sony Group Corporation, Tokyo, Japan) equipped with Feima D2000 aerial survey system is 6000 × 4000 pixels and the lens focal length is 25 mm (Figure 3b) [36].

(**a**) (**b**)

Figure 3. (**a**) The Feima UAV platform; (**b**) the SONY a-6000 digital camara.

The flight time was selected from 12:00 to 14:00 on 4 December 2020, which is the best time for solar illumination. All aerial images were obtained through two flights. The flight adopts the ground-simulating flight mode, with a relative height of 200 m from the ground. The flight parameters were defined off the shelf (Table 1) and the preset flight range and route were imported into the remote controller before the flight.

Table 1. The flight parameters of the aerial image acquisition.

Parameter	Value
Flight altitude (above take-off)	946 m
Ground resolution	4 cm
Number of flights	2
Flight duration	50 min
Course overlap	80%
Lateral overlap	60%
Relative height	314 m
Area covered	3.3 km^2

In Table 2, the well-known UAV systems used in the AMD detection of mines in the recently published literature are listed. The UAV system used in this study has advantages of light weight and long duration.

Table 2. The instrumental parameters of the UAV aerial photography systems.

Parameter	Feima D2000	DJI Matrice 210 V2 [19]	Texo DSI [15]	Tholeg THO-R-PX8 [21]
Duration	60 min	24 min	30 min	20 min
Sensor weigh	200 g	508.8 g	680 g	720 g
Cruise speed	72 km/h	61 km/h	36 km/h	40 km/h
Maximum relative height	±1000 m	±500 m	±500 m	±500 m

2.3. Pre-Processing

This study used Context Capture software version 10.20.0.4117 (Bentley Engineering Software System Co., Ltd., Exton, PA, USA) to compute aerial triangulation of UAV aerial data. The software automatically generates digital surface model (DSM) by matching the same image points on the stereo relative and calculates the elevation of the ground points using the collinear equation. True digital orthophoto map (TDOM) is a resampling based on DSM to correct the geometric distortion of the original image. Since DSM contains elevation information of buildings, bridges, trees, and other ground objects, the resulting image not only corrects the terrain, but also corrects ground objects such as surface buildings, maintains the vertical angle of the surface landscape, and resolves the difficulties of splicing large-scale orthophoto images [37]. The geometric resolution of the UAV orthophoto images collected in this study is 6 cm × 6 cm. The map projection used CGCS2000 ellipsoidal and Gauss Kruger projection (Ministry of Natural Resources, Beijing, China). The true-color image contains three channels of red, green, and blue.

2.4. Ground Truth Data

This study collected the latest land-use map of the study area from Xi'an Geological Survey Center of China Geological Survey, and combined similar land types according to recent land cover. Since dry land, tea plant, shrubbery, orchard, woods, bamboo, and vegetables are all vegetation types, and abandoned land, unpaved road, rock dumps, and rock cuts are bare-ground types, a total of five land-cover types were displayed. The five surface-cover types of mine acid drainage, water body, roof, vegetation, and bare land are most accurately classified as the ground truth data (Figure 4).

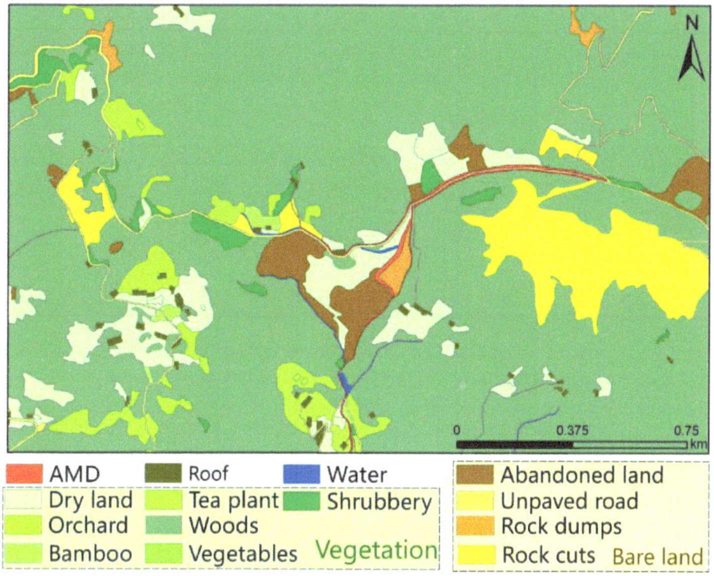

Figure 4. The latest land-use map of the stone coal mining area.

2.5. Methods

2.5.1. Selection of Study Samples

The learning image is an aerial image taken by an unmanned aerial vehicle in the vicinity of the study area, covering an area of 87,680 km^2 (Figure 5a). The learning image contains five types of land cover: AMD, roof, water, vegetation, and bare land (Figure 5b). These learning labels are interpreted by human visual interpretation according to the land-use-type map of the mining area. Because the roads in the mining area are non-paved and the surface is consistent with the bare-land type, the bare-land type includes traditional bare land and the roads in the mining area.

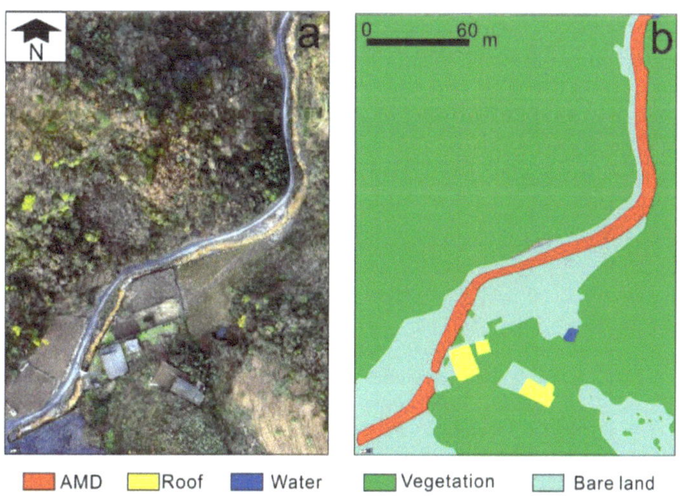

Figure 5. (**a**) The UAV image of the training area; (**b**) the ground truth of the training area.

2.5.2. Support Vector Machine

This study used Support Vector Machine (SVM) to extract landslide information. This method is a machine-learning model algorithm based on statistical theory. This method uses optimization problem solving to determine the optimal classification hyperplane in the high-dimensional feature space of the data, which can deal with complex data classification problems [38–40]. SVM classifies by mapping the set of low-dimensional space vectors to the high-dimensional space vectors, and then constructs a kernel function, which minimizes the classification error and maximizes the generalization ability of the classifier [40].

The SVM optimal classification function expression is as follows:

$$S(x) = sign\left[\sum_{i=1}^{n} a_i y_i K(x_i, x_j) + b\right] \quad (1)$$

In the formula, a_i is a non-negative Lagrange multiplier; y_i is the category; $K(x_i, x_j)$ are kernel functions; b is the classification threshold.

At present, there are four kinds of kernels commonly used in SVM, mainly Sigmoid kernels, Polynomial kernels, Radial basis functions (RBF), and Linear kernels. In this study, the polynomial is 2nd-order polynomial and all the expressions of the core functions are as follows:

Linear kernel function is:

$$K(x_i, x_j) = x_i^T x_j \quad (2)$$

Polynomial kernel function is:

$$K(x_i, x_j) = \left(g x_i^T x_j + r\right)^c \quad (3)$$

Radial basis function is:

$$K(x_i, x_j) = exp\left(-g\|x_i - x_j\|^2\right), g > 0 \quad (4)$$

Sigmoid kernel is:

$$K(x_i, x_j) = \tanh\left(g x_i^T x_j + r\right) \quad (5)$$

where, $g > 0$; c is a natural number, c = 2, r is hyperparameter.

2.5.3. Random Forest

Random forest algorithm was originally proposed by Breiman as a machine-learning algorithm [41]. Random forest is a bagging-based decision tree method. Its core is the bagging algorithm. Given a training set D of size n, the bagging algorithm selects n′ size subsets D_i in a number of m from the training set evenly, and playback (using self-service sampling) is used as a new training set. Using classification, regression, and other algorithms on this m training set, a number of m models can be obtained, and then bagging results can be obtained by averaging and majority votes. That is, given a training set $X = X_1$, ..., X_n and target $Y = Y_1$, ..., Y_n, the bagging method repeats back (B times) samples from the training set and then trains the tree model on these samples. After the training, the prediction of the unknown sample X can be achieved by averaging the predictions of all the single regression trees on X [42].

2.5.4. U-Net

The U-Net network structure, proposed by Olaf Ronneberger, Philipp Fischer, and Thomas Brox at the IEEE International Symposium on Biomedical Imaging (ISBI) Competition 2015, is a U-shaped structure consisting of shrinking subnetworks and expanding subnetworks. Therefore it is named U-Net [43]. The structure of the U-Net network is shown in Figure 6. U-Net first extracts feature information by downsampling through convolution and pooling, and then fuses lower feature maps before upsampling and clipping through transposed convolution for accurate positioning. The U-Net method repeats this process until the output signature map is obtained, and then the segmented map is finished through the activation function.

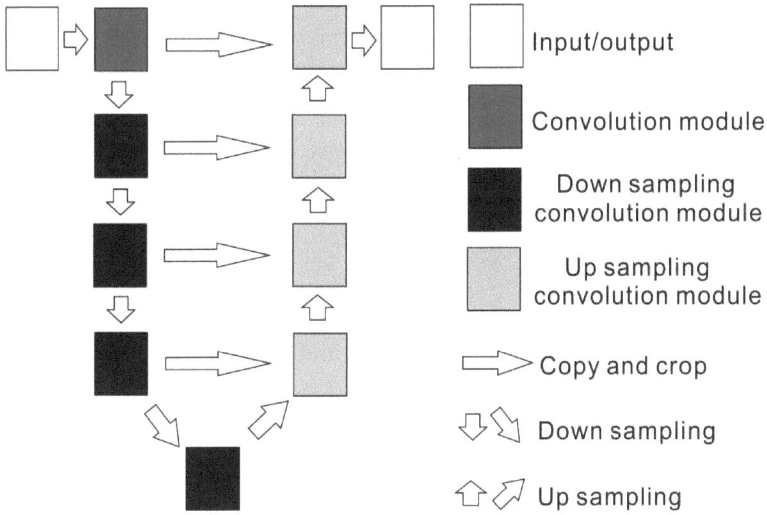

Figure 6. The structure of the U-Net algorithm.

2.5.5. Confusion Matrix

In this study, the confusion matrix method was used to evaluate the accuracies of AMD and land-cover classification in this mining area. The confusion matrix is calculated by comparing the location and category of each real pixel on the surface with the corresponding location and category in the classification image [44]. The evaluation includes overall accuracy and accuracy of each category. For AMD and other land-cover categories in mining area, the overall accuracy is the percentage produced by dividing the sum of the correct classification inspection points by the total selected inspection points. The accuracy of each category refers to the probability that the results of each classification are comparable with the actual reference data, to calculate the correct classification of each category.

3. Results

In general, all three methods have been used to classify RGB aerial images from UAVs effectively and have identified the distribution of AMD successfully. The SVM and RF methods classified all the pixels of the entire image, while the U-Net method classified some pixels poorly matched with the five types of land cover as unclassified. Among the SVM classification results, bare land covered the largest area (62.77% of the entire area) of the study area, followed by vegetation (22.39% of the entire area). Roof class contained threadiness such as non-paved roads in mining areas (Figure 7). AMD and water bodies were mainly distributed along ponds and streams in the study area with a small coverage percent of 0.14% (Figure 7). Because the study area is located in the Qinling Mountains, the earth's surface is mainly covered with dense forest and vegetation. Therefore, the SVM classification results distorted the classification of vegetation and bare land. The results of RF method showed the largest distribution area is bare land (39.89% of the entire area) and followed by vegetation (38.99% of the entire area) (Figure 8). Most pixels of roof class were real roofs of buildings, a few were unpaved roads in the mining area. Unpaved roads in the mining area were mainly incorporated into bare-land class due to their similar color. AMD and water bodies were still distributed along ponds and rivers with a smaller area (5.84% of the entire area) in the results of the RF method (Figure 8). However, some of the vegetation distinguished by the SVM classification method were misclassified into the AMD category, possibly because some of the withered leaf color in autumn was close to the orange color of AMD. The U-Net method showed the largest vegetation area (39.54% of the entire area), followed by bare land (8.92% of the entire area). There were almost no misclassified mining roads in the roof class. Water bodies and AMD were still distributed along ponds and rivers with a smaller area in the result of the U-Net method (Figure 9). The distribution of U-Net classification results was better than that of the SVM and RF methods with fewer discrete classification points. This feature greatly improves the aesthetics of the next mapping step.

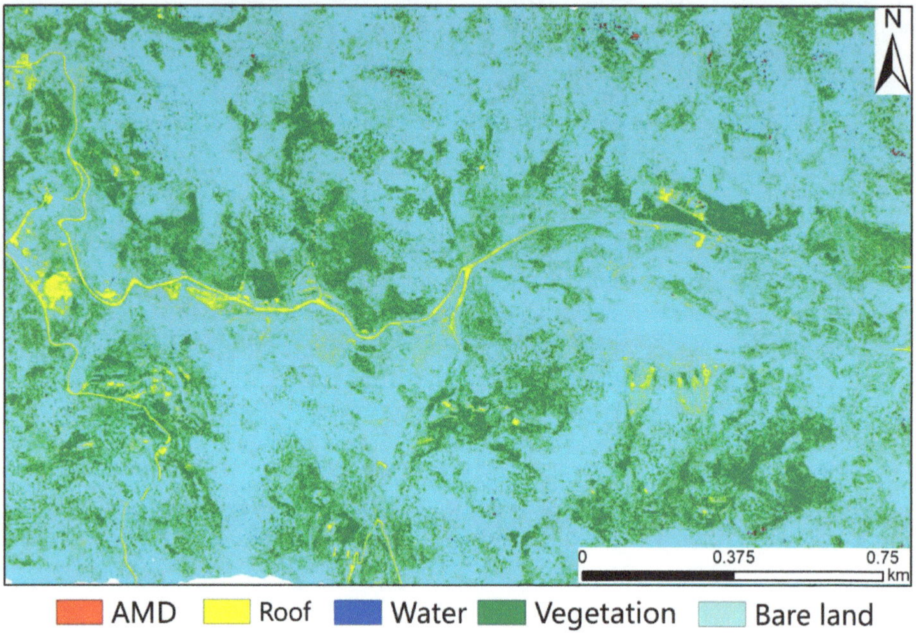

Figure 7. The classification map of SVM method.

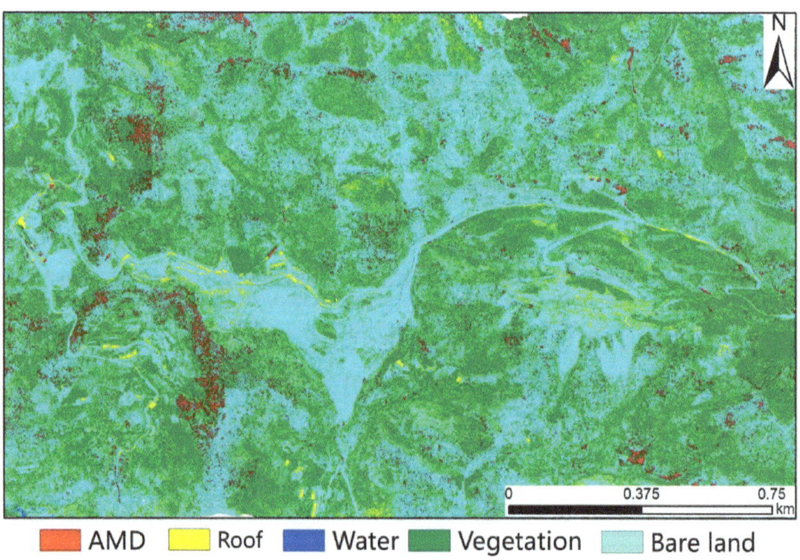

Figure 8. The classification map of RF method.

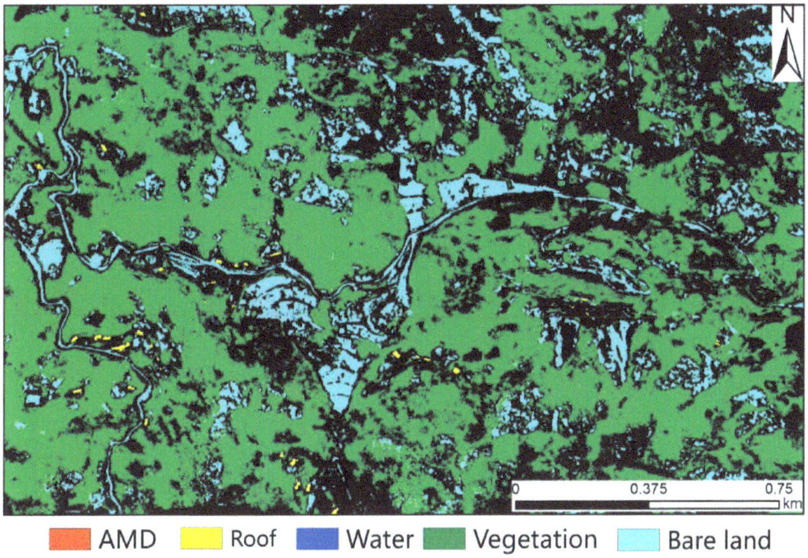

Figure 9. The classification map of U-Net method.

A small scale of the classification area was selected for detail comparisons among the results of the different methods and the classes (Figure 10a). Among the vegetation classification results of all the three methods, the SVM method mapped a smaller vegetation land cover than the other two methods (Figure 10b). The RF and the U-Net methods identified a similar vegetation distribution, but the RF vegetation recognition results are more discrete than those of the U-Net (Figure 10c,d). Among the bare-land classification results, the SVM method identified the largest bare-land area. Some vegetation pixels were misidentified as bare land, while some unpaved roads were also misclassified as roof by the SVM (Figure 10b). The RF method can more accurately identify bare land, and fewer areas of bare land were misclassified as roof (Figure 10c). The bare-land areas incorrectly identified by the U-Net method were mainly the road in the mining area and the surface of the mining gangue piles (Figure 10d). The formation of AMD in stone coal mines in the study area is due to the peeling off of rock and soil during open-pit mining, resulting in the exposure of fresh rock surfaces and loose mining gravel to an oxidative environment. Under the action of oxygen, atmospheric precipitation, and microorganisms, minerals such as pyrite (FeS) are oxidized to form sulfate-containing AMD. AMD flows downstream along the drainage channel built on the mining waste dump and forms a confluence into the Xiaomi stream, so acidic water in the study area is mainly distributed in the drainage channel and the Xiaomi stream of the mining area. As shown in Figure 10b, the AMD identified by the SVM method appeared intermittently along the drainage channel in the mining area. As shown in Figure 10c, the AMD identified by the RF method was more contiguous along drainage channels in mining areas, but withered leaves are also misidentified as AMD in vegetation areas. As shown in Figure 10d, the continuity of the AMD distribution identified by the U-Net method was close to RF, and there was no misclassification in vegetation areas.

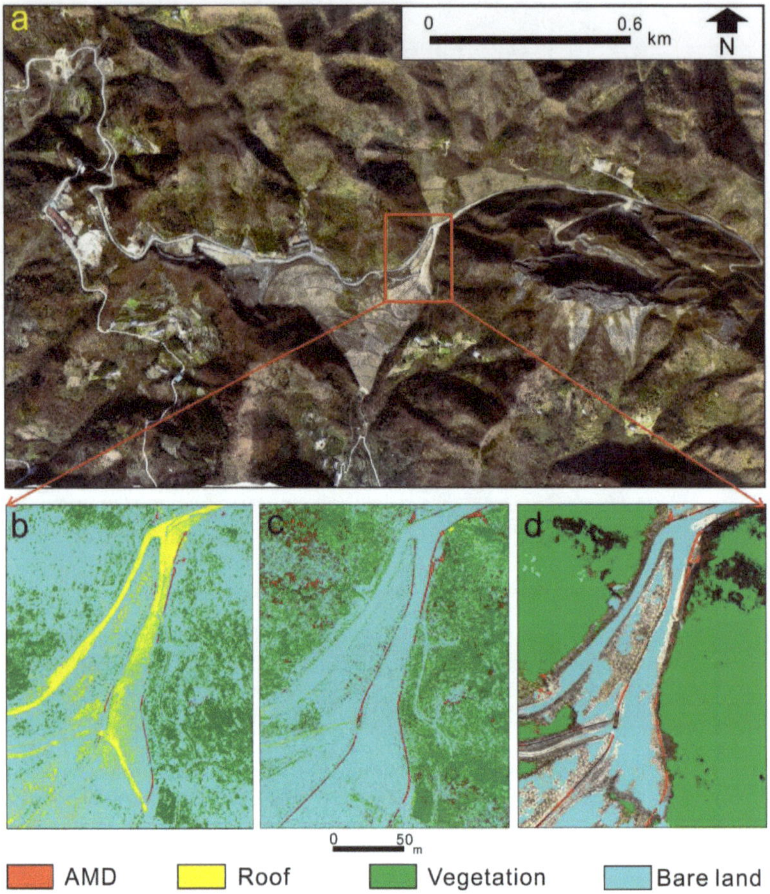

Figure 10. The classification map of U-Net method.((**a**) the UAV image of the study area, (**b**) SVM classification, (**c**) RF classification, (**d**) U-Net classification).

4. Discussion

This study used SVM, RF, and U-Net models combined with land-use classification maps to identify AMD, bare land, roof, vegetation, and water bodies in a stone coal mine area in Ziyang county with over 6 cm high-resolution RGB images obtained using UAVs. The overall classification accuracies of the SVM, RF, and U-Net methods were 30.35%, 49.84%, and 56.26%, respectively, (Tables 3–5). Compared with traditional machine-learning classification methods such as RF and SVM, the U-Net method improves the classification accuracy by 6.42–25.91% using the same RGB image and the ground truth land-cover map. This result is in agreement with previous studies that applied the deep learning network model to remote-sensing land-use classification [45–47]. In this study, to form a reciprocity in data with traditional machine-learning classification methods, without too many manual steps to extract and select features [47,48], using 6 cm very-high-resolution UAV color images for AMD and mine land-cover classification, the overall classification accuracies of traditional machine-learning classification methods SVM and RF were lower than that of U-Net.

Table 3. The confusion matrix of the SVM results (%).

Class	Bare Land	Vegetation	AMD	Roof	Water
Unclassified	0.01	0.01	0.00	0.00	0.00
Bare land	70.85	71.87	75.47	50.63	79.57
Vegetation	8.91	27.33	2.34	0.37	19.01
AMD	0.15	0.55	13.60	0.18	0.00
Roof	20.08	0.25	8.59	48.82	1.43
Water	0.00	0.00	0.00	0.00	0.00
Total	100.00	100.00	100.00	100.00	100.00

Note: Overall Accuracy = 30.35%.

Table 4. The confusion matrix of the RF results (%).

Class	Bare Land	Vegetation	AMD	Roof	Water
Unclassified	0.01	0.01	0.00	0.00	0.00
Bare land	77.43	42.80	47.49	37.61	48.49
Vegetation	12.18	47.85	12.76	7.36	34.93
AMD	4.31	7.15	38.65	12.47	2.85
Roof	6.06	2.16	1.09	42.53	13.30
Water	0.01	0.03	0.00	0.04	0.42
Total	100.00	100.00	100.00	100.00	100.00

Note: Overall Accuracy = 49.84%.

Table 5. The confusion matrix of the U-Net results (%).

Class	Bare Land	Vegetation	AMD	Roof	Water
Unclassified	43.44	37.91	26.28	37.33	47.49
Bare land	49.80	4.56	1.76	16.75	1.62
Vegetation	6.58	57.50	2.97	1.51	2.46
AMD	0.00	0.00	68.97	0.00	5.91
Roof	0.18	0.02	0.01	44.35	0.00
Water	0.00	0.01	0.00	0.05	42.51
Total	100.00	100.00	100.00	100.00	100.00

Note: Overall Accuracy = 56.26%.

Among the confusion matrices of the five types of ground cover distinguished by the three methods, the AMD classification accuracy of the SVM method was the lowest at 13.60% and a large number pixels were misclassified into bare land (Table 3). The AMD classification accuracy of the RF method was 38.65%, and the misclassification was also mainly bare land (Table 4). The AMD classification accuracy of the U-Net method was 68.97% and the highest (Table 5). The classification accuracy of bare land by the SVM method was 70.85% (Table 3). The bare-land classification accuracy of the RF method was the highest at 77.43% (Table 4). The classification accuracy of bare land by the U-Net method was 49.80%, with the lowest classification accuracy, and a large number of pixels were marked as unclassified (Table 5). The vegetation classification accuracy of the SVM method was the lowest at 27.33% and many pixels were misclassified as bare land (Table 3). The precision of vegetation classification by the RF method was 47.85% (Table 4). The classification accuracy of bare land by the U-Net method was 57.50% and the highest (Table 5). The roof classification accuracy of the SVM method was the highest at 48.82% (Table 3). The roof classification accuracy of the RF method was the lowest at 42.53% with many pixels incorrectly classified as bare land (Table 4). The roof classification accuracy of the U-Net method was 44.35% and some pixels were classified as bare land (Table 5). The water classification accuracy of the SVM method was the lowest at 0.00%. In total 79.57% and 19.01% pixels of water were misclassified as bare land and vegetation, respectively, (Table 3). The roof classification accuracy of the RF method was 0.42% and the classification accuracy was also very low. Approximately 48.49% and 34.93% pixels of the roof were misclassified as bare land and vegetation (Table 4). The roof classification accuracy of the U-Net method was 42.51% and many pixels failed to be classified (Table 5). In some previous studies, the accuracy of RF land-cover classification is generally higher than that of SVM classification [49].

Meanwhile, due to the limitation of flight time and spectrum of the RGB images collected using UAVs, non-paved roads, open pits, and waste dumps of stone coal mines all belong to mining land. The color and texture of these cover types are similar to those of natural bare land, which results in failure to classify these similar land-cover types by machine-learning algorithm. However, along the pathway of AMD, the water bottom exhibited orange color due to the deposition of richer iron oxides on the water bottom. This color can be distinguished from most other objects and clean water bodies. Only some dry vegetation in autumn and winter show brown-red color which leads to incorrect classification of AMD identification. Therefore, follow-up research should consider the use of growing season images to refine other land-cover classifications for in-depth learning fine classification.

The data used in this study are true-color remote-sensing data with 6 cm very high spatial resolution of the UAV, which contains richer texture information. In particular, AMD performs a strong sense of granularity on the water bottom of the high-resolution UAV image, and the fine features of the water bottom of AMD are more significant. Extracting geometric texture features of AMD underwater stones combined with current edge information detection algorithm could better identify the distribution of AMD, which may provide direction for further research [50].

5. Conclusions

In this study, very-high-resolution images acquired by a UAV system with a spatial resolution of 6 cm were used to classify AMD and land-cover types in a stone coal mining area of Ziyang County. Two conventional machine-learning classification methods including SVM and RF, and additionally the currently emerging U-Net deep artificial neural network (DANN) method, were used. All three classification methods have successfully identified five land-cover types: AMD, roof, water, vegetation, and bare land. The overall classification accuracy of the SVM method is the lowest, and the U-Net method the highest. The identification results for AMD also resulted in the lowest classification accuracy of 13.60% for the SVM method and the highest identification accuracy of 68.97% for the U-Net method. The results illustrated the capability of a UAV system equipped with an RGB camera to perform its advantages of high resolution and fast data acquisition in mine environmental monitoring. The very-high-resolution imagery acquired by a UAV system could enable comparatively higher accuracy of AMD identification and mine environment investigation. In terms of identification algorithms, the DANN methods, especially U-Net, generally perform with higher classification accuracy than general machine-learning methods in the identification of environmental problems such as AMD. In addition, UAV systems equipped with RGB cameras are lighter and cheaper than those equipped with multispectral or hyperspectral sensors, and RGB color images are easier to process than multispectral and hyperspectral data. Therefore, the UAV systems and RGB color images are particularly suitable for monitoring mining environmental problems by mining corporations and local environmental protection administration with nonspecialist remote-sensing technology. In the next step, a serious attempt might be taken to improve U-Net or other DNN classification algorithms to achieve better precision in mining environmental surveys using UAV aerial photography systems with RGB color cameras.

Author Contributions: Conceptualization, X.K.; methodology, D.H.; software, Y.C.; validation, H.S. and M.Y.; formal analysis, X.Z.; investigation, M.Y.; writing—original draft preparation, M.Y.; writing—review and editing, M.Y.; supervision, H.L.; project administration, X.K.; funding acquisition, Y.C. All authors have read and agreed to the published version of the manuscript.

Funding: This research was funded by Natural Science Basic Research Program of Shaanxi (Program No. 2021JLM-57, 2021JLM-56 and 2021JM-350) and the APC was funded by Northwest Engineering Corporation Limited Major Science and Technology Projects, grant number XBY-ZDKJ-2020-08.

Data Availability Statement: The data presented in this study are available on request from the corresponding authors.

Acknowledgments: We are thankful to Xi'an Geological Survey Center of China Geological Survey for their provision of the land-use map. The authors would like to thank the reviewers for their very helpful and constructive reviews of this manuscript.

Conflicts of Interest: The authors declare no conflict of interest.

References

1. Lottermoser, B. *Mine Wastes;* Springer: Berlin/Heidelberg, Germany, 2010; Volume 44, p. 085201.
2. Liu, Q.; Chen, B.; Haderlein, S.; Gopalakrishnan, G.; Zhou, Y. Characteristics and environmental response of secondary minerals in AMD from Dabaoshan Mine, South China. *Ecotoxicol. Environ. Saf.* **2018**, *155*, 50–58. [CrossRef] [PubMed]
3. Ferrier, G. Application of imaging spectrometer data in identifying environmental pollution caused by mining at Rodaquilar, Spain. *Remote Sens. Environ.* **1999**, *68*, 125–137. [CrossRef]
4. Kemper, T.; Sommer, S. Estimate of heavy metal contamination in soils after a mining accident using reflectance spectroscopy. *Environ. Sci. Technol.* **2002**, *36*, 2742–2747. [CrossRef] [PubMed]
5. Zvereva, V.P.; Frolov, K.R.; Lysenko, A.I. Formation of mine drainage in the Far Eastern region and its impact on the ecosphere and public health. *Min. Sci. Technol.* **2022**, *7*, 203–215. [CrossRef]
6. Arefieva, O.; Nazarkina, A.V.; Gruschakova, N.V.; Skurikhina, J.E.; Kolycheva, V.B. Impact of mine waters on chemical composition of soil in the Partizansk Coal Basin, Russia. *Int. Soil Water Conserv. Res.* **2019**, *7*, 57–63. [CrossRef]
7. Lazareva, E.; Myagkaya, I.; Kirichenko, I.; Gustaytis, M.; Zhmodik, S. Interaction of natural organic matter with acid mine drainage: In-situ accumulation of elements. *Sci. Total Environ.* **2019**, *660*, 468–483. [CrossRef]
8. Banfield, J.F.; Welch, S.A. Microbial controls on the mineralogy of the environment. In *Environmental Mineralogy*; Vaughan, D.J., Wogelius, R.A., Eds.; Mineralogical Society of Great Britain and Ireland: London, UK, 2000; Volume 2, ISBN 978-963-463-133-0.
9. Singer, P.C.; Stumm, W. Kinetics of the oxidation of ferrous iron. In Proceedings of the Second Symposium on Coal Mine Draiage Research, National Coal Association/Bituminous Coal Research, Pittsburgh, PA, USA, 14–15 May 1968; Wiley: New York, NY, USA, 1968; pp. 12–34.
10. Bowles, J.F.W. Hydroxides. In *Encyclopedia of Geology*, 2nd ed.; Alderton, D., Elias, S.A., Eds.; Academic Press: Oxford, UK, 2021; pp. 442–451. ISBN 978-0-08-102909-1.
11. Schwertmann, U. Effect of pH on the formation of goethite and hematite from ferrihy-drite. *Clays Clay Miner.* **1983**, *31*, 277–284. [CrossRef]
12. Das, S.; Hendry, M.J.; Essilfie-Dughan, J. Transformation of two-line ferrihydrite to goethite and hematite as a function of pH and temperature. *Environ. Sci. Technol.* **2011**, *45*, 268–275. [CrossRef]
13. Desborough, G.A.; Smith, K.S.; Lowers, H.A.; Swayze, G.A.; Hammarstrom, J.M.; Diehl, S.F.; Leinz, R.W.; Driscoll, R.L. Mineralogical and chemical characteristics of some natural jarosites. *Geochim. Cosmochim. Acta* **2010**, *74*, 1041–1056. [CrossRef]
14. Fraser, S. Discrimination and identification of ferric oxides using satellite thematic mapper data: A Newman case study. *Int. J. Remote Sens.* **1991**, *12*, 614–635. [CrossRef]
15. Gopinathan, P.; Parthiban, S.; Magendran, T.; Al-Quraishi, A.M.F.; Singh, A.K.; Singh, P.K. Mapping of ferric (Fe^{3+}) and ferrous (Fe^{2+}) iron oxides distribution using band ratio techniques with aster data and geochemistry of Kanjamalai and Godumalai, Tamil Nadu, South India. *Remote Sens. Appl. Soc. Environ.* **2020**, *18*, 100306.
16. Jackisch, R.; Lorenz, S.; Zimmermann, R.; Möckel, R.; Gloaguen, R. Drone-borne hyperspectral monitoring of acid mine drainage: An example from the Sokolov Lignite District. *Remote Sens.* **2018**, *10*, 385. [CrossRef]
17. Mielke, C.; Boesche, N.K.; Rogass, C.; Kaufmann, H.; Gauert, C.; De Wit, M. Spaceborne mine waste mineralogy monitoring in South Africa, applications for modern push-broom missions: Hyperion/OLI and EnMAP/Sentinel-2. *Remote Sens.* **2014**, *6*, 6790–6816. [CrossRef]
18. Chalkley, R.; Crane, R.A.; Eyre, M.; Hicks, K.; Jackson, K.-M.; Hudson-Edwards, K.A. A multi-scale feasibility study into acid mine drainage (AMD) monitoring using same-day observations. *Remote Sens.* **2023**, *15*, 76. [CrossRef]
19. Bacova, D.; Khairutdinov, A.M.; Gago, F. Cosmic geodesy contribution to geodynamics monitoring. *IOP Conf. Ser. Earth Environ. Sci.* **2021**, *906*, 12074. [CrossRef]
20. Golik, V.I.; Klyuev, R.V.; Martyushev, N.V.; Brigida, V.; Efremenkov, E.A.; Sorokova, S.N.; Mengxu, Q. Tailings utilization and zinc extraction based on mechanochemical activation. *Materials* **2023**, *16*, 726. [CrossRef]
21. Swayze, G.A.; Smith, K.S.; Clark, R.N.; Sutley, S.J.; Pearson, R.M.; Vance, J.S.; Hageman, P.L.; Briggs, P.H.; Meier, A.L.; Singleton, M.J.; et al. Using imaging spectroscopy to map acidic mine waste. *Environ. Sci. Technol.* **2000**, *34*, 47–54. [CrossRef]
22. Isgró, M.A.; Basallote, M.D.; Caballero, I.; Barbero, L. Comparison of UAS and sentinel-2 multispectral imagery for water quality monitoring: A case study for acid mine drainage affected areas (SW Spain). *Remote Sens.* **2022**, *14*, 4053. [CrossRef]

23. Montero, I.C.; Brimhall, G.H.; Alpers, C.N.; Swayze, G.A. Characterization of waste rock associated with acid drainage at the Penn Mine, California, by ground-based visible to short-wave infrared reflectance spectroscopy assisted by digital map-ping. *Chem. Geol.* **2005**, *215*, 453–472. [CrossRef]
24. Flores, H.; Lorenz, S.; Jackisch, R.; Tusa, L.; Contreras, I.C.; Zimmermann, R.; Gloaguen, R. UAS-based hyperspectral environmental monitoring of acid mine drainage affected waters. *Minerals* **2021**, *11*, 182. [CrossRef]
25. Richter, N.; Staenz, K.; Kaufmann, H. Spectral unmixing of airborne hyperspectral data for baseline mapping of mine tailings areas. *Int. J. Remote Sens.* **2008**, *29*, 3937–3956. [CrossRef]
26. Choe, E.; van der Meer, F.; van Ruitenbeek, F.; van der Werff, H.; de Smeth, B.; Kim, K.W. Mapping of heavy metal pollution in stream sediments using combined geo-chemistry, field spectroscopy, and hyperspectral remote sensing: A case study of the Rodalquilar mining area, SE Spain. *Remote Sens. Environ.* **2008**, *112*, 3222–3233. [CrossRef]
27. Shang, J.; Morris, B.; Howarth, P.; Lévesque, J.; Staenz, K.; Neville, B. Mapping mine tailing surface mineralogy using hyperspectral remote sensing. *Can. J. Remote Sens.* **2009**, *35*, S126–S141. [CrossRef]
28. Davies, G.E.; Calvin, W.M. Mapping acidic mine waste with seasonal airborne hyper-spectral imagery at varying spatial scales. *Environ. Earth Sci.* **2017**, *76*, 1–14. [CrossRef]
29. Sares, M.; Hauff, P.; Peters, D.; Coulter, D. Characterizing sources of acid rock drain-age and resulting water quality impacts using hyperspectral remote sensing—Examples from the Upper Arkansas. In Proceedings of the Advanced Integration of Geospatial Technologies in Mining Reclamation, Atlanta, GA, USA, 7–9 December 2004.
30. Crowley, J.; Williams, D.; Hammarstrom, J.; Piatak, N.; Chou, I.-M.; Mars, J. Spectral reflectance properties (0.4–2.5 μm) of secondary Fe-oxide, Fe-hydroxide, and Fe-sulphate-hydrate minerals associated with sulphide-bearing mine wastes. *Geochem. Explor. Environ. Anal.* **2003**, *3*, 219–228. [CrossRef]
31. Davies, G.E.; Calvin, W.M. Quantifying iron concentration in local and synthetic acid mine drainage: A new technique using handheld field spectrometers. *Mine Water Environ.* **2016**, *36*, 299–309. [CrossRef]
32. Booysen, R.; Gloaguen, R.; Lorenz, S.; Zimmermann, R.; Nex, P.A. Geological remote sensing. In *Reference Module in Earth Systems and Environmental Sciences*, 2nd ed.; Elsevier: Amsterdam, The Netherlands, 2020; Volume 64, pp. 267–274.
33. Jia, Z.; Lv, T. Geological characteristics and stone-bearing coal analysis of dove formation in Shuanglong Area, Ankang City. *Coal* **2014**, *23*, 52–53.
34. Liu, X.; Lian, X.; Yang, W.; Wang, F.; Han, Y.; Zhang, Y. Accuracy assessment of a UAV direct georeferencing method and impact of the configuration of ground control points. *Drones* **2022**, *6*, 30. [CrossRef]
35. He, K.; Liu, B.; Hu, X.; Zhou, R.; Xi, C.; Ma, G.; Han, M.; Li, Y.; Luo, G. Rapid characterization of landslide-debris flow chains of geologic hazards using multi-method investigation: Case study of the Tiejiangwan LDC. *Rock Mech. Rock Eng.* **2022**, *55*, 5183–5208. [CrossRef]
36. Liu, X.; Zhu, W.; Lian, X.; Xu, X. Monitoring mining surface subsidence with multi-temporal three-dimensional unmanned aerial vehicle point cloud. *Remote Sens.* **2023**, *15*, 374. [CrossRef]
37. Zhao, Z. Researching on 3D modeling of small object with tilt photogrammetry and Context Capture. *Digit. Technol. Appl.* **2020**, *38*, 29–31.
38. Mazzoni, D.; Garay, M.J.; Davies, R.; Nelson, D. An operational MISR pixel classifier using support vector machines. *Remote Sens. Environ.* **2007**, *107*, 149–158. [CrossRef]
39. Zhang, R.; Ma, J. State of the art on remotely sensed data classification based on support vector machines. *Adv. Earth Sci.* **2009**, *24*, 555–562.
40. Long, Y.; Qiao, W.; Sun, J. Change detection of remote sensing images in Datun Mining Area based on support vector machine. *Geomat. Spat. Inf. Technol.* **2020**, *43*, 107–115.
41. Breiman, L. Random forests. *Mach. Learn.* **2001**, *45*, 5–32. [CrossRef]
42. Zhang, J.; Yao, Y.; Cao, N. Prediction of whether precipitation based on decision tree. *J. Geomat.* **2017**, *42*, 107–109.
43. Ronneberger, O.; Fischer, P.; Brox, T. U-net: Convolutional networks for biomedical image segmentation. Medical image computing and computer-assisted intervention–MICCAI 2015. In Proceedings of the 18th International Conference, Munich, Germany, 5–9 October 2015; Proceedings, Part III 18. Springer International Publishing: Berlin/Heidelberg, Germany, 2015; pp. 234–241.
44. Ye, R.; Niu, R.; Zhang, L.; Yi, S. Mineral contents determination and accuracy evaluation based on classification of petrographic images. *J. China Univ. Min. Technol.* **2011**, *40*, 810–822.
45. Fang, X.; Wang, G.; Yang, H.; Liu, H.; Yan, L. High resolution remote sensing image classification combining with mean-shift segmentation and fully convolution neural network. *Laser Optoelectron. Prog.* **2018**, *55*, 22802. [CrossRef]
46. Cao, Q.; Zhong, Y.; Ma, A.; Zhang, L. Urban land use/land cover classification based on feature fusion fusing hyperspectral image and LiDAR data. In Proceedings of the IGARSS 2018—2018 IEEE International Geoscience and Remote Sensing Symposium, Valencia, Spain, 22–27 July 2018; IEEE: Piscataway Township, NJ, USA, 2018; pp. 8869–8872.
47. Xia, M.; Cao, G.; Wang, G.; Shang, Y. Remote sensing image classification based on deep learning and conditional random fields. *J. Image Graph.* **2017**, *22*, 1289–1301.
48. Pang, B.; Huang, Z.; Wu, Y.; Lu, Y. Mapping of impervious surface extraction of high resolution remote sensing imagery based on improved fully convolutional neural network. *Remote Sens. Inf.* **2020**, *35*, 47–55.

49. Ma, L.; Fu, T.; Blaschke, T.; Li, M.; Tiede, D.; Zhou, Z.; Ma, X.; Chen, D. Evaluation of feature selection methods for object-based land cover mapping of unmanned aerial vehicle imagery using random forest and support vector machine classifiers. *ISPRS Int. J. Geo-Inf.* **2017**, *6*, 51. [CrossRef]
50. Chen, H.; Bian, Z.; Zhao, X. Research on attribute morphological profiles based on multifeature ultra-high resolution remote sensing image classification. *Geomat. Spat. Inf. Technol.* **2019**, *42*, 115–119.

water

MDPI

Article

Geochemical Response of Surface Environment to Mining of Sn-Pb-Zn Sulfide Deposits: A Case Study of Dachang Tin Polymetallic Deposit in Guangxi

Bo Li [1], Tao Yu [2,3], Wenbing Ji [4], Xu Liu [1], Kun Lin [1], Cheng Li [1], Xudong Ma [1] and Zhongfang Yang [1,3,*]

1 School of Earth Sciences and Resources, China University of Geosciences, Beijing 100083, China
2 School of Science, China University of Geosciences, Beijing 100083, China
3 Key Laboratory of Ecological Geochemistry, Ministry of Natural Resources, National Research Center for Geoanalysis, Beijing 100037, China
4 Nanjing Institute of Environmental Sciences, Ministry of Ecology and Environment, Nanjing 210042, China
* Correspondence: yangzf@cugb.edu.cn

Abstract: The rational development of mineral resources provides necessary materials for economic development, but environmental pollution caused by mining activities is an inevitable consequence. Here, we present a case study of Chehe Town in Guangxi, an area with integrated metals mining and smelting. The geochemical distribution, migration, and transformation behaviors of Cd and other heavy metals were studied in detail by systematically collecting surface media such as atmospheric dust, surface water and stream sediments, ores, tailings, mine drainage, soil, and crops in and around the mining area. We used these data to explore the geochemical response of the surface environment to mining and smelting of metal sulfide deposits. The annual flux of Cd and other heavy metals near the mining and smelting sites was high. Due to the topography, heavy metals in the atmosphere are mainly transported via vertical deposition, influencing areas downwind for 25 km. The mine drainage exceeded As and Zn standards but had little impact on the surface water. The surface water quality was good, without acidification. Risks due to ore were much higher than that for tailings. Heavy metals buffered by surrounding carbonate rocks and secondary minerals mainly migrated as solid particles, resulting in the contamination of stream sediment by heavy metals. In mountainous areas, rivers are mainly affected by topography, flowing fast and dominated by downcutting, which caused heavy metal pollution in the sediment have a limited effect on the soil near the river. Heavy metal concentrations in the cultivated soil were greatly influenced by external input such as substantial atmospheric dust. However, only Cd accumulated in the crops, with very high concentrations in rice, but safe and edible levels in corn. Thus, in the mining area, the most sensitive to heavy metals was the atmospheric environment. High concentrations of heavy metals beyond the ore district are mainly concentrated in the sediment, with distant impacts. Therefore, it is necessary to monitor and control risks associated with sediment transport, conduct treatment, and adjust crop planting. The soil, river, and agriculture respond differently to mining activities, but the risk is low and can be managed as needed.

Keywords: sulfide deposit; surface environment; heavy metals; migration; transformation; Chehe Town; Guangxi

Citation: Li, B.; Yu, T.; Ji, W.; Liu, X.; Lin, K.; Li, C.; Ma, X.; Yang, Z. Geochemical Response of Surface Environment to Mining of Sn-Pb-Zn Sulfide Deposits: A Case Study of Dachang Tin Polymetallic Deposit in Guangxi. *Water* **2023**, *15*, 1550. https://doi.org/10.3390/w15081550

Academic Editor: Domenico Cicchella

Received: 6 March 2023
Revised: 31 March 2023
Accepted: 13 April 2023
Published: 14 April 2023

1. Introduction

Environmental disturbance inevitably results from ore deposit development; long-term mining operations exert enormous pressure on the surrounding ecosystem. While mining is an important resource extraction activity, dumping of solid wastes and dust and effluent discharges result in environmental pollution [1–3]. The environmental effects of mining non-metallic deposits are dominated by physical disturbance, including geological hazards, noise, and dust [4,5]. The intensity of the hazards depends on factors such as the scale of the ore deposit and the mining technology. On the basis of physical disturbance, metal mine areas undergo strong chemical reactions, of which acidification and heavy metal pollution are frequently encountered. Sulfide deposits are mainly formed by endogenesis. Due to anthropogenic disturbance from deposit exploration, mining, smelting, etc., deposits originally formed under anaerobic conditions are exposed to air, water, and microbial activity, resulting in oxidation and leaching of minerals [6–8]. Under the dual effects of physical migration and chemical release, acidic discharges and heavy metals can enter the environment, causing potentially serious threats to the atmosphere, water, soil, and biological systems.

Many previous studies have been conducted on the environmental effects of sulfide mining, but most of this research has focused on a single medium such as acid mine drainage, tailings ponds, or soils, as well as discussion of surficial reaction mechanisms of sulfide minerals [9–14]. However, few of these have carried out multimedia studies of geochemical characteristics.

The overall response of the environment to mines is affected by basic factors such as contaminant concentrations in the mine, the type of ore, and mining operations, as well as characteristics of the surrounding rock, topography, hydrology, climate, etc., [15]. Thus, the distribution and migration of heavy metals in surface media such as the atmosphere dust, tailings, soil, surface water, and stream sediment are mutually influenced and closely related. Recently, as tension between the demand for mineral resources and restoration of mining environments has gradually sharpened, green mining has put forward higher requirements on the analysis of pollution sources and exploration of pollution pathways in the ore districts [16]. Only by systematically identifying the environmental pollution characteristics of multiple elements in multiple media, as well as migration pathways and constraints on heavy metals in the environment based on the geochemical behavior of sulfide minerals, can we achieve the purpose of classification management and precise policy to ensure adequate supply of resources and simultaneously maintain the quality of the ecosystem.

We studied a typical Sn-Pb-Zn polymetallic sulfide mining area in Nandan County, Guangxi Zhuang Autonomous Region, including surface media such as atmospheric dust, ore, tailings, mine drainage, surface water, stream sediments, soil, and crops. Multimedia samples were systematically collected to conduct a comprehensive investigation of the migration and contamination of heavy metals in the surface environment under the influence of mining operations. Geochemical transformations of heavy metals in various environmental media in the ore district were analyzed in detail, taking into account the characteristics of secondary minerals formed by sulfide weathering, the lithology of the surrounding rocks, climatic conditions, landforms, and geotectonic background. This information provides a sufficient scientific basis for the accurate identification of ecological risks, safe utilization of land resources, and pollution remediation in areas affected by polymetallic sulfide mines.

2. Materials and Methods

2.1. The Study Area

Nandan County is located in the northern area of Hechi City, Guangxi Zhuang Autonomous Region, at the southern foot of the Yunnan-Guizhou Plateau, and is a branch of the Fenghuang Mountains. The terrain is higher in the north and lower in the south and the topography is complex and mainly mountainous. The climate has the characteristics and seasonal patterns of the plateau area. The temperatures are generally low with abundant rainfall and little sunlight. The area is rich in water resources and developed karst landforms, forming many intermittent open rivers and undercurrents; most stream segments have rapid water flow or waterfalls due to steep slopes and high rocky ridges. Nandan County is rich in mineral resources, with more than 20 non-ferrous metals such as tin, lead, zinc, silver, and manganese. Tin reserves are the largest in the country and lead and zinc reserves are also among the most abundant. The area is known as "The Hometown of Non-Ferrous Metals in China" and "Tin Capital".

The Dachang tin-sulfide polymetallic ore field is located in the central section of the Danchi (Nandan-Hechi) metallogenic belt in northwest Guangxi. Its mineral resources are mainly concentrated in the Dachang and Chehe anticlines, where the exposed strata are clastic rock-carbonate rock-siliceous assemblages of the Devonian–Triassic system [17]. This area is one of the important non-ferrous metal industrial bases in China and is globally known for its long mining history and rich mineral resources. Mining and smelting activities are mainly concentrated at the head of the Diaojiang River. In addition to a number of state-owned mining enterprises, it also accommodates hundreds of private beneficiation sites at peak times. With an annual raw ore processing volume of millions of tons, this area makes outstanding contributions to local economic development. However, it also contributes to a series of environmental problems along the Diaojiang River, resulting in abandonment of large areas of farmland and abnormal blood lead levels among residents. The ore mainly consists of cassiterite, pyrite, pyrrhotite, marmatite, galena, etc. The mine recovers Sn, Pb, Zn, Sb, and other resources from the ore for a long time while discharging a large number of sulfides such as pyrite, pyrrhotite, arsenopyrite, and gangue as tailings. Currently, there are as many as dozens of tailing ponds above medium size in the mining area, including Chehe, Changpo, Lutang, etc. Chehe Tailings Pond is circled with mountains in three sides and has a valley outlet in one side, which allows precipitation to quickly flow through the tailings pond along the hillside and merge into the river while carrying a large number of alluvium and endangering the water environment. Generally, the drainage of the tailings pond is mostly treated with lime, which to some extent alleviates the discharge of acidic water.

The Diaojiang River Basin is located in the sloping transition zone from the Yunnan–Guizhou Plateau to the Guangxi Basin. The topography is elevated in the northwest and lower in the southeast, with typical karst geomorphic development. Due to the many mountains in the area and few plains, the area of paddy fields is small, only about 20% of the total arable land area. Chehe town is located along the upper reaches of the Diaojiang River and is in the subtropical monsoon climate zone. The wind direction is mainly southeast and rainfall is concentrated from May to July annually. The primary crop is corn, mainly distributed in the hilly areas on both sides of the Diaojiang River. Rice crops are mainly planted in the karst depressions, sporadically distributed, and account for a very small percentage of the area. The town also has a variety of mining enterprises, such as concentrators, smelters, etc., and is one of the most important towns in Guangxi Zhuang Autonomous Region integrating industry, mining, commerce, and agriculture.

2.2. Sampling Methods

To evaluate the geochemical response of the surficial environment to the mining, dressing, and smelting activities at the Dachang tin-polymetallic ore field in Guangxi, Chehe town was selected as the sampling center as it has relatively concentrated mining and smelting activities. We studied mining activities, climate, topography, and the direction of wind and water flow, and systematically sampled atmospheric dust, surface water, stream sediment, ore, tailings, mine drainage, soil, and crops in and around the mining area. The sampling plan is shown in Figure 1. Ten atmospheric dust samples were collected at locations along the southeasterly wind direction during the dry season and the rainy season. Atmospheric dust samples are collected using high-density polyethylene bottles (25 cm in diameter) and placed at a height of 5–10 m from the ground to prevent soil resuspension. Six sets of co-located surface water and stream sediment samples were collected along the stream segment near Chehe Town in the Diaojiang River, with one sample upstream of the mine site as a background sample, three sampling locations within the ore district, and two sampling locations downstream of the mining area (Figure 1). The surface water samples were collected at the centerline of the river, 30–50 cm below the surface. Composite stream sediment samples were collected at the same locations as the surface water samples, composed of 3–5 subsamples uniformly mixed. The concentrators and tailings ponds are adjacent to each other. Ore samples were collected from the concentrator and mine drainage and tailings were collected from the tailings pond. Mine drainage is stored in an enclosed depression of the tailings dam and the ore and tailings are stored in open-air piles. Soil and crop samples were collected in agricultural fields, concentrated in mountainous areas. A total of 62 samples of corn and root soil and 3 samples of rice and root soil were collected.

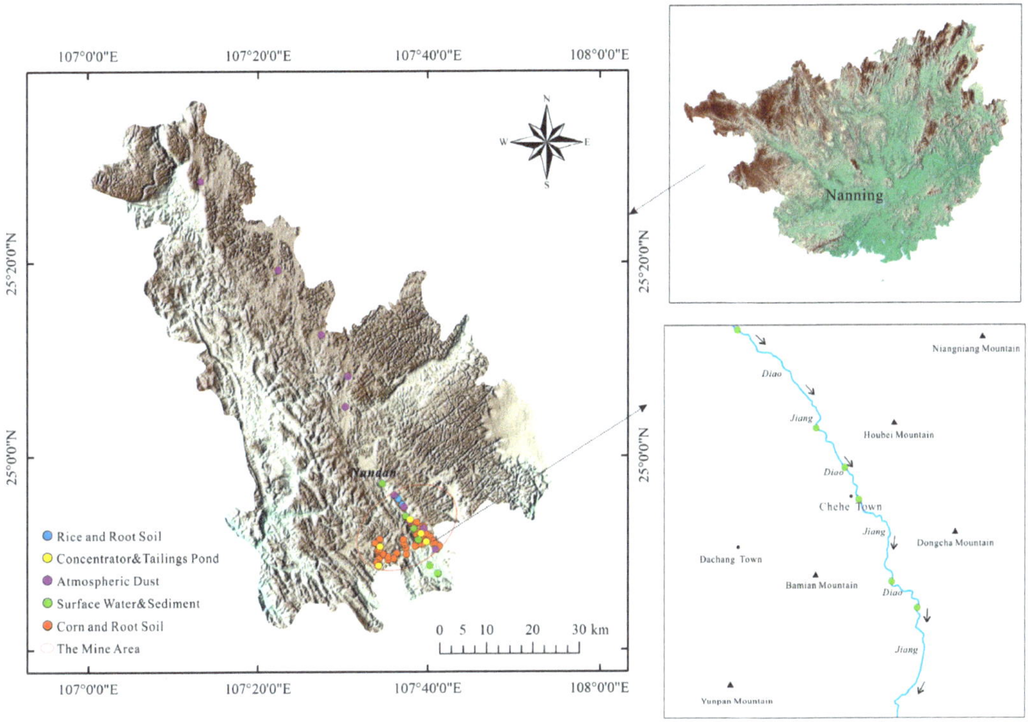

Figure 1. Distribution of sample locations.

2.3. Analysis Methods

After the atmospheric samples were collected, they left the deposition jars for 2–3 d. The supernatant was siphoned and the sediment was transferred into a beaker. We recorded the volume and total amount of the supernatant and sediment, respectively. The supernatant was added with HNO_3 (Guaranteed Reagent), and the precipitate was filtered and dried at 65 °C for testing.

The total volume of each water sample sent for analysis was 5 L. No preservatives were added to 2 L, and this original sample was used to analyze the pH, acidic ions, As, and other elements. The remaining 3 L of each sample was acidified with HCl (Guaranteed Reagent) to analyze concentrations of Cd, Pb, Zn, and other heavy metals.

Soil and sediment samples were dried under natural conditions and ground to 200 mesh. The samples were digested in a polytetrafluoroethylene tube (a mixture of $HClO_4$, HNO_3, and HF) and then dissolved in aqua regia. After stewing, transferred the liquid and diluted with HNO_3 (3%) for analysis.

After washing the crop samples with deionized water, they were ventilated and dried at 60–70 °C for 24–48 h. The dried samples were placed in an agate mortar and ground to 0.42–0.25 mm and mixed evenly for testing.

Sample analysis was conducted by Geology & Mineral Analysis & Test Research Center of Guangxi Zhuang Autonomous Region, which mainly used inductively coupled plasma mass spectrometry (ICP-MS, X-SERIES, Thermo Electron, Waltham, MA, USA; incident power 1400 W), atomic fluorescence spectrometry (AFS, Model AFS-230E, Kechuang Haiguang Instrument Co., Ltd., Beijing, China), inductively coupled plasma emission spectrometry (ICP-AES, IRIS Advantage, Thermo Fisher, Waltham, MA, USA; incident power 1150 W), X-ray fluorescence spectrometry (XRF, PW2440, Philips Co., Eindhoven, The Netherlands), and a pH meter to analyze heavy metals such as As, Cd, Pb, and Zn and other physicochemical indices in the multimedia samples. The measurement errors for the reference materials were within 5% and the errors for the coded samples were <5%. Thus, the data quality was considered reliable. X-Ray Diffraction (XRD, DMAX RAPID II, Rigaku Co., Tokyo, Japan) was completed in the Key Laboratory of Surficial Geochemistry of Ministry of Education of Nanjing University. The test used Cu targets and capillary projection method with a scanning angle of 0–40°.

2.4. Data Processing

2.4.1. Deposition Ratio

The deposition coefficient $R_{deposition}$ is calculated as shown in equations (1), where C_i is average annual atmospheric dust flux density in the mine site and C_j average annual atmospheric dust flux density outside the mine site. $R_{deposition}$ is used to assess the proportion of longitudinal deposition of elements within the mining area.

$$R_{deposition} = (C_i - C_j)/C_i \tag{1}$$

2.4.2. Content Variation Coefficient

Content variation coefficient (ΔC) is used to characterize the relative changes in the content of elements before and after geochemical migration. The formula is shown in (2). C is the concentration of the element, and i, j represents the post-migration and pre-migration assignment media of the element, respectively. $\Delta C > 0$ indicates that during geochemical migration, the element is enriched, and the greater its value, the higher the degree of enrichment. $\Delta C < 0$ indicates that during geochemical migration, the element is depleted, and the greater its absolute value, the higher the degree of dilution.

$$\Delta C_{i-j} = (C_i - C_j)/C_j \times 100\% \tag{2}$$

3. Results and Discussion

3.1. Atmospheric Dust

Heavy metal releases to the atmosphere include dust created by transportation of uncovered ore and smelting emissions. The study area is surrounded by mountains and the dust is not easily dispersed, resulting in significant concentrations of heavy metals in the atmosphere (Figure 2). Annual deposition flux density characteristics of atmospheric heavy metals in the study area are shown in Table 1. There were significant seasonal differences in the deposition flux density of heavy metals such as As, Cd, Pb, and Zn, significantly influenced by rainfall. The deposition flux density of heavy metals in the rainy season was generally higher than that in the dry season, with lower concentrations and high deposition, while the dry season has the opposite characteristics (Figure 3). Over the entire year, the annual deposition flux density for each element was high in the mining area. The annual deposition flux density of Cd was two-times higher than the standard (30 $g/hm^2 \cdot a$) of DZ/T 0295-2016 "Specification of Land Quality Geochemical Assessment [18]", indicating that the concentrations of heavy metals in atmospheric dust is sufficient to produce significant harm to surface environments such as soil.

Figure 2. Smelter emission in the research area.

Table 1. Geochemical characteristics of heavy metal flux in atmospheric dust (g/hm^2/a; n = 10).

	Dry Season				Rainy Season			
	As	**Cd**	**Pb**	**Zn**	**As**	**Cd**	**Pb**	**Zn**
Min	1.27	0.78	22.55	52.12	4.65	1.24	24.08	74.75
Max	552.15	58.70	197.71	2632.15	1628.54	76.19	1161.17	3173.91
Mean	89.18	13.68	87.95	577.33	196.22	27.86	421.26	1123.85

Annual deposition flux density of heavy metals in the atmosphere of different regions [19–24]

	Nandan	**Guixi**	**Dinghu Mountain**	**Dabaoshan**	**Changchun**	**Daqing**	**Chengdu**
As	285	-	-	-	47.9	8.1	27.7
Cd	41.55	65.6	5.7	1.97	2.5	1.7	17.7
Pb	509	700	428.4	-	123.1	157.1	459.5
Zn	1701	2250	1857.8	-	481.5	788.1	1478.3

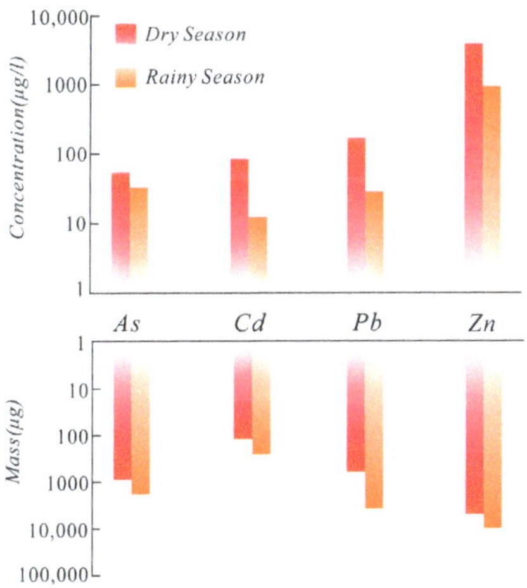

Figure 3. Comparison of characteristics of heavy metals in the atmosphere in different seasons. Log-rithmic coordinates. Comparison of differences in concentration and quality of heavy metals in atmospheric deposition during the dry season and rainy season.

The annual deposition flux densities of heavy metals in the ore district were much higher than those of typical mining agglomerations (Dabaoshan mine), heavy industrial areas (Dinghushan, Guangdong; Daqing; Changchun), economic development zones (Chengdu), and other high-emissions functional areas, and equivalent to those of copper smelting areas (Guixi) (Table 1). The effects of smelting operations on air quality are far greater than those of ore mining, industrial production, and other activities, and there is a risk of secondary pollution by deposition of dust onto the surface environment [19–24].

Emissions characteristics and migration of heavy metals through atmospheric dust transport in the study area were preliminarily evaluated (Figure 4). Due to isomorphism, the migration of Cd, Zn, and other heavy metals in the atmosphere showed similar geochemical behaviors and variations along the wind direction. Namely, the annual deposition flux densities were highest in the mine area and gradually decreased with increasing distance from the ore district. Pb and Zn show similar trends to some extent, but their distribution is also different due to ore type and grade limitations. As is the main component of gangue minerals and also exhibits characteristics with high concentrations in the mining area, but its fluctuations are relatively low. The annual precipitation flux densities of heavy metals in the mine area were closely related to the distribution of ore deposits and mining activities, with significant fluctuations along the wind direction. Heavy metal fluxes in mining areas with strong human disturbance were higher than those in areas with weak human disturbance. Under the combined influences of topography, climate, and the specific gravity of the metal sulfide or metal oxide, diffusion in the atmosphere demonstrated hysteresis. Longitudinal deposition was greater than lateral migration; the proportion of longitudinal deposition for each element was calculated according to (1): As (97.49%), Cd (92.26%), Zn (89.64%), and Pb (88.36%). Beyond about 25 km away from the mine area, the annual atmospheric deposition flux density of heavy metals decreased rapidly and the impact of mining activities on the atmosphere gradually disappeared.

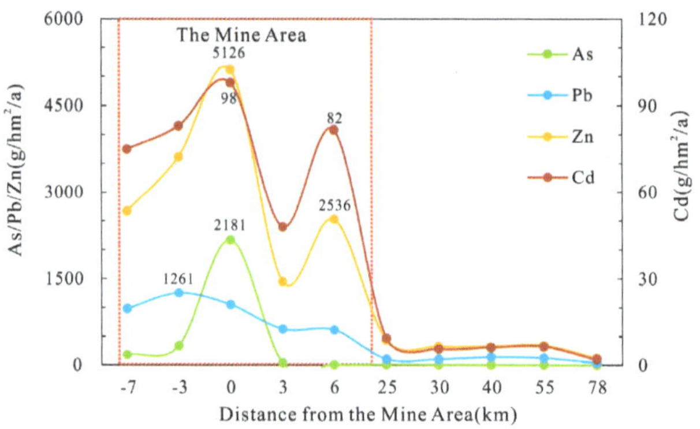

Figure 4. Spatial migration of heavy metals in the atmosphere. (Changes in the concentration of heavy metals migrating along the wind).

3.2. Mine Drainage and Surface Water

Environmental and geochemical characteristics of the mine drainage and surface water within the mine site were analyzed (Table 2). The mine drainage from the tailings pond/dam showed various degrees of exceedance for As and Zn over the limits specified in GB 25466-2010 "Emission Standard of Pollutants for Lead and Zinc Industry", while the concentrations of Cd and Pb were relatively low and did not exceed the limits [25]. Due to oxidation of sulfide minerals in the ore, concentrations of SO_4^{2-} in the mine age were high, with a maximum value of 1472 mg/L. However, SO_4^{2-} readily reacts with Ca^{2+} in the surrounding carbonate rocks to precipitate gypsum, so, after discharge into the river, SO_4^{2-} concentrations in the surface water appear to have been significantly reduced through precipitation and dilution.

Heavy metal concentrations in the river water were much lower than those in the mine drainage. Cd, Pb, and Zn concentrations complied with the Class III limits of GB 3838-2002 "Environmental Quality Standards for Surface Water", indicating low risk, while As exceeded the standard [26]. According to previous data, in 1992, the Diaojiang River was affected by mining activities. The contamination with heavy metals such as As (114 mg/L),

Cd (0.64 mg/L), Pb (54.8 mg/L), and Zn (50.5 mg/L) was very high [27]. In 1996, the river remained contaminated by multiple chemicals and the area of pollution had significantly expanded [27]. By 1998, the mine area had implemented large-scale corrective measures, the water quality had significantly improved and dissolved heavy metals concentrations met the national surface water Class III standards [27]. The present study demonstrates that the physical and chemical properties of the surface water in the ore district have remained stable over the past 20 years, with neutral to alkaline pH and good water quality.

Table 2. Geochemical properties of water in the study area (mg/L).

Medium	Sampling Site	As	Cd	Pb	Zn	SO_4^{2-}	pH
Mine Sewage (n = 3)	Tailings pond/dam	0.33 0.007 1.61	0.0005 0.019 0.016	0.0007 0.0007 0.04	0.037 2.45 0.079	1372 215 1472	6.61 7.29 7.28
GB 25466-2010	Direct Discharge	0.3	0.05	0.5	1.5	-	6–9
	Indirect Discharge	0.3	0.05	0.5	1.5	-	6–9
	Concentration	**As**	**Cd**	**Pb**	**Zn**	SO_4^{2-}	**pH**
Surface Water (n = 6)	Range	0.018–0.18	0.0001–0.0032	0.0006–0.0007	0.0095–0.047	87.4–250	7.75–8.06
	Mean	0.07	0.0011	0.0007	0.0218	155	-
GB 3838-2002	Class III	0.05	0.005	0.05	1.0	250	6–9

3.3. Solid Wastes

As the primary solid waste resulting from mining and the source of heavy metal pollution, tailings have been the focus of much research on mine pollution. However, in our study area, in addition to large accumulations of tailings, there is also ore that has not been smelted after grinding. The ore is more sensitive to the supergene environment than tailings due to its much higher sulfide content prior to smelting. As a source of pollution, it presents a greater risk than tailings (Table 3). Stream sediment is an important medium that is both a "source" and "sink" for heavy metals. The geochemical characteristics of the sediment are the result of the combined effects of chemical weathering and physical migration of pollutants from the upstream mines. The sediment also acts as a carrier of heavy metals in the river and can be a source of secondary pollution. Heavy metal concentrations in the sediment of the study area were high (Table 3).

Table 3. Statistics of heavy metals characteristics of solid wastes.

		Fe_2O_3	S	As	Cd	Pb	Zn	pH
		%	%	μg/g	μg/g	μg/g	μg/g	-
Tailings (n = 2)	Min	3.57	2.51	5484	22.6	328	2833	7.71
	Mean	6.64	3.45	5908	47	2203	6397	-
	Max	9.70	4.39	6331	71	4078	9960	7.85
Ore (n = 4)	Min	21.42	21.87	32,440	85.90	3395	10,310	3.20
	Mean	21.52	24.57	65,073	91.98	3759	10,663	-
	Max	21.69	27.14	99,080	99.30	4010	11,210	4.88
Sediment (n = 6)	Min	5.00	0.06	32.60	1.60	34	243	7.32
	Mean	7.98	0.73	2321	35.70	218	3710	-
	Max	10.86	1.34	3788	64.70	334	5850	8.80
$\Delta C_{sediment\text{-}tailings}$	-	20.25%	−78.77%	−60.71%	−23.72%	−90.10%	−42.01%	-
$\Delta C_{sediment\text{-}ore}$	-	−62.93%	−97.02%	−96.43%	−61.19%	−94.20%	−65.21%	-

Based on the mineral compositions of various media (Figure 5), the ore is mainly composed of primary sulfide minerals such as sphalerite, pyrite, and pyrrhotite. It has a strongly acidic pH and very high concentrations of various metals related to mineralization, such as S, Fe, As, Cd, Pb, and Zn, which present a great potential risk. Tailings and sediment that are exposed to hypergene conditions for a long time, in addition to retaining some primary sulfide minerals, quartz and calcite, are mainly composed of secondary minerals such as kaolinite, goethite, illite, and gypsum. The pH becomes neutral to alkaline and the concentrations of S and metals precipitously decline compared to those in the ore. The tailings had undergone the processes of mining, smelting, and open storage, the heavy metals were mainly in the form of oxides, and the potential risk was greatly reduced compared with that of ore. Heavy metal concentrations in sediment are the result of the combined action of chemical weathering and physical migration of ore and tailings. Although the concentrations of As, Cd, Pb, and Zn in sediment had decreased significantly compared to the source material, it was still elevated relative to other surface media such as soil and crops, and poses a threat to the safety of the environment.

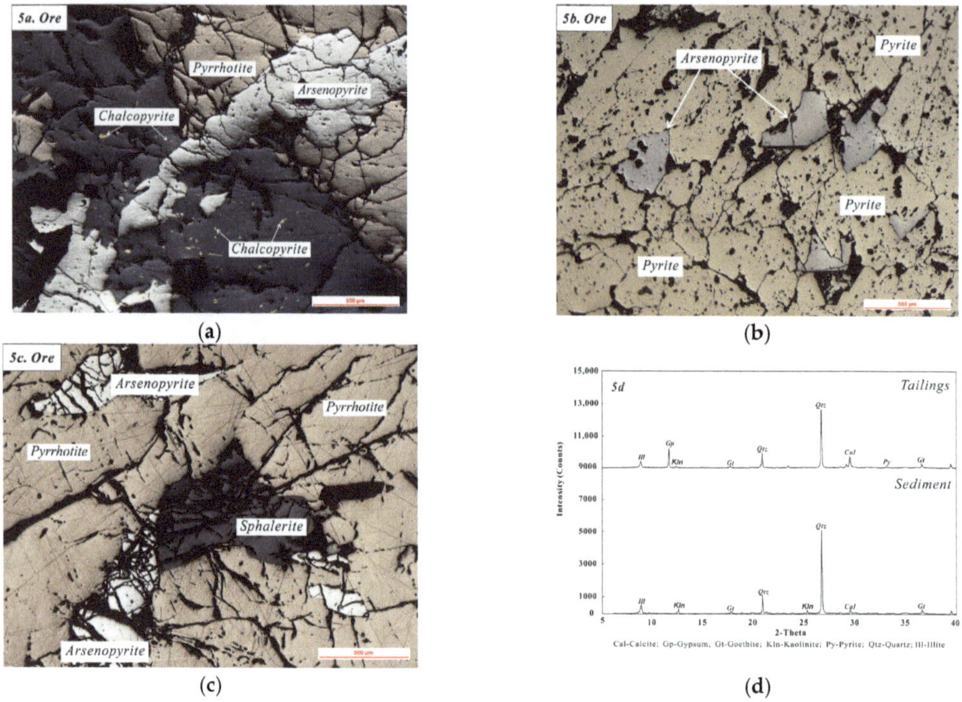

Figure 5. Mineral composition of ore, tailings, and stream sediment. (**a**–**c**) Show the microscopic results of rock slices, and (**d**) shows the analysis of XRD of tailings and sediments. Figure 5 reflects the mineral composition of the samples.

The migration and transformation of elements during the supergene process can be characterized by ΔC. The heavy metals in the river sediments, driven by both physical and chemical dynamics, underwent different degrees of leaching compared to the ore and tailings samples. As and Pb were leached more heavily, followed by Zn and Cd with a similar ΔC. The tailings had gone through the dressing and smelting process and, thus, the heavy metals were relatively stable, with leaching rates far lower than those of the ore. One of the primary sulfide minerals constituting the deposit is pyrrhotite, which has a defect structure. Some of the Fe^{2+} in the crystal structure is replaced by Fe^{3+} and the cation position has some vacancies, resulting in instability of the crystal structure. Thus, it oxidizes before other minerals, forming a dissolution-acid supply system based on Fe^{2+} oxidation and Fe^{3+} hydrolysis [28–30]. As oxidation continues, pyrite also begins to weather and dissolve, and the low-pH pore water continuously leaches and releases additional Fe^{3+} to the elemental cycle. This process greatly increases the decomposition rate of primary sulfides such as arsenopyrite, sphalerite, and galena [28–30].

The stability of sphalerite is intermediate between that of pyrrhotite and pyrite, and the weathering and dissolution of sphalerite results in release of large amounts of Zn^{2+} and Cd^{2+} into the environment. The released heavy metals either co-precipitate or adsorb into the early-formed iron oxides/hydroxides or metasomatize with carbonate rocks to form precipitates, thus reducing harm to the environment [6,30]. Galena has greater resistance to weathering than sphalerite, because during weathering of galena, insoluble anglesite ($PbSO_4$) is formed and adheres to the mineral surface, preventing further oxidation. However, coexistence with pyrite greatly enhances the dissolution of galena, and Pb^{2+} released in a carbonate environment metasomatizes with Ca^{2+} in the surrounding rock to form insoluble cerussite [28,29,31]. Studies have shown that smithsonite has a much higher ability to migrate in supergene fluids than cerussite, which is largely retained in situ after formation [32]. In contrast, Zn readily migrates with fluids [32].

Migration of mineralization-related elements such as S, Fe, As, Cd, Pb, Zn, etc., in the ore-tailings-sediment system are somewhat correlated due to interactions, synergistic release, adsorption, and precipitation, and the regulation of carbonate rocks and secondary minerals (iron oxide/hydroxide). To a certain extent, these processes reduce the harmful effects of deposit mining on the environment. In addition to the release and migration of heavy metals due to chemical weathering, physical migration also affects heavy metal concentrations in the sediment, resulting in a high-concentration area of heavy metals downstream of the mine site (Figure 6). These results are consistent with those of Jian (2010), whose research results indicate that the impact of the mine on sediment extends up to 200 km downstream [33].

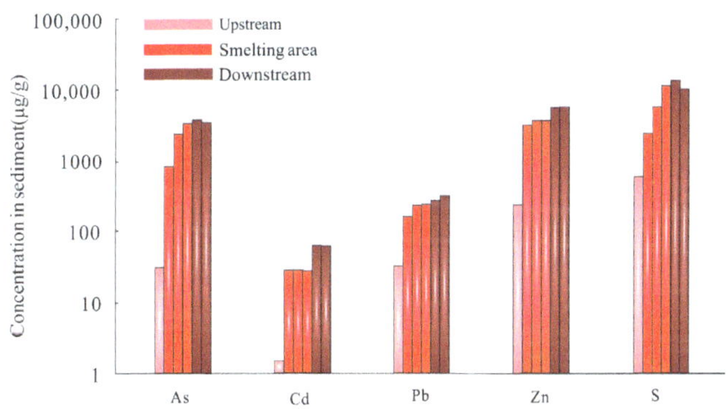

Figure 6. Migration characteristics of different elements in stream sediment. Logarithmic coordinates. The color from light to dark indicates the concentration characteristics of heavy metals in the upstream, midstream, and downstream in order.

3.4. Risk Assessment for Soil and Crops in Mining Area

Soil is the end receptor of geochemical processes such as atmospheric deposition and weathering of ore/tailings, and its environmental quality directly affects the safe production of crops and human health. Heavy metal concentrations in soil in the study area are shown in Table 4. It is apparent that mining and smelting in the study area have resulted in serious contamination in the soil. According to the risk screening values of GB 15618-2018 "Soil Environmental Quality Risk Control Standard for Soil Contamination on Agricultural Land [34]", the dryland soils in the study area are seriously contaminated with heavy metals, with exceedance rates for metals such as As, Cd, and Zn of >90% and an exceedance rate for Pb of 70.97%. However, overall acidification is not high; the soil samples were mainly neutral to alkaline, which promotes fixation of heavy metals. Exceedances for heavy metals in paddy fields were greater than for dryland. The soil pH was slightly acidic, which is conducive to mobilization of heavy metals, increasing risks. However, due to a limited planting area, the sample quantity for the paddy fields was not statistically significant and the results are usable only for reference. The high concentrations of heavy metals in the topsoil near the mine site are mainly influenced by contributions from the mining activities, including atmospheric deposition and river transport. The surface water environmental quality is good and its contribution of heavy metals to the soil was much lower than that of the atmosphere. Thus, contamination with heavy metals of the cultivated soils was mainly due to atmospheric deposition.

Table 4. Characteristics of heavy metals content in topsoil (μg/g).

Paddy Field (n = 3)	**As**	**Cd**	**Pb**	**Zn**	**pH**
Min	49.3	1.92	61.5	222	5.68
Mean	61	4.04	138.50	420	-
Max	67.2	5.41	196	548	6.2
Exceedance rate	100%	100%	66.67%	100%	-
Dryland (n = 62)	**As**	**Cd**	**Pb**	**Zn**	**pH**
Min	23.3	0.15	28	67.9	4.47
Mean	685	12.53	569	1562	-
Max	5273	108	5003	14,715	8.48
Exceedance rate	90.32%	98.39%	70.97%	93.55%	-

The heavy metal concentrations in corn and rice samples from the study area are shown in Table 5. According to the provisions of GB 2762-2017 "National Standard for Food Safety Maximum Levels of Contaminants in Foods [35]", heavy metals in corn do not exceed these standards and are safely edible, consistent with the results of Lu et al. (2017) [36]. In contrast, the Cd concentrations in rice were all higher than the national standard. Although the sample quantity is small, the potential for high risk cannot be ignored.

Table 5. Characteristics of heavy metals in crops (μg/g).

Rice (n = 3)	**As**	**Cd**	**Pb**	**Zn**	**Corn (n = 62)**	**As**	**Cd**	**Pb**	**Zn**
Min	0.25	0.26	0.078	19.1	Min	0.03	0.0075	0.07	19.9
Mean	0.27	0.60	0.09	21.03	Mean	0.10	0.02	0.09	31.10
Max	0.3	1.13	0.1	24.8	Max	0.32	0.085	0.2	46.2
Exceedance rate	-	100%	0%	-	Exceedance rate	0%	0%	0%	-

The differences in risks between the crops are not only by differences in absorption, but also by the terrain and landforms in the study area, the areas of cultivation, and the degree to which they are influenced by the supergene media. Nandan County is located at the southeast edge of the Yunnan-Guizhou Plateau. After crustal uplift, the riverbed was elevated, resulting in a relative lowering of the groundwater level and a difference in hydraulic head. Because the river channel is relatively narrow, the river flow is rapid and the area was dominated by downcutting. These conditions are not conducive to sediment deposition or the formation of a floodplain. Therefore, the study area is mainly planted with dryland crops, and paddy crops are sparsely distributed. The dryland crops are mainly planted on the slopes of both banks of the river, which are less affected by surface water and sediment. Corn is also covered in leaves and has little contact with the atmosphere; thus, it is less affected by atmospheric deposition. Therefore, despite the high risk associated with heavy metals in the soil, absorption of heavy metals by corn is low, and the edible portions are safe. Paddy fields are scattered in karst depressions on both sides of the river where irrigation is easy. Irrigation with river water carries a large amount of sediment with high concentrations of heavy metals. These metals are more bioavailable under acidic pH conditions, which results in unsafe concentrations in rice.

3.5. Comprehensive Environmental Response and Pollution Prevention Suggestions

3.5.1. Mechanism of Heavy Metal Migration and Transformation

Ore and tailings exposed to the surface will undergo strong oxidation under epigenetic conditions. The dissolution of major minerals such as marmatite, galena, pyrite, and pyrrhotite can cause the release of large amounts of heavy metals, and cause acidification of surrounding rivers and soils. The reaction is as follows [8]:

$$\text{Pyrite: } FeS_{2(s)} + 15/4O_2 + 7/2H_2O \rightarrow 2SO_4{}^{2-} + Fe(OH)_3 + 4H^+$$

$$\text{Pyrrhotite: } Fe_{(1-x)}S_{(s)} + (2 - x/2)O_2 + xH_2O \rightarrow (1 - x)Fe^{2+} + SO_4{}^{2-} + 2xH^+$$

$$\text{Marmatite: } (Zn_{(1-x)}Fe_x)S_{(s)} + 2O_2 \rightarrow (1 - x)Zn^{2+} + xFe^{2+} + SO_4{}^{2-}$$

$$\text{Galena: } PbS + O_2 \rightarrow Pb^{2+} + SO_4{}^{2-}$$

$$\text{Arsenopyrite: } 4FeAsS(s) + 11O_2 + 6H_2O \rightarrow 4Fe^{2+} + 4H_3AsO_3 + 4SO_4{}^{2-}$$

Generally, the released heavy metals migrate as cations with high biological activity and risk. However, under the background of karst, metal cations will combine with $CO_3{}^{2-}$ to form precipitation, and the H^+ released by the weathering of sulfide minerals will also be buffered to a certain extent by $CO_3{}^{2-}$-$HCO_3{}^-$, as shown below [37]. Therefore, lime addition is also used in many mining areas to reduce the ecological hazards of sulfide deposit mining. In addition to co-precipitation with carbonate ions, metal cations can also be adsorbed by clay minerals such as illite and kaolinite, as well as iron and manganese oxides. Therefore, heavy metals in water mainly accumulate in sediments as particulate, which is consistent with the research results of this paper. It is worth noting that the buffering effect of carbonate rocks on heavy metals is limited, and serious acidification and heavy metals pollution can still occur in ore districts with long mining time, high grade and large scale.

$$HCO_3{}^- + H^+ \rightarrow H_2O + CO_2$$

$$CaCO_3 + H^+ \rightarrow Ca^{2+} + HCO_3{}^-$$

$$Zn^{2+} + HCO_3{}^- \rightarrow H^+ + ZnCO_3 \text{ (Smithsonite)}$$

$$Pb^{2+} + HCO_3{}^- \rightarrow H^+ + PbCO_3 \text{ (Cerussite)}$$

According to the previous results, in addition to the heavy metals released by chemical dissolution, smelting is also one of the important ways for heavy metal emissions. The heavy metals discharged from smelting undergo high-temperature oxidation and are mostly emitted into the environment in the form of oxides. The reaction process is shown below. Therefore, although the annual flux density of heavy metals in the atmosphere is relatively high, their activity is low and they are less harmful to crops. However, long-term accumulation in the soil can affect soil functions and lead to ecological imbalances.

$$2ZnS + 3\,O_2 \text{ (high-temperature)} \rightarrow 2ZnO + 2SO_2$$

$$2PbS + 3\,O_2 \text{ (high-temperature)} \rightarrow 2PbO + 2SO_2$$

3.5.2. Potential Risk Assessment

Based on the above investigation results and mechanism, the atmosphere is the most sensitive to mining activities within the mining area, while the pollution downstream of the mining area is mainly concentrated in sediments.

Atmospheric heavy metals have a high proportion of vertical deposition and the risk receptors are mainly soil and human. (1) Soils: the accumulation of heavy metals in soil is long-term, cumulative, concealment, latency, and irreversibility, which have adverse effects on the physical and chemical properties of soil and microbial structure [1]. The further intensification of pollution will also pose a threat to the survival of biological communities [38]. (2) Humans: atmospheric particulates enter the human trachea, lungs, and even the blood system through the respiratory pathway, leading to cardiovascular and other diseases [39]. Nandan County has been reported that human lead poisoning in smelting areas, and developmental delays and immune deficiencies are common in children. It can be seen that heavy metals ingested through respiration, although they are trace elements, are extremely harmful to human health. Therefore, it is important to control emissions and deposition of atmospheric heavy metals in the mine area to maintain a safe soil environment and protect human health.

The stream sediment receives high concentrations of heavy metals from the ore and tailings. However, due to the influence of the mountainous area, the anthropogenic disturbance of the water is relatively low, making the sediment in the mining area either carried away by the current and deposited downstream, or deposited in the river bottom, less harmful. Sediments with high heavy metals concentration pose a great threat to the downstream floodplain because, in open areas, downcutting gradually decreases and erosion increases. With the development of floodplains, sediments will gradually accumulate on convex banks. Under specific conditions (e.g., low pH), accumulated heavy metals will be activated and released, leading to high-risk secondary pollution, endangering soil, crops, and even human beings along the food chain. Therefore, risk abatement in the downstream area should focus on the river sediment.

3.5.3. Suggestions on Prevention and Control of Pollution

In view of the above environmental results, the following suggestions are provided for risk management and control in the study area:

(1) The area of concentrated mining activities is mainly contaminated through atmospheric deposition, of which smelting emissions are the main source, supplemented by dust from transportation. Therefore, it is necessary to control atmospheric heavy metal deposition within the mine area. Rainfall has a significant impact on the concentration and flux of atmospheric deposition. Therefore, the smelting operation should be properly adjusted in the rainy season to prevent flux increases due to the driving force of rainfall, and the mining and transportation plan can be improved to minimize dust emissions.

(2) Downstream of the mining area, the main focus should be on sediment management. Through the ore district, the river is fast flowing and downcutting the surrounding geology. The middle and lower reaches are more prone to lateral erosion and sedimentation as the flow velocity decreases and the river channel widens. Therefore, special attention should be paid to the risk of secondary pollution caused by heavy metals in sediment where the tributaries converge into the main river, river convex banks, and floodplains. There are differential hazards to corn and rice crops. Consequently, we recommend adjusting the planting structure accordingly, gradually converting paddy fields to dryland crops to reduce the risks associated with paddy soils and irrigation, and planting mainly dryland crops that are safe for consumption.

4. Conclusions

1. The heavy metals pollution caused by atmospheric deposition in the mine area is relatively high. The annual deposition flux density of Cd is two-times higher than the relevant standards. Influenced by climate and topography, heavy metals contamination from atmospheric deposition migrates about 25 km along the wind direction and then decreases. About 90% of the heavy metals migrate in the form of vertical deposition in the mine area.
2. Effluent drainage from the ore district contains individual heavy metals exceeding the standards, but the water quality of the river is less affected by the mining activities. After treatment, only As slightly exceeded the standard at the river bend, the pH of

the water was neutral–alkaline, and the concentrations of soluble heavy metal ions were low.

3. Weathering and migration of ore and tailings contribute to high concentrations of heavy metals in river sediments, mainly downstream of the mine. Risks associated with migration of sulfide particles are high, while heavy metals migrating in chemical form are more stable after co-precipitated with carbonates or adsorbed by secondary iron oxides.
4. The soil in the mine area is greatly affected by mining activities. The surface soil is significantly enriched in heavy metals and greatly exceeds the standards. It is mainly influenced by atmospheric deposition. The risks of the soil environment to various crops differ. Cd in rice greatly exceeds the standard, while corn does not exhibit heavy metals exceedances and can be safely consumed.

Author Contributions: Conceptualization, writing manuscript draft, and supervision, B.L.; writing—reviewing and editing, T.Y.; investigation, W.J. and X.L.; data curation writing—reviewing and editing, K.L.; investigation, C.L. and X.M.; resources, project administration, conceptualization, data curation, supervision, visualization, and writing—reviewing and editing, Z.Y. All authors have read and agreed to the published version of the manuscript.

Funding: This study was financially supported by the Geological Survey Project of China Geological Survey (Granted No. DD20211414), the project of Geochemical study on selenium and heavy metal elements in central—eastern area of Guangxi, China (2015–2016), Study on the genesis and ecological effect of Se, Ge and Cd in soil of Guangxi, China (2017–2019), Ecological and geochemical survey and study on the heavy metals in typical soil of Guangxi, China (2018–2019). The authors sincerely thank all the participants for their efforts.

Data Availability Statement: Data used in this study are available from the first authors on reasonable request.

Conflicts of Interest: The authors declare no conflict of interest.

References

1. Li, Z.; Ma, Z.; Kuijp, T.J.v.d.; Yuan, Z.; Huang, L. A Review of Soil Heavy Metal Pollution From Mines in China: Pollution and Health Risk Assessment. *Sci. Total Environ.* **2014**, *468–469*, 843–853. [CrossRef] [PubMed]
2. Han, Y.-S.; Youm, S.-J.; Oh, C.; Cho, Y.-C.; Ahn, J.S. Geochemical and Eco-Toxicological Characteristics of Stream Water and Its Sediments Affected by Acid Mine Drainage. *Catena* **2017**, *148*, 52–59. [CrossRef]
3. Tian, S.; Liang, T.; Li, K. Fine Road Dust Contamination in a Mining Area Presents a Likely Air Pollution Hotspot and Threat to Human Health. *Environ. Int.* **2019**, *128*, 201–209. [CrossRef] [PubMed]
4. Hong, Y. Analysis of Environmental Impact and Prevention and Protection Measures of Small Non-Metallic Mines. *Inn. Mong. Sci. Technol. Econ.* **2013**, *298*, 50–51.
5. Liu, S.; Guo, A.; Li, L. Discussion on Geological Environment Problems and Recovery of One Limestone Deposit in Guilin City of Guangxi Province. *Shandong Land Resour.* **2017**, *33*, 65–70.
6. Li, B.; Yang, Z.; Ji, W.; Yu, T.; Hou, Q.; He, H.; Zhang, Q.; Wu, T.; Qin, J. Ecological Effect of A Typical Sulfide Deposit in Carbonate Area—Xijikeng Lead-Zinc Mine in Guigang, Guangxi. *Geoscience* **2020**, *34*, 957.
7. Zhou, S.; Shi, Z. Study on the Environment of Metal Sulfide Mines. *Acta Mineral. Sin.* **2013**, *33*, 733–734.
8. Lindsay, M.B.J.; Moncur, M.C.; Bain, J.G.; Jambor, J.L.; Ptacek, C.J.; Blowes, D.W. Geochemical and Mineralogical Aspects of Sulfide Mine Tailings. *Appl. Geochem.* **2015**, *57*, 157–177. [CrossRef]
9. Shu, X.; Zhang, Q.; Lu, G.; Yi, X.; Dang, Z. Pollution Characteristics and Assessment of Sulfide Tailings from the Dabaoshan Mine, China. *Int. Biodeterior. Biodegrad.* **2018**, *128*, 122–128. [CrossRef]
10. García-Giménez, R.; Jiménez-Ballesta, R. Mine Tailings Influencing Soil Contamination by Potentially Toxic Elements. *Environ. Earth Sci.* **2017**, *76*, 51. [CrossRef]
11. Lei, L.; Song, C.; Xie, X.; Li, Y.; Wang, F. Acid Mine Drainage and Heavy Metal Contamination in Groundwater of Metal Sulfide Mine at Arid Territory (BS Mine, Western Australia). *Trans. Nonferrous Met. Soc. China* **2010**, *20*, 1488–1493. [CrossRef]
12. Zheng, K.; Li, H.; Wang, L.; Wen, X.; Liu, Q. Pyrite Oxidation Under Simulated Acid Rain Weathering Conditions. *Environ. Sci. Pollut. Res.* **2017**, *24*, 21710–21720. [CrossRef] [PubMed]
13. Zhang, Y.; Zhao, H.; Qian, L.; Sun, M.; Lv, X.; Zhang, L.; Petersen, J.; Qiu, G. A Brief Overview on the Dissolution Mechanisms of Sulfide Minerals in Acidic Sulfate Environments at Low Temperatures: Emphasis on Electrochemical Cyclic Voltammetry Analysis. *Miner. Eng.* **2020**, *158*, 106586. [CrossRef]

14. Battistel, M.; Stolze, L.; Muniruzzaman, M.; Rolle, M. Arsenic Release and Transport during Oxidative Dissolution of Spatially-Distributed Sulfide Minerals. *J. Hazard. Mater.* **2021**, *409*, 124651. [CrossRef] [PubMed]
15. Li, J.; Zhang, T.; Yang, W.; Zhang, Y. The Environmental Impact of Mining and Its Countermeasures. *MATEC Web Conf.* **2016**, *63*, 04010. [CrossRef]
16. Wu, W.; Wang, L.; Song, Z.; Zhou, J. Discussions on the Contradiction between Development of Mineral Resources and Ecological Environment Protection in the New Era in China. *China Min. Mag.* **2020**, *29*, 6–10.
17. Xu, M.; Cai, M.; Peng, Z.; Wang, X.; Chen, Y. Research on Metallogenic Zoning Characteristics and Control Mechanism of Dachang Ore Field. *Miner. Resour. Geol.* **2011**, *25*, 29–33.
18. *DZ/T 0295-2016;* Specification of Land Quality Geochemical Assessment. Ministry of Ecology and Environment: Beijing, China, 2016.
19. Tang, Q.; Yang, Z.; Zhang, B.; Feng, H.; Wang, H. A study of Elements Flux and Sources from Atmospheric Bulk Deposition in the Chengdu Economic Region. *Earth Sci. Front.* **2007**, *14*, 213–222.
20. Tao, M.; Zhou, J.; Liang, J.; Cui, H.; Xu, L.; Zhu, Z. Atmospheric Deposition of Heavy Metals in Farmland Area Around a Copper Smelter. *J. Argo-Environ. Sci.* **2014**, *33*, 1328–1334.
21. Ye, L.; Huang, M.; Zhong, B.; Wang, X.; Tu, Q.; Sun, H.; Wang, C.; Wu, L.; Chang, M. Wet and Dry Deposition Fluxes of Heavy Metals in Pearl River Delta Region (China): Characteristics, Ecological Risk Assessment, and Source Apportionment. *J. Environ. Sci.* **2018**, *70*, 106–123. [CrossRef]
22. Chen, Q.; Ma, M.; You, Y.; Yang, D. Cadmium Input Flux in Farmland Soil Near the Dabaoshan Mining Area of Guangdong Province. *Geol. Miner. Resour. South China* **2020**, *36*, 147–152.
23. Tang, J.; Li, N.; Li, H.; Bian, J.; Li, Z.; Cui, Y. Flux and Source Appointment of Heavy Metals from Atmospheric Dry and Wet Deposition in Daqing City, China. *J. Jilin Univ. Earth Sci. Ed.* **2012**, *42*, 507–513.
24. Yang, Z.; Lu, W.; Long, Y. Atmospheric Dry and Wet Deposition of Heavy Metals in Changchun City, China. *Res. Environ. Sci.* **2009**, *22*, 28–34.
25. *GB 25466-2010;* Emission Standard of Pollutants for Lead and Zinc Industry. Ministry of Ecology and Environment of the People's Republic of China: Beijing, China, 2010.
26. *GB 3838-2002;* Environmental Quality Standards for Surface Water. Ministry of Ecology and Environment of the People's Republic of China: Beijing, China, 2002.
27. Zhou, Y.; Song, S.; Zhang, C.; Yang, X.; Liu, C. Water Environmental Geochemical Response of Rivers to Mines and Mining Activity—A Case Study of the Diaojiang River Drainage System, Guangxi, China. *Geol. Bull. China* **2005**, *24*, 940–944.
28. Moncur, M.C.; Jambor, J.L.; Ptacek, C.J.; Blowes, D.W. Mine Drainage from the Weathering of Sulfide Minerals and Magnetite. *Appl. Geochem.* **2009**, *24*, 2362–2373. [CrossRef]
29. Zheng, K.; Li, H.; Wang, L.; Wen, X.; Liu, Q. Galena Weathering Under Simulated Acid Rain Conditions: Electrochemical Processes and Environmental Assessments. *Environ. Sci. Process Impacts* **2018**, *20*, 822–832. [CrossRef]
30. McKibben, M.A.; Tallant, B.A.; del Angel, J.K. Kinetics of Inorganic Arsenopyrite Oxidation in Acidic Aqueous Solutions. *Appl. Geochem.* **2008**, *23*, 121–135. [CrossRef]
31. Wu, P.; Liu, C.; Yang, Y.; Zhang, G. Release and Transport of (Heavy) Metals and Their Environmental Effect in Mining Activities. *Acta Mineral. Sin.* **2001**, *21*, 213–218.
32. Wu, Z.; Song, Y.; Hou, Z.; Liu, Y.; Zhuang, L. The World-Class Huoshaoyun Nonsulfide Zinc-Lead Deposit, Xinjiang, NW China: Formation by Supergene Oxidization of a Mississippi Valley-Type Deposit. *Earth Sci.* **2019**, *44*, 1987–1997.
33. Jian, L. Research on Spatial Distribution and Formation Mecharism of Heavy Metals and Arsenic Species in Guangxi Diaojiang River. Master's Thesis, Northwest A&F University, Xianyang, China, 2010.
34. *GB 15618-2018;* Soil Environmental Quality Risk Control Standard for Soil Contamination on Agricultural Land. Ministry of Ecology and Environment of the People's Republic of China: Beijing, China, 2018.
35. *GB 2762-2017;* National Standard for Food Safety Maximum Levels of Contaminants in Foods. China's National Health Commission (NHC): Beijing, China, 2017.
36. Lu, S.; Zhang, Y.; Yu, Y.; Zhong, X.; Tian, M.; Huang, Y.; Song, B. Characteristics of Heavy Metal Accumulation in Soil-Corn System Contents and Their Health Risks in Nandan, Guangxi. *J. Ecol. Rural. Environ.* **2017**, *33*, 706–714.
37. Berger, A.C.; Bethke, C.M.; Krumhansl, J.L. A process model of natural attenuation in drainage from a historic mining district. *Appl. Geochem.* **2000**, *15*, 655–666. [CrossRef]
38. Huang, C.-C.; Liang, C.-M.; Yang, T.-I.; Chen, J.-L.; Wang, W.-K. Shift of bacterial communities in heavy metal contaminated agricultural land during a remediation process. *PLoS ONE* **2021**, *16*, e0255137. [CrossRef] [PubMed]
39. Fiordelisi, A.; Piscitelli, P.; Trimarco, B.; Coscioni, E.; Iaccarino, G.; Sorriento, D. The mechanisms of air pollution and particulate matter in cardiovascular diseases. *Heart Fail. Rev.* **2017**, *22*, 337–347. [CrossRef] [PubMed]

MDPI AG
Grosspeteranlage 5
4052 Basel
Switzerland
Tel.: +41 61 683 77 34

Water Editorial Office
E-mail: water@mdpi.com
www.mdpi.com/journal/water

www.ingramcontent.com/pod-product-compliance
Lightning Source LLC
Chambersburg PA
CBHW042011260926
51850CB00020B/3162